Lecture Notes in Mechanical Engineering

Lecture Notes in Mechanical Engineering (LNME) publishes the latest developments in Mechanical Engineering—quickly, informally and with high quality. Original research reported in proceedings and post-proceedings represents the core of LNME. Volumes published in LNME embrace all aspects, subfields and new challenges of mechanical engineering. Topics in the series include:

- Engineering Design
- Machinery and Machine Elements
- Mechanical Structures and Stress Analysis
- Automotive Engineering
- Engine Technology
- Aerospace Technology and Astronautics
- Nanotechnology and Microengineering
- Control, Robotics, Mechatronics
- MEMS
- Theoretical and Applied Mechanics
- Dynamical Systems, Control
- Fluid Mechanics
- Engineering Thermodynamics, Heat and Mass Transfer
- Manufacturing
- Precision Engineering, Instrumentation, Measurement
- Materials Engineering
- Tribology and Surface Technology

To submit a proposal or request further information, please contact the Springer Editor of your location:

China: Dr. Mengchu Huang at mengchu.huang@springer.com
India: Priya Vyas at priya.vyas@springer.com
Rest of Asia, Australia, New Zealand: Swati Meherishi at swati.meherishi@springer.com
All other countries: Dr. Leontina Di Cecco at Leontina.dicecco@springer.com

To submit a proposal for a monograph, please check our Springer Tracts in Mechanical Engineering at http://www.springer.com/series/11693 or contact Leontina.dicecco@springer.com

Indexed by SCOPUS. The books of the series are submitted for indexing to Web of Science.

More information about this series at http://www.springer.com/series/11236

Mahavir Singh · Yasser Rafat
Editors

Recent Developments in Acoustics

Select Proceedings of the 46th National Symposium on Acoustics

Springer

Editors
Mahavir Singh
Department of Physico-Mechanical Metrology
CSIR-National Physical Laboratory
New Delhi, Delhi, India

Yasser Rafat
Mechanical Engineering Department
Aligarh Muslim University
Aligarh, Uttar Pradesh, India

ISSN 2195-4356 ISSN 2195-4364 (electronic)
Lecture Notes in Mechanical Engineering
ISBN 978-981-15-5775-0 ISBN 978-981-15-5776-7 (eBook)
https://doi.org/10.1007/978-981-15-5776-7

This Springer imprint is published by the registered company Springer Nature Singapore Pte Ltd.
The registered company address is: 152 Beach Road, #21-01/04 Gateway East, Singapore 189721, Singapore

Contents

Editors and Contributors

About the Editors

Dr. Mahavir Singh is the Head of the Acoustics, Ultrasonics and Vibration (AUV) section of the CSIR - National Physical Laboratory, New Delhi, India. He has over 25 years of experience in acoustics and vibration. He received his Ph.D in acoustics from IIT, Delhi. His main research interests are in the field of building acoustics, noise and vibration control, calibration and testing of electro-acoustic equipment, acoustical products and devices, maintenance, up-gradation and realization of national primary standards for acoustics and vibration. Dr. Singh has published over 125 research papers in journals and conference proceedings as well as authored a book on noise control in buildings. He is a fellow of the Acoustical Society of India (ASI), Ultrasonics Society of India (USI), Metrology Society of India (MSI), the Institution of Engineers (IEI) and the International Institute of Acoustics and Vibration (IIAV-USA).

Dr. Yasser Rafat is a faculty in the Mechanical Engineering department of the Aligarh Muslim University, India. He received his B. Tech. in Mechanical Engineering from the Aligarh Muslim University, India. For further studies, he moved to Canada where he pursued his M. Tech. from Concordia University and his Ph.D from McGill University. Dr. Rafat has authored several publications and delivered talks in the area of thermofluids and thermoacoustics. He has served as a reviewer for POMA (Acoustical Society of America). He has also filed and published three patents in the area of charging infrastructure of xEVs and cloud based health management system. His research interests include thermoacoustic refrigerators and engines, building acoustics, ride vibration analysis, sports projectile aerodynamics, and heat management of Li-ion batteries. He is a life member of the Acoustical Society of India (ASI).

Contributors

Ajish K. Abraham Department of Electronics, All India Institute of Speech and Hearing, Mysuru, Karnataka, India

Ravinder Agarwal Thapar University, Patiala, India

Deepak C. Akiwate Indian Institute of Technology Hyderabad, Telangana, India

V. Akshatha Department of Speech-Language Pathology, All India Institute of Speech and Hearing, Mysore, India

Ferasat Ali Department of Kulliyat, Ajmal Khan Tibbiya College, A.M.U, Aligarh, UP, India

Mohd Masoom Ali Computational Aerodynamics Lab, MED, ZHCET, AMU, Aligarh, India

Syed Fahad Anwer Computational Aerodynamics Lab, MED, ZHCET, AMU, Aligarh, India

Rajendar Bahl Centre for Applied Research in Electronics, Indian Institute of Technology Delhi, New Delhi, India

Archi Banerjee Sir C.V. Raman Centre for Physics and Music, Jadavpur University, Kolkata, India;
Rekhi Centre of Excellence for the Science of Happiness, IIT Kharagpur, Kharagpur, India

P. Bangaru Babu National Institute of Technology, Warangal, Telangana, India

Shobha Bhatt U.S.I.C.T, Guru Gobind Singh Indraprastha University, Dwarka, Delhi, India

Chandrashekhar Chauhan BITS Pilani K K Birla Goa Campus, Goa, India

Mahendra D. Date Indian Institute of Technology Hyderabad, Telangana, India

Sagar Deshmukh Department of Mechanical Engineering, Indian Institute of Technology Bombay, Mumbai, India

Amita Dev Indira Gandhi Delhi Technical University for Women, New Delhi, India

Shahab Fatima Center of Automotive Research & Tribology (Formerly ITMMEC), Indian Institute of Technology Delhi, New Delhi, India

P. V. S. Ganesh Kumar Naval Science and Technological Laboratory, Visakhapatnam, India

Swati Gautam ITMMEC, IIT Delhi, New Delhi, India

Dipak Ghosh Sir C.V. Raman Centre for Physics and Music, Jadavpur University, Kolkata, India

K. S. Girish Department of Electronics, All India Institute of Speech and Hearing, Mysuru, Karnataka, India

T. Gunasekhar JNTU Kakinada-University College of Engineering, Vizianagaram, India

Anurag Jain U.S.I.C.T, Guru Gobind Singh Indraprastha University, Dwarka, Delhi, India

T. Jayakumar Department of Speech-Language Sciences, All India Institute of Speech and Hearing, Mysuru, India

Vimala J. Kasturi Department of Speech Language Pathology, All India Institute of Speech and Hearing, Mysuru, India

V. Kavya Department of Speech-Language Pathology, All India Institute of Speech and Hearing, Mysore, India

Ahmad Ali Khan Department of Mechanical Engineering, ZHCET, AMU, Aligarh, Uttar Pradesh, India

Sidra Khanam ZHCET, Aligarh Muslim University, Aligarh, Uttar Pradesh, India;
Department of Mechnical Engineering, Aligarh Muslim University, Aligarh, India

Arun Kumar Centre for Applied Research in Electronics, Indian Institute of Technology Delhi, New Delhi, India

Nishant Kumar Thapar University, Patiala, India

Sanjay Kumar Government Senior Secondary Boys School, Phagwara, Punjab, India

S. R. Mahadeva Prasanna Department of Electrical & Electronics Engineering, IIT, Guwahati, Guwahati, Assam, India

V. Mani kumar JNTU Kakinada-University College of Engineering, Vizianagaram, India

Santosh Maruthy Department of Speech-Language Sciences, All India Institute of Speech and Hearing, Mysuru, India

A. R. Mohanty Mechanical Engineering Department, IIT Kharagpur, Kharagpur, India

Sayan Nag Department of Electrical Engineering, Jadavpur University, Kolkata, India

V. Namitha Department of Speech-Language Sciences, All India Institute of Speech & Hearing, Mysuru, India

V. Namratha Department of Speech-Language Sciences, All India Institute of Speech & Hearing, Mysuru, India

Vijay Kumar Narne Department of Clinical Research, Facility of Health, South Denmark University, Sønderborg, Denmark

Huma Noor Department of Kulliyat, Ajmal Khan Tibbiya College, A.M.U, Aligarh, UP, India

Soumyadeep Pal Department of Electrical Engineering, Jadavpur University, Kolkata, India

Debasish Panda Mechanical Engineering Department, IIT Kharagpur, Kharagpur, India

Surojit Poddar Centre for Automotive Research & Tribology (formerly ITMMEC), Indian Institute of Technology Delhi, New Delhi, India

Prashanth Prabhu All India Institute of Speech and Hearing, Mysore, Karnataka, India

V. Priyadharshini All India Institute of Speech and Hearing, Mysore, Karnataka, India

M. Pushpavathi Department of Speech-Language Pathology, All India Institute of Speech and Hearing, Mysore, Karnataka, India

Yasser Rafat Department of Mechanical Engineering, Aligarh Muslim University, Aligarh, Uttar Pradesh, India

Manish Raj Center of Automotive Research & Tribology (Formerly ITMMEC), Indian Institute of Technology Delhi, New Delhi, India

R. Rajasudhakar Department of Speech-Language Sciences, All India Institute of Speech & Hearing, Mysuru, India

V. Ramakrishna Scientist 'E', Naval Science and Technological Laboratory, Visakhapatnam, India

Sripriya Ramamoorthy Department of Mechanical Engineering, Indian Institute of Technology Bombay, Mumbai, India

D. A. Rao Naval Science and Technological Laboratory, Visakhapatnam, India

Udhayakumar Ravirose All India Institute of Speech and Hearing, Mysore, Karnataka, India

Manoj Kumar Riyal Department of Physics, Government Post Graduate College, Kotdwara, Pauri Garhwal, Uttarakhand, India

Ch. Sankara Rao Naval Science and Technological Laboratory, Visakhapatnam, India

Shankha Sanyal Sir C.V. Raman Centre for Physics and Music, Jadavpur University, Kolkata, India;
School of Languages and Linguistics, Jadavpur University, Kolkata, India

Uddalok Sarkar Department of Electrical Engineering, Jadavpur University, Kolkata, India

Ranjan Sengupta Sir C.V. Raman Centre for Physics and Music, Jadavpur University, Kolkata, India

Rashika Sharma Speech Language Pathologist, Jawahar Lal Nehru Medical College and Hospital, Ajmer, Rajasthan, India

Mahavir Singh A & V Metrology, PMM Division, CSIR-National Physical Laboratory, New Delhi, India

Shashank Shekhar Singh Department of Mechanical Engineering, JECRC, Jaipur, Rajasthan, India

P. M. Singru BITS Pilani K K Birla Goa Campus, Goa, India

Kirti Soni A & V Metrology, PMM Division, CSIR-National Physical Laboratory, New Delhi, India

Neetu Srivastava Department of Mathematics, Amrita School of Engineering, Amrita Vishwa Vidyapeetham, Bengaluru, India

S. Suryakumar Indian Institute of Technology Hyderabad, Telangana, India

Ch. Suryanarayana Scientist 'E', Naval Science and Technological Laboratory, Visakhapatnam, India

N. Tandon Centre for Automotive Research & Tribology (formerly ITMMEC), Indian Institute of Technology Delhi, New Delhi, India

Prithivi Thanikaiarasu All India Institute of Speech and Hearing, Mysore, Karnataka, India

Rajeev Kumar Upadhyay Department of Physics, V. A. Government Degree College, Atrauly District Aligarh, India

M. Vasupradaa All India Institute of Speech and Hearing, Mysore, Karnataka, India

Radhika Vathsan BITS Pilani K K Birla Goa Campus, Goa, India

B. Venkatesham Indian Institute of Technology Hyderabad, Telangana, India

Mohd Wajid Department of Electronics Engineering, ZHCET, Aligarh Muslim University, Aligarh, India;
Centre for Applied Research in Electronics, Indian Institute of Technology Delhi, New Delhi, India

Yudhisther K. Yadav A & V Metrology, PMM Division, CSIR-National Physical Laboratory, New Delhi, India

K. Yeshoda All India Institute of Speech and Hearing, Mysore, Karnataka, India

Speech Hearing and Perception

Voice as a Parameter to Assess *Mizāj*

Huma Noor, Ferasat Ali, and Yasser Rafat

Abstract The need for this article is to strengthen the observational hypothesis and to find out the correlation between voice and Mizāj (Temperament). Voice is one of the most important physiological functions of the body which has some characteristic features for the differentiation of human body on the basis of Mizāj. Science needs continuous progression in every filed. Hence, many civilizations have been working to understand human body almost on every aspect of it, namely, anatomy, physiology, biochemistry, etc. to keep them alive and healthy. In search of the same, Unani scholars found out seven basic constituents of human body, where Mizāj comes on second number. There are some parameters to diagnose temperament; physiological function is one of them. Voice is a physiological function of the body and has various characteristics which are present in all the individuals according to their Mizāj. Mizāj could be either Ḥa*r* (hot) or Barid (cold) associated with passive properties Ruṭūbat and yubusāt (moistness and dryness). To see the history and course of Mizāj with respect to voice, ancient classical Unani literature was reviewed on this topic. Voice samples were collected, and time taken to speak a specific sentence in healthy female volunteers having different types of temperament was determined. With the help of the literature, it can be concluded that Ḥar Mizāj (Hot Temperament) has stronger characteristics than Barid Mizāj (Cold Temperament) as various previous works done on different parameters of temperament proved it. Here, with respect to voice, also results are in favour of the above conclusion. The results show significant difference between different temperaments with respect to time given to them to complete the sentence for the assessment purpose.

Keywords Unani · Voice · Temperament · Ajnās 'Ashara · Voice parameters

H. Noor (✉) · F. Ali
Department of Kulliyat, Ajmal Khan Tibbiya College, A.M.U, Aligarh, UP, India
e-mail: humabums08.hn@gmail.com

Y. Rafat
Department of Mechanical Engineering, AMU, Aligarh, UP, India

M. Singh and Y. Rafat (eds.), *Recent Developments in Acoustics*, Lecture Notes in Mechanical Engineering, https://doi.org/10.1007/978-981-15-5776-7_1

Unani system of medicine has been one of the most accepted systems of medicine throughout the world until the introduction of allopath. The role of Unani medicine in the progress of medical sciences is well known. The government of India has been extending its support and funds for the multidirectional progress of Unani medicine along with other indigenous systems of medicine. India is the world leader in the Unani System of Medicine, having the widest network of educational, research, health care and pharmaceutical institutions of the system. The central council for research in Unani medicine (CCRUM), an autonomous system of government of India, under the ministry of health and family welfare is engaged in developing scientific research on various fundamental and applied aspects of Unani medicine. The growth of the council has attracted attention of the scientific community in both the country and abroad. It is heartening to note that WHO since the declaration of Alma Ata (1978) has been taking great interest in the revival of traditional medicine including Unani medicine.

1 Introduction

Unani medicine is based upon the theory of humours given by Hippocrates, which presupposes the existence of four humours in the body, namely, blood, phlegm, yellow bile and black bile. The temperament of an individual is expressed as sanguine, phlegmatic, choleric and melancholic according to predominance of the respective humours. The humours have their respective temperaments. Blood is hot and moist; phlegm is cold and moist; yellow bile is hot and dry; and black bile is cold and dry. Categorization of the subjects for any kind of study or research is the necessity of work for easiness. In various civilizations, human body and its functions were studied. The Greeks also studied the human body, with respect to its structure and functions. There are seven primary factors, viz., Arkān (Elements), Mizāj (Temperament), Akhlāt (Humours), A'ḍa (Organs), Arwāḥ (Life spirit, pneuma), Quwā (Faculty) and Afa'l (Action) which comes under Umūr Ṭabī'yya; no one can escape even a single from these seven fundamentals. Mizāj occupies second place among these factors. Human body is divided into four types of Mizāj (Temperament): Ḥar Raṭab (hot and wet), Ḥar Yabis (hot and dry), Barid Raṭab (cold and wet) and Barid Yabis (cold and dry). In these four qualities, two are active, Ḥararāt and Būrudāt (hotness, and coldness), while two are passive Ruṭūbat and yubusāt (moistness and dryness). Temperament is a vast term which encompasses all matchless traits of an individual. Every individual has a unique temperament that acquires distinctive characters. Specific temperament is inherited by every human being and encounters environmental assaults while interacting with transformed atmosphere in the second phase of life, i.e. post-natal. The significance of this predestined interaction results in acquired temperament. Every cell/tissue/organ possesses their own temperament and thus works distinctively. Functionally, organs of all human beings are comparable, yet their capacities are not identical. This functional variability results in physiological extent of variations [1].

2 The Human Temperament

About hundred trillion cells organised into different functional structures that work together as a single unit in a human body. Each cell, tissue, organ, and the whole body is governed by an innate power called Tabi'at which helps in Tadeel-e-Mizāj (homeostasis) through thousands of mechanisms for the preservation of the individual as well its species [2]. Every species has a specific temperament with the range that is confined in all the members of that species. So, human being also possesses Mizāj which is just equable to their required normal functions within a normal limit. Diseases are the result of alteration or derangement in the equable temperament. It is obvious if the temperament gets altered, and then the body functions get disturbed too.

3 Determination of Human Temperament

Ancient Unani physician described certain parameters in their manuscripts which are related to the morphological, physiological and psychological conditions of the human body. ***Jalinoos*** **(130–200 A.D)** mentioned five determinants of temperament, i.e. Head configuration, qualities of his sense of perception, how a person acts/reacts, reasoning power and function of the body [3]. ***Rabban Tabri*** **(770–850 A.D.)** had described the following five parameters: body colour, body hairs, muscles and fat, touch, and functions of five parameters: touch, complexion, hairs, physique and body function [5]. ***Mohammad Bin Zakariya Razi*** **(841–925 A.D.)** describes the following five parameters: complexion, physique, touch, functions and excreta [6]. ***Jurjani*** had described the following five determinants: touch, flesh and fat, state of organs, texture and colour and distribution of hairs, complexion [7]. According to ***Arzani***, the determination of temperament was done through sense of touch, body muscles and fat, hairs and skin complexion [8]. ***Ibn-e-Rushd*** had described the following parameters: functions of organs, body or skin complexion and physique [9]. The most accepted one is that of ***Ibn-e-Sina*** **(980–1037 A.D.)**, who had given ten parameters known as Ajnās 'Ashara. [10–12]. These are as follows: **(1)** *Malmas* (Touch): The hotness, coldness, softness and hardness of the body are to be checked by this parameter. The hotness of the skin is an indicator of hot temperament, whereas the colder skin is of cold temperament people. The quality of having soft and smooth skin indicates the moistness of temperament and vice versa. The combination of warm as well as soft skin is found in the hot andwet temperament. Those who are cold and moist in their temperament have cold and soft skin, while cold and rough skin points towards the cold and dry temperament [12]. **(2)** *Laḥm-wa-Shaḥm* (Muscles and Fat): The high amount of flesh in a person indicates heat while that of fat indicates cold. It is described in Unani literature that the excess of muscle is present in hot and moist temperament, while fat in excess amount is an indication of cold and moist temperament [13]. **(3)** *Sha'ar* (Hairs): Growth rate, i.e. rapid or slow, hair colour, texture (smooth or rough), distribution and density are seen for assessing temperament [12]. **(4)** *Lawn-e-Badan* (Body Complexion): ***Tabri*** has mentioned the dominance of a particular humour (Khilṭ) being real cause of body colour and its diversion towards the skin at the time of embryo formation. If the Ṣafrā (yellow bile) is the dominating humour, then the body colour would be yellowish; blackish colour of skin is due to Sawdā (black bile), reddish colour is due to Dam (blood) and whitish colour is due to Balgham (phlegm) [4]. Climate and zones always have their impact on skin and hair colour. One should keep this factor also in mind while assessing their colour. That is why one should not expect a person of central Africa origin to have fair skin colour. **(5)** *Haiyat-e-A'ḍā* (Physique): Indicators of hot temperament are chest to be broad, veins and joints being prominent; muscles are well-developed pulse found rapid and strong, larger extremities and tall in height. Opposite qualities of it are found in cold temperament that is narrow and small chest, shorter extremities, slow and weak pulse, hidden joints and veins, shorter stature and excessive fat deposition [12, 14]. **(6)** *Kayfiyāt-e-Infi'al* (Responsiveness of organs): Reaction of organs as well as whole body is seen through this parameter. How the body reacts towards the states of hotness, coldness, moistness or dryness? Quickness of the body in reacting to a certain state tells about the preponderance of that particular quality in the body [15]. If an organ possesses hot temperament in itself, then considerable response will be given by that organ to the stimuli which is of hot quality. Similar reaction will not be given by the cold organ [16]. **(7)** *Nawm-o-yaqẓa* (Sleep and Wakefulness): Balanced sleep and wakefulness show equability of temperament, especially of the brain. Moisture and cold results in excessive sleep, while wakefulness is due to increased heat and dryness in the brain and body, and thus results in hot and dry temperament of a person. This is because the individuals of phlegmatic temperaments have excess of sleep and people with bilious temperament having less sleep [17]. **(8)** *Afa'l-ul-A'ḍā* (Functions of the body organs): Organs work with the pace as demanded by the physiology, and then they are considered to be normal. Similarly, assessment of perfection in somatic functions indicates balanced temperament. Acceleration of functions and actions of the body beyond physiological limits becomes indicative of hot temperament of the organ and vice-versa. [17]. **(9)** *Fuzlat-e-Badan* (Excreta of the body): The body excreta are one of the transparent windows in the diagnosis of temperament in both healthy and in diseased state. The body excreta are urine, stool, sweat, various body discharges, etc. The characteristics like colour, consistency, odour, taste, etc. help in assessing temperament as they are different in different individuals [1]. Preponderance of hot qualities in the body makes excreta acrid, strong or foul smelling and deep in colour and hence is an indication of hot temperament, while less smell and dull colour indicate coldness. Due to increased body metabolic rates, there is excess sweating which also signifies body hotness. (10) *Infi'ālāt-e-Nafsāniyya* (Psychic reactions): The nervous functions are taken into consideration through this determinant. This gives information about the psychological and mental setup of an individual. The intensity of anger, depth as well as persistence of sadness and sorrow, retentive power or memory status, decision-making power, over trust, confidence, insensitivity, attention/alertness, kindness and active habits all indicates the hot temperament. A cold temperament is characterised by the qualities opposite to those mentioned before. If there is persistence of happiness, anger and good memory, dryness of temperament is the cause, while forgetfulness is due to excessive moisture [17, 18].

4 Voice as One of the Body Functions *(Afa'l-e-Badan)* and Temperament *(Mizāj)*

Afa'l-ul-A'ḍā' are important determinants of temperament. Unani physicians have always been trying to extract clues about the individual temperament by observing the functional state of the body. It is believed that the individuals having hot temperament have more rapid physiological functions because of the dominance of hotness which symbolises the active property in comparison to the body functions of the individuals of cold temperament. Classical Unani literature includes many quotes and philosophical facts about the voice in different temperaments, which shows that voice is an important body function and it had been used as a determinant of temperament in previous time. In this regard, while describing the qualities of hot temperament ***Majoosi*** mentioned that the signs of hot temperament with respect to voice are that they ***talk rapidly*** and ***do not pause during conversation.*** Similarly, regarding the cold temperament, it was mentioned that the individuals having cold temperament ***owns a heavy tongue and pauses during conversation*** [5, 9–11]. Unani Philosophers have mentioned that voice is one of the strong aspects among the functions which can be used for characterisation of the temperament. ***Strong and loud voice (within normal range), frequent speaking pattern, can speak for long***, short-tempered, rapid body movements, rapid blinking of eyes are all in support of hot temperament [5, 9, 19].

5 Voice in Unani System of Medicine

The following quotations are mentioned in Unani literature, showing voice characteristics with respect to temperament as mentioned above.

(1) "….one whose temperament is hot, the voice of that person will be loud or high and clear….." ***(Kamil-us-Sana)*** [5]. In addition, it is also mentioned in the literature that in wet temperament (*Raṭab Mizāj*) the person will be unable to make his/her speech louder. In dry temperament (*Yabis ul Mizāj*), voice would have harshness; if this kind of a person is asked to make his/her speech loud, its resemblance will be with a bird whose name is *tehu* [5, 9–11].

(2) ***RAZI*** "……loud voice indicates hot temperament while soft and low voice is the indication of clod temperament….." At another place, Razi has mentioned that "….fluent speech shows the hotness of the temperament…." ***(Kitab-ul-Mansoori)*** [6].

(3) ***GHULAM HUSAIN KANTOORI*** "……when a human body has cold temperament then the walking style of such a person is lazy, poor intelligence, he will own a heavy tongue which will make him to take pauses while conversation, his activities will be slow and he will feel a burden to do any work……" [10].

(4) ***HKM.AZAM KHAN*** "……hotness makes voice loud while coldness makes it dull or low. Dryness makes voice clear or in other words harsh, while wetness makes it quite unclear and soft….." ***(Akseer-e-AA'ḍām)*** [20].

(5) ***IBN RUSHD—"….***if the temperament of lungs is equable/moderate (moatadil), then the respiration is neither too fast nor too slow but it is normal and the voice of such a person is also moderate, neither too high (loud) nor too low (dull)……" ***(Kitab-ul- Kullīyat)*** [5, 6, 9].

The elaboration in support of the above quotes is given in literatures with some detail: Soft and low voice indicates wet temperament, while loud, clear and sharp voice indicates dry temperament. Person having wet temperament of lungs; if they start speaking with little efforts, then their voice would not be clear due to wetness of the lungs which disrupts the movement and action of trachea because of the phlegmatic secretions. Dry temperament of lungs helps to produce clear voice. Only hotness and coldness are not responsible to make voice high and low. The anatomy of trachea is also responsible for this, so if lumen of trachea is wide then it will help to produce high pitch, and the narrow lumen is responsible for low pitch quality. Coldness and hotness of trachea are temporary and come under the influence of lungs; hot temperament of lung which is given by *Ṭabī'at* widens the trachea, because hotness dilates and widens the lumen. Cold temperament of lungs contracts the lumen of trachea because *Burūda*t has the property to contract the lumen. Soft voice is due to wetness of trachea, and rough voice is a result of dryness of trachea [5].

According to *Ali Ibne Abbas Majoosi,* there are five parameters to diagnose the individual's temperament. One of them is body function: Under this parameter, he stated that *Ḥar Mizāj* person's speech will be fluent with lesser pause, and speech would be loud and clear. *Barid Mizāj* person's speech will be less fluent with more number of pauses [5]. If the functions of the body are slower than normal, it indicates coldness. Rapidity in functions of the body indicates hotness. There is a strong relationship between movement and hotness (Ḥarārat) [12].

6 View of Voice in Ayurvedic System of Medicine

Unani as well as Ayurvedic system of medicines is based on ancient principles. So, there is similarity between Ayurvedic and Unani medicine as they share similar ideas based on the basic principles. As in Unani System of medicine there are four humours in the body which are responsible for assigning the unique temperament to the person, similarly in the Ayurvedic System of medicine there are three kinds of humours, these are Vata, Pitta and Kapha which are responsible for different body types [15]. They have described the voice features according to the type of humours which are as follows:

i. **Vata**—Voice of people having vata as the prominent humour among all the tridoshas have a dry, hoarse and high-pitched or loud voice. They are fast talker, jumps from topic to topic with erratic rhythm and focus. Vata is very sensitive to excess noise.
ii. **Pitta**—People having pitta as the major humour have a loud and sharp voice. They are well focused, persuasive, dominates conversation, argumentative, challenging and due to their impatience they tend to finish other people's sentences.
iii. **Kapha**—Those who have kapha as the dominating humour in their body, they have a moist, soft, deep and calm voice. Their voice is sweet and melodious which is pleasant to listen.

Ayurveda shares the Vedic concepts like Prana, Panchakoshas, Chakras, Pancha Mahabhootas, Mind and nadis. They have described four types of Voice from gross to subtle—Vaikhari, Madhyama, Pashyanti and Para. Sound occurs in four levels and dimensions. These four levels of sound relate to frequency, fine quality, perceiving level and power as follows:

1. **Vaikhari**—The coarse voice which is the ordinary, audible and material sound. It is this level of Nada that is used in singing.
2. **Madhyama**—The mental sound with even little practice of music or Yoga. A common man can hear, understand and feel this level of Nada. To understand the concept of shruti, one should practice inner awareness to reach this level.
3. **Pashyanti**—The visualised sound.
4. **Para**—The magnificent sound [21].

On the basis of octaves and registers for voice, Ayruveda has defined six limbs or qualities of good voice. A balanced voice in all the three octaves and registers is one which is well heard (loud enough), well tuned and richly textured and smooth and that which is not harsh is sweet and harmonious. These six limbs are as follows:

i. ***Shravaka***—Voice which is loud enough, well heard by the audience even at a long distance when necessary is called *shravaka.* Even today the importance of such voice is not less, but the advanced sound projection technologies are available nowadays which has minimised its use.
ii. ***Ghana***—Voice which is very pleasing and tuned. This voice has a rich texture (especially which has a 'bass' effect).
iii. ***Snigdha***—The voice which sounds smooth, soft and sweet that which is not harsh is *snigdha.*
iv. ***Madhura***—Musical and pleasant voice is called as *madhura.*
v. ***Avadhanavan***—A voice that is sweet and harmonious, a voice which knows how to tune itself perfectly is termed as *avadhanavan.*
vi. ***Tristhanashobhi***—This voice is properly balanced in all the octaves and registers. The quality of this voice has stability in all three registers.

Sangeet Ratnakar: Sharangadeo in his text 'Sangeet Ratnakar' gives more than 30 qualities of the voice. He says that basically voice is of three types, which bear the qualities of *kapha, pitta* and *vata.* Further, he says, they can be combined into infinite qualities of *mishraka* (combination) like ***Khahula***—derived from *kapha*—having qualities of *snigdha, madhura* and *komala,* i.e. soft, sweet and tender. This type of voice in lower (*mandra*) and middle (*madhya*) octave is called as *adilla.* ***Narata***—derived from pitta—has qualities of *ghana, gambhira* and *Lina* (*asphuta*), i.e. rich textured, having good bass effect and humble. ***Bombaka***—derived from *vata*—has qualities of *nihssara, kathora* and *tara,* i.e. dry, harsh and high-pitched. ***Mishraka*** is any combination of the above three qualities [22].

The various concepts about the voice and its qualities in different traditional sciences, i.e. Unani, Ayurveda and other sciences, have been reviewed in the previous sections. The whole literature was compared and deeply analysed to look at the base for an experiment. After going through theories and concepts of the different traditional sciences, it was observed that the uniqueness of Unani system of medicine is its theory of temperament which is based on four qualities Ḥar, Barid, Raṭab and Yabis. From the above-compiled literature, it can be easily understood that the voice is an important parameter to assess temperament of a person. It has a lot of characteristics on which experimental and scientific researches should be done for the benefit of human being and such an attempt is being done by this study.

7 Methodology

A meeting was held by Institutional Ethics Committee (IEC), Ajmal Khan Tibbiya College, Faculty of Unani Medicine, Aligarh Muslim University, Aligarh on 21 June 2016, with the reference to dispatch number 215/FUM. Prof. Qayyum Husain was the chairman along with other members of the committee. The committee did not find anything objectionable/unethical as the study was totally noninvasive and did not harm any kind of ethical issue. The proposal is therefore awarded ethical clearance. Voice parameters were analysed scientifically and experimentally. In the study presented here time taken to read a sample sentence chosen as one of the parameters to assess the quality of voice among different temperaments. For this purpose, a study is carried out on healthy volunteers. Firstly, all the volunteers were categorised into four groups by assessing their temperament through a well-designed pro-forma which includes all the ten determinants essential for temperament differentiation. The criteria to assign volunteer, a specific temperament, were based on scores. The pro-forma and consent form shown in Table 4. Then, the study was done by cross-checking their temperament on the basis of their voice qualities as mentioned in literature. A sample of 69 females with no history of any voice disorders was taken. A self-made sentence was recorded (two-time repetition) in an acoustically treated room by condenser microphone and then analysed for different voice parameters by PRAAT software. The sentence was made according to the points of articulation for getting all the possible physiological human sounds during conversational speech.

After obtaining data, it was statistically interpreted by applying unpaired t-test, and significance was noticed among different groups of temperaments.

8 Statistical Analysis of Time as One of the Parameters of Voice to Assess Temperament

The data regarding time taken by female volunteers of different temperaments to speak a sentence is given below:

Among 69 females, there were 19 bilious, 23 sanguine, 22 phlegmatic and 5 melancholic, respectively, as shown in Table 1.

Table 2 shows mean and standard deviation of the time variation among different groups. Significance between groups is shown in Fig. 1.

Mean and S.D. of time in bilious, sanguine, phlegmatic and melancholic volunteers were 4.54 ± 0.43, 4.30 ± 0.52, 5.19 ± 0.74 and 4.80 ± 0.48, respectively.

The above table shows the significance of time in different temperaments. The result shows that there is statistically no significant difference between bilious to sanguine, bilious to melancholic, sanguine to melancholic and phlegmatic to melancholic, as time (which shows the rate of speech of the volunteers) taken by volunteers when compared was very much close to each other. But the difference is highly significant in case of bilious to phlegmatic and sanguine to phlegmatic because as mentioned in the literatures the people of phlegmatic temperament speaks slowly, so

Table 1 Number of females in different groups of temperaments

Sex	Bilious	Sanguine	Phlegmatic	Melancholic	Total
Female	19	23	22	05	69

Table 2 Mean and standard deviation of time in females of different temperaments

TIME	Bilious	Sanguine	Phlegmatic	Melancholic
Mean ± *S.D*	4.54 ± 0.43	4.30 ± 0.52	5.19 ± 0.74	4.80 ± 0.48

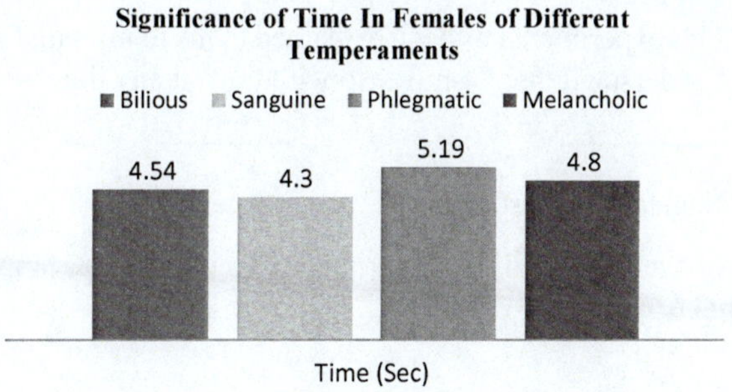

Fig. 1 Significance of time in different groups of temperament

Table 3 Comparison of time to check the significance between different groups of temperament

Comparison between	t-value	p-value	Significance
Bilious to Sanguine	1.533	0.1332	Insignificant
Bilious to Phlegmatic	3.396	0.0016	Highly significant
Bilious to Melancholic	1.188	0.2475	Insignificant
Sanguine to Phlegmatic	4.676	0.0001	Highly significant
Sanguine to Melancholic	1.957	0.0612	Insignificant
Phlegmatic to Melancholic	1.125	0.2711	Insignificant

the time taken by them was longer in comparison to others. The result is in accordance with the logic-based theory (Table 3).

Consent of The Volunteer *(The participant should complete the whole of this sheet himself/herself)*

Title of experiment: **A study to assess Different Temperaments Through Voice Analysis**

Name of the Experimenter: **Huma Noor**

Please tick boxes

1. I confirm that I have read and understood the information sheet for the above experiment. ☐
2. I have had opportunities to ask questions and my questions have fully been answered. ☐
3. I understand that my participation is voluntary and that I am free to withdraw at any time, without giving any reason. ☐
4. I have received enough information about the experiment. ☐
5. I agree to take part in the above experiment. ☐

This experiment has been explained to me to my satisfaction, and I agree to take part. I understand that I am free to withdraw at any time.

__________ __________ __________

Name of the Participant Date Signature

I have explained the experiment to the above participant and he/she has agreed to take part.

Table 4 Mizāj assessment pro-forma

Parameter (Evidence)	Damwi (Sanguineous)	Balghami (Phlegmatic)	Ṣafrawi (Bilious)	Saudāwi (Melancholic)
		MORPHOLOGICAL		
Skin texture/temperature Score:01	Warm & Smooth ☐	Soft & Moist ☐	Hard & Hot ☐	Rough & Cold ☐
Complexion Score:05	Reddish ☐	Whitish ☐	Pale ☐	Blackish ☐
Body built Score:5	Muscular ☐	Fatty ☐	Moderate ☐	Lean & Thin ☐
Texture of hairs Score:0.5	Thick & Lusty ☐	Thin & Smooth ☐	Curly ☐	Straight ☐
Growth and distribution of hairs score:0.5	Rapid, Average ☐	Slow, Scanty ☐	Moderate/profuse ☐	Excessive ☐
Colour of hairs Score:0.5	Blackish ☐	Brownish ☐	Yellow-Black (Golden) ☐	Black & White (Mixed) ☐
		PHYSIOLOGICAL		
Urine Score:01	Moderate in Quantity ☐	White, More in Quantity ☐	Yellow, Less in Quantity ☐	Turbid, Less in Quantity ☐
Tolerate well Score:01	Dryness ☐	Summer ☐	Cold ☐	Dampness ☐
Remains well in Score:03	Spring ☐	Summer ☐	Winter ☐	Autumn ☐
Appetite Score:01	Strong Appetite (Can Skip a meal) ☐	Less Appetite (feel heaviness after eating) ☐	Strong Appetite (Can't Skip A Meal) ☐	Irregular Appetite ☐
Thirst Score:01	Average (++) ☐	Poor (+) ☐	Increased (++++) ☐	Low (+++) ☐
Digestion Score:01	Average ☐	Slow ☐	Strong ☐	Irregular ☐
Movements and activities Score:03	Average in Physical Activity ☐	Dull Laziness ☐	Brisk, Hyperactive ☐	Less ☐
Sleep Score:01	Average ☐	Excess ☐	Disturbed Sleep ☐	Insomnia ☐
		PSYCHOLOGICAL		
Dream Score:01	Blood, Red Objects ☐	Water, Snow ☐	Fire, Yellow Objects ☐	Black, Fearful Dreams ☐
Anger /joy Score:01	Comes On Easily & Easily Lost ☐	Comes On Hardly ☐	Frequent, Severe & Persists For Long ☐	Infrequent But Persist ☐
Response to external stimuli in adverse condition Score:01	Aggressively Respond ☐	Weekly Respond ☐	Bravely Respond ☐	Cowardly Respond ☐
Decision taking power Score:01	Take Boldly ☐	Hesitate in Taking Decisions ☐	Take Quickly ☐	Afraid in Taking Decisions ☐
Memory Score:01	Good, Retention Also Good ☐	Not Good ☐	Good, But Can't Retain For Long ☐	Can't Learn Easily But Excellent Retention ☐

Huma Noor

__________ __________ __________

Name of the Experimenter Date Signature

Name of the volunteer ……….. Father's name …….. Age….. Gender……… Height …..

Weight ……….. Pulse …… B. P ……. Occupation……. Mobile no :…… Address ………..

Total Collection:——— Damwi:——— Safrawi:——— Balghami:——— Sawdāwi:———

Diagnosed Temperament:——— **Signature of the Investigator:**———

9 Conclusion

This study is done to ascertain some updated parameters to diagnose the human temperament with respect to voice. Those parameters of voice were taken into consideration which are measurable and can be obtained digitally with maximum ease. At the same time, it was also an objective that these all parameters must be very precise and easily applicable; also they must be free of bias and errors. Human temperament is a very unique aspect of Unani system of medicine. A false diagnosis of temperament will be misleading or may be of no use as it is not helpful in curing diseases or bringing back the misbalanced temperament to the equable temperament. For this, above result of the digital examination of voice is giving direction to go to quantitative aspect of voice instead of qualitative which is more correct and validated.

References

1. Zaidi IH, Zulkifle M (1999) Temperamentology: a scientific appraisal of human temperament, 1st edn. Aligarh, pp 6–13, 18–23, 64–65
2. Mehdi HS (1973) A Rational interpretation of the four cosmic element says operating in Alchemy, theories and philosophy of medicine, 2nd edn. IHMMR, New Delhi, pp 209
3. Jalinoos (YNM) Kitab Fil Mizāj, edited and translated by Prof. Syed Zillur Rahman, Published by Ibne Sina Academy, Aligarh, U.P., Printed At: International Printing Press, Aligarh, U.P., P 137
4. Tabri R (2002) Firdaus ul Hikmat. (Urdu translation by Hakeem Mohd Awwal Shah Sambhali). New Delhi. Idara Kitab ul Shifa, pp 58–59, 86, 251
5. Majoosi AIA (2010) Kamil-us-Sana, Urdu translation by Kantoori GH, Idara Kitab-ul-Shifa. New Delhi, pp 24, 25, 41–42, 45–46, 110, 190–191
6. Zakariya R (1991) Kitab Al Mansoori, Urdu translation, CCRUM. New Delhi, pp 59–80
7. Jurjani AH (1902) Zakhira Khwarizam Shahi, vol 1, Urdu translation by Hadi Hasan Khan. Lukhnow, Matba Nawal Kishore. pp 14, 18, 25, 235, 242

8. Azrani A (1939) Ikseerul Quloob, Urdu translation of Mufarrehul Quloob by Baqar, MS, Matba Munshi Nawal Kishore. Lukhnow, pp 54, 308, 317
9. Rushd I (1980) Kitab-Al-Kullīyat, Urdu Translation, CCRUM, New Delhi, pp 35, 156, 157, 159
10. Sina I (2010) Al Qanoon Fil Tibb, Urdu translation by Kantoori, GH, Idara Kitab Ul Shifa. New Delhi, pp 83, 133–138
11. Sina I (1993) Al Qanoon Fil Tibb, Book 1, English translation of critical Arabic Text, Jamia Hamdard. New Delhi, pp 7–13, 65, 190–197
12. Nafees B (1954) Kullīyat-e-Nafisi, vol 1. (Urdu Translation By HKM. Kabiruddin), Idara Kitabus Shifa. New Delhi, pp 136, 266–280
13. Gruner OC (1983) The four temperament Avicenna's tract on cardiac drugs and essays on arab cardio therapy. IHMMR, New Delhi, pp 119–123
14. Shah MH (2007) The general principles of Avicenna's Canon of medicine, Idara Kitab ul Shifa. New Delhi, pp 24–29, 156–192
15. Ahmed SI (1980) Introduction To Al-Umur Al- Tabiyah, Nuzhat Ishtiaq, pp 27–40
16. Hock RC, The four temperament, The Angelicum Academy, Cited From http://Angelicum.Net/Newsroom/The-Four-Temperament-By-Revcarrad-Hock
17. Shah MH (1972) Temperament, explanation and interpretation philosophy of medicine and science problems and perspective, Compiled by department of philosophy of medicine and science. New Delhi, IHMMR, pp 123–128, 109–111
18. Azmi AA (1993) Basic concepts of unani medicine-a critical study, 1st edn. Department of history of medicine, Jamia Hamdard, New Delhi, pp 57–59, 73–79
19. Ahmed SI (1983) Kullīyate Asri, Mrs Nuzhat Ishtiaq, Staff Quarter, Tibbiya College, Karol Bagh, New Delhi, pp 78
20. Khan A (2010) Ikseere A'ḍām, Published by Idara Kitab ul Shifa. New Delhi, pp 246–247
21. Gokhale MK, Applied yoga and ayurveda for voice and singing
22. Sreelekshmi G (2015) Udana Vayu-physiological understanding, review article. Int Ayurvedic Med J 3(8). ISSN 2320–5091

Neuro-Physiological Correlates of Voice Onset Time in Kannada and Tamil Speaking

T. Jayakumar and Vijay Kumar Narne

Abstract Voice Onset Time (VOT) refers to the time difference between two events—one articulatory and the other laryngeal. VOT is measured as the time difference between the onset of the burst (articulatory event) and the onset of vocal fold vibration (laryngeal event) Lisker [14]. The cross-linguistic variation with respect to voicing perception is more in multilingual countries like India. For example, in Tamil language, there is no good distinction between voiced and unvoiced stops; however, in Kannada, Telugu, and Hindi, there exists a difference. To verify the cross-linguistic difference, the behavioral response for VOT continuum in native Kannada and Tamil speakers, and the neuro-physiological changes for VOT continuum in native Kannada and Tamil speakers were taken as objective of the study. Two groups of subjects were participated in the study. Group I consisted of ten male Kannada speakers in the age range of 20–35 years. All the subjects were native speakers of Kannada language, and they belong to Mysore dialect and they are fluent speakers in Kannada only. All though they were exposed to English, they were not fluent speakers of English or Hindi. Group II consisted of ten male Tamil speakers in the age range of 20–35. All the subjects were native speakers of Tamil language, and they belong to Coimbatore/Chennai dialect and they are fluent speakers in Tamil only. Although they were exposed to English, they were not fluent speakers of English or Kannada or Hindi. From the naturally recorded speech sounds, /da-ta/ continuum was created using copy past synthesis method. This continuum had 10 tokens. Using this stimulus, behavioral identification curve was generated. Also electrophysiological, N100 potential was recorded using Neuro-Scan instrument (Compumedies, AUS) with five tokens, which covers the entire dynamic range of the /da-ta/ continuum. The recording was done with 16 channels. The behavioral result showed there was difference between Tamil and Kannada languages; however, electrophysiological results showed that there was a change with N100 latency with changing VOT but there was no language

T. Jayakumar (✉)
Department of Speech-Language Sciences, All India Institute of Speech and Hearing, Mysuru, India
e-mail: jayakumar82@gmail.com

V. K. Narne
Department of Clinical Research, Facility of Health, South Denmark University, Sønderborg, Denmark

M. Singh and Y. Rafat (eds.), *Recent Developments in Acoustics*, Lecture Notes in Mechanical Engineering, https://doi.org/10.1007/978-981-15-5776-7_2

difference found. The author concluded that N100 may not be an effective measure to indicate to represent the voice /voiceless categorical perception. May be higher potential might give better information.

1 Introduction

The term voice onset time was coined and described by Abramson and Lisker [2]. Voice Onset Time (VOT) refers to the time difference between two events—one articulatory and the other laryngeal. VOT is measured as the time difference between the onset of the burst (articulatory event) and the onset of vocal fold vibration (laryngeal event) Lisker and Abramson [1]. In case of voiced stop consonants, voicing starts before the onset of the burst, and hence VOT is negative or leads VOT. In case of unvoiced stop consonants, voicing starts after the burst, and hence VOT is positive or lags VOT. It is known that VOT is one of the cues in the perception of voicing in stop consonants Liberman et al. (1982), [3], Shankweiler (1961), Williams [4, 5], Savithri et al. [6], Satya [7]. However, the change in the percept from voiced to unvoiced stop consonant depends upon the classification of stop consonants in a language. For example, in English, the percept changes at +20 ms VOT, whereas in Kannada and Telugu, the percept change at lead VOT and in Tamil the voicing contrast is negligible in spoken language.

Kannada language is one of the important Dravidian languages of India, spoken majorly in the state of Karnataka. Kannada language is native language of people who live in the State of Karnataka, India. Their population is roughly 38 million, making it the 27th most spoken language in the world. It is one of the scheduled languages of India and the official and administrative language of the state of Karnataka (http://www.en.wikipedia.org/wiki/Kannada_language), and Tamil is a major Dravidian language spoken by Tamil people of the Indian subcontinent. It has official language status in the Indian state of Tamil Nadu and in the union territory of Puducherry. Their population is roughly around 66 millions of people. (http://www.en.wikipedia.org/wiki/Tamil_language).

Few researchers investigated the neural encoding of VOT in the brain through auditory evoked potentials. Sharma and Dorman [8] measured cortical auditory evoked potential (CAEP) in conjunction with behavioral perception of /da- ta/ continuum. Neuro-physiological correlates of VOT encoding were investigated using the N1 CAEP which reflects sensory encoding of stimulus features, and they found distinct changes in N1 morphology which was related to VOT encoding. For the stimuli that were behaviorally identified as /da/, a single negativity (N1) was apparent; however, for stimuli identified as /ta/, two distinct negativities (N1 and N1') were apparent. This difference in N1 morphology seen in the region of the /da-ta/ phonetic boundary appears to provide neuro-physiological correlates of categorical perception for VOT.

Since speech perception abilities are altered by experience with a particular language and the lack of experience with a particular phonetic contrast has the effect of reducing sensitivity to contrast, Sharma and Dorman [9] investigated neurophysiologic correlates of VOT perception of native and nonnative phonetic categories. Using Hindi and English listeners in response to a /ba-ta/ continuum of naturally produced bilabial CV stimuli that were different in VOT from –90 to 0 ms. The changes in N1 latency which reflected the duration of pre-voicing across the stimulus continuum were not significantly different for Hindi and English listeners. On the other hand, in response to the /ba–pa/ stimulus contrast, a robust MMN was seen only in Hindi listeners and not in English listeners. These results suggested that neurophysiologic levels of stimulus processing reflected by the MMN and N1 are differentially altered by linguistic experience. The authors also checked the reliability of N1 onset response from a single to double peak using /ga-ka/ and /ba-pa/ continuum [9]. Behavioral identification result from ten subjects revealed a mean category boundary at a VOT of 46 ms for the /ga-ka/ continuum and at a VOT of 27.5 ms for the /ba-pa/ continuum. ERP showed single N1 component for stimuli with VOT of 40 ms and more for both continua. That is, the change in N1 morphology (from single to double) coincided with the change in the perception from voiced to unvoiced for stimuli from /ba-pa/ continuum, but not for stimuli from the /ga-ka/ continuum. Horev et al. [10] also researched on the behavioral measures and auditory event-related potentials (ERPs) were measured from 14 normal-hearing Hebrew speakers. This speaker voicing distinction is different from English, during identification and discrimination stimuli: a VOT continuum, created by editing natural productions of /ba/ and /pa/. The above research results supported the voicing perception, and voicing boundaries are controlled mainly by linguistic experience rather than by innate temporal sensitivity.

Above studies indicate the evidence for neuro-physiological change concurrent to change in phonetic perception. However, first the reliability of the N1 double peak (change in N1 morphology) still needs investigation. Second, the differences in the properties of stop consonants could bring about a change in the behavioral and neuro-physiological response. Third, the cross-linguistic variation with respect to voicing perception is more in multilingual countries like India.

In the present study, Kannada and Tamil languages were selected. These two languages are specifically selected because Kannada is different from English in that it is a four-way language. Therefore, any differences in the results could be attributed to the difference in language especially with respect to any speech continuum. Second, Tamil is selected because the voicing contrast is poor in this language.

The objective of the present study will be to investigate (a) the behavioral response for VOT continuum in native Kannada and Tamil speakers, (b) the neurophysiological changes for VOT continuum in native Kannada and Tamil speakers, and (c) the existence and the reliability of double peak (morphological change (N1, N1’) with the change in stimulus property.

2 Method

The present study got clearance from AIISH ethical committee (All India Institute of Speech and Hearing, Mysuru), and written consent was taken from individual participants of the study before starting the study.

I. **Behavioral experiment**

Subjects: Two groups of subjects were participated in the study.

Group I consisted of ten male Kannada speakers in the age range of 20–35 years (Mean age 26.8 years). All the subjects were native speakers of Kannada language, and they belong to Mysore dialect and they are fluent speakers in Kannada only. Although they were exposed to English, they were not fluent speakers of English or Hindi.

Group II consisted of ten male Tamil speakers in the age range of 20–35 years (Mean age 25.3 years). All the subjects were native speakers of Tamil language and they belong to Coimbatore/Chennai dialect and they are fluent speakers in Tamil only. Although they were exposed to English, they were not fluent speakers of English or Kannada or Hindi.

Common criteria for Group I and II

- All Subjects had completed minimum of 10+2 or PUC.
- All subjects had completed their schooling (till 10th) in their native languages (Group I is in Kannada and group II is in Tamil).
- None of the subjects had history of speech and hearing disorders which was verified through the oral interview.
- Ling sound test was used to screen their hearing by the audiologist and speech language pathologist.

Stimuli: Stop consonants, retroflex voiced /da./, retroflex unvoiced unaspirated /ta/, were selected to check the voicing contrast in both the groups. These stop consonants were uttered five times by a normal female Hindi speaker who has audio-recorded and stored into the computer memory. Authors had selected more intelligible and clearly articulated speech sounds based on the perpetual judgment and spectrogram. Figure 2 shows the waveform of the selected speech sounds for the purpose of making speech continuum.

Using the waveform display of the PRAAT software, voice onset time (VOT) was measured. VOT was measured as the time difference between the onset of burst and the onset of voicing. The measured VOT was used to make the continuum for behavioral experiment. VOT was truncated from the voiced stops. However, it was made sure that VOT always starts with zero cross only. When VOT was 0 ms or it reaches the burst, silence was introduced between burst and onset of voicing for the following vowel. Each modified stimulus was called as token. Each continuum had 10 number of tokens. In this manner, there were two continua. The tokens in each continuum were equated in intensity and iterated thrice. An inter-stimulus interval

Table 1 VOT details of the /da-ta/ continuum

/da-ta/ continuum	
Token	VOT
Token1	−60
Token2	−45
Token3	−25
Token4	−10
Token5	0
Token6	+10
Token7	+20
Token8	+30
Token9	+40
Token10	+50

+ indicates lag VOT and – indicates lead VOT

of 500 ms was maintained using Adobe Audition Version 3 software. Table 1 shows the VOT details of the /da-ta/ continuum and procedure.

Procedure

All subjects were tested individually. They were instructed to listen to the edited tokens (CV syllables) through headphones and identify each token as a voiced or unvoiced; otherwise, voice or unvoice aspirated in a binary-forced choice format. These experiments were carried out in the sound-treated room. Subjects were given an initial practice session. Each continuum token was randomized, and three sets were made. Each subject listened to each token three times in a random order for each continuum.

Analysis

The response of the subject for all the binary-forced choice listening task was averaged, and percent response was calculated using the following formula:

$$\text{For example, } /\text{ga-ta}/\text{continuum} = \frac{\text{Obtained}/\text{ga}/\text{or}/\text{ta}/\text{response for each token}}{\text{Total no. of iteration ofthe tokens}} \times 100$$

The percent identification response for the target syllables was tabulated for each subject for VOT continuum and based on this, identification curve was drawn. Fifty percent crossover, lower limit of phoneme boundary (LLPB), upper limit of phoneme boundary (ULPB), and phoneme boundary width (PBW) were obtained (Doughty 1949) from identification curve. Fifty percent crossover is the point at which 50% of the subject's response corresponds to the voiced (unvoiced or unvoiced aspirated) category. Lower limit of phoneme boundary width is the point along the acoustic cue continuum where an individual identified voiced (voiceless or voiceless aspirated)

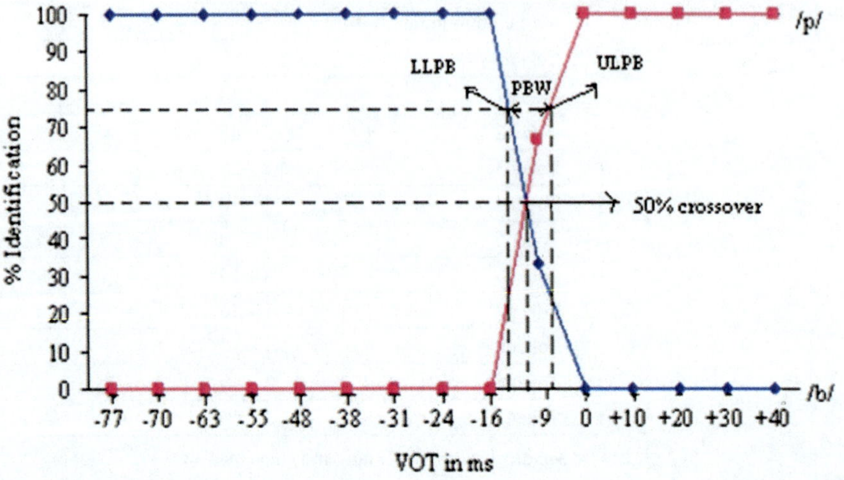

Fig. 1 Illustration of 50% crossover, lower limit, and upper limit of phoneme boundary on an identification curve

75% of the time and upper limit of phoneme boundary width defined as the corresponding point of the identification of the unvoiced or unvoiced aspirated (voiced) 75% of the time. The VOT value corresponding to 75% identification, immediately before/after the 50% crossover, was considered as LLPB/ULPB for both single and multiple crossovers. Phoneme boundary width was determined by subtracting the lower limit from upper limit of boundary width. Figure 1 illustrates the identification curve with x-axis representing VOT and y-axis representing the percentage of identification.

II. **Electrophysiologic experiment**

Subjects: The same subjects who participated in the behavioral experiment were involved in electrophysiologic experiment.

Stimuli: The same stimulus used for behavioral experiment (/da-ta/ continuum) was used for electrophysiological experiment. However, in behavioral experiment, each speech stimuli continuum had 10–13 tokens but for the electrophysiological experiment only five tokens were selected which covered the entire dynamic range of continuum. Similarly, for non-speech stimuli, only three tokens were selected. Repeated presentation of each token separated by an inter-stimulus interval of 800 ms was used to elicit the N1. The order of presentation of stimuli will be counterbalanced across subjects. Tables 2 and 3 show the details of the tokens used for each continuum in electrophysiological experiment, and Figs. 2, 3, 4, 5, 6, 7, 8, 9 show the waveform of the tokens used for electrophysiological experiment.

Table 2 Details of the tokens used for speech continuum in electrophysiological experiment

/da-ta/ continuum	
Token	VOT
Token 1	−60
Token 2	−25
Token 3	0
Token 4	+30
Token 5	+50

Table 3 Comparison of /da-ta/ continuum response across Kannada and Tamil

Language	Identification curve for /da-ta/ continuum			
	50% crossover	Phoneme boundary width (PBW)	Lower limit of phoneme boundary (LLPB),	Upper limit of phoneme boundary (ULPB)
Kannada	−09 VOT	31 ms	−27 VOT	04 VOT
Tamil	03 VOT	36 ms	−25 VOT	11 VOT

Recording Procedure

N1-evoked potential component was recorded using a Neuro-Scan (Compumedics, USA). The responses were recorded using Scan 4.3 software. Quick Cap silver chloride electrodes were used to obtain 16 channel recordings. ERPs were recorded from standard electrode sites over both hemispheres (International 10–20 System sites Cz, C4, C3, T8, T7, C5, C6, FCz, Fz, Pz, CP4, CP3, FC4, FC5, FT7, and FT8), referenced to the left mastoid during recording. Impedance was 5 kΩ or less at all sites. The stimuli were presented binaurally through calibrated ER 3A insert receiver at 75 dB SPL. Figure 3 shows the 16-channel montage of electrodes on scalp-based 10-20 system.

Subjects were seated comfortably in a sound-treated room. They were asked to fix their eyes at one position and were instructed to refrain from making large eye movements. Evoked responses were bandpass filtered online from 0.1 to 30 Hz. The recording window includes a 100-ms pre-stimulus period and 600 ms post-stimulus time. To control for arousal, subjects were made to watch silent videotaped movies with subtitles of their choice. The stimulus (tokens) was presented binaurally through an insert earphone at 75 dB SPL. Evoked responses were bandpass filtered online from 0.1 to 100 Hz. Responses that are judged noisy or that are greater than 100 mV was rejected offline. For the N1 recordings, 300 sweeps elicited in response to each stimulus (tokens) from the continuum (voiced to unvoiced and unvoiced aspirated) were collected.

Data analysis

Data analysis was carried out individually for each subject. For individual subjects, sweeps were averaged to compute an individual average waveform. All the continuous waveforms (.cnt) recorded through Scan 4.3 were first preprocessed through

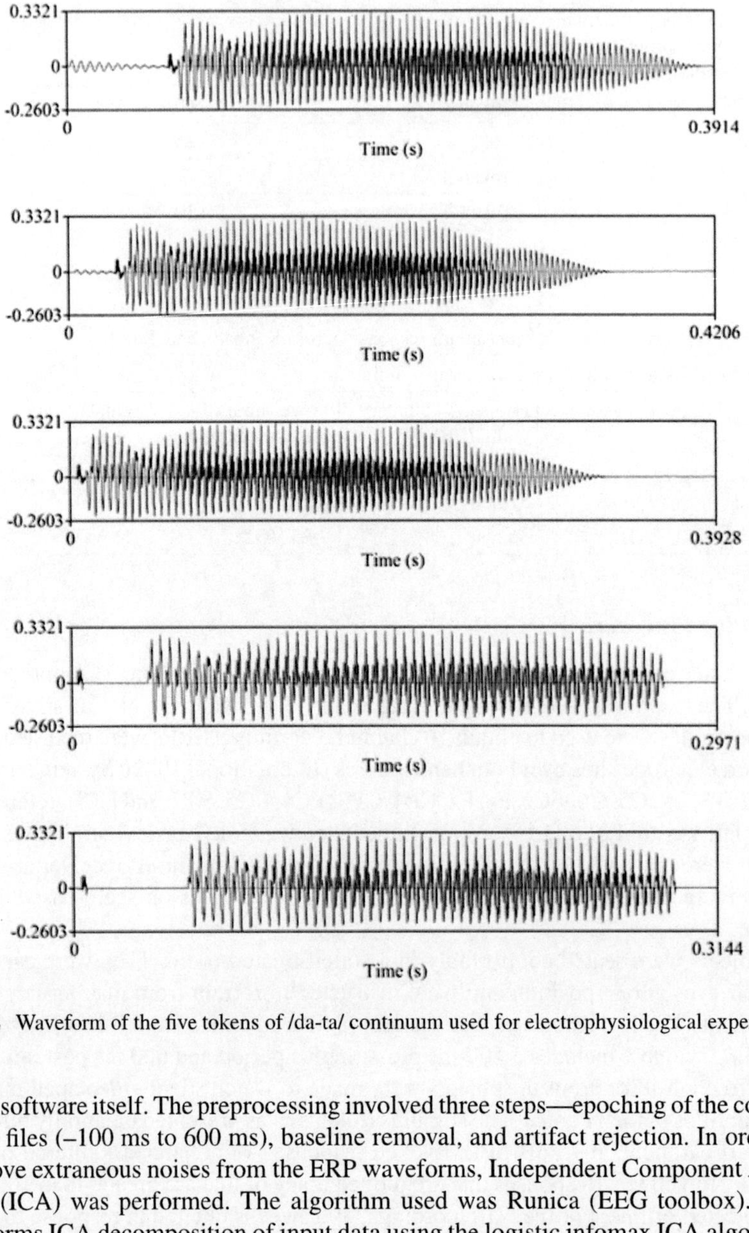

Fig. 2 Waveform of the five tokens of /da-ta/ continuum used for electrophysiological experiment

scan software itself. The preprocessing involved three steps—epoching of the continuous files (–100 ms to 600 ms), baseline removal, and artifact rejection. In order to remove extraneous noises from the ERP waveforms, Independent Component Analysis (ICA) was performed. The algorithm used was Runica (EEG toolbox). This performs ICA decomposition of input data using the logistic infomax ICA algorithm by Bell and Sejnowski (1995). The components which were noisy were removed from the ERP including eye and scale muscle artifices. All 16 waveforms (potentials) were digitally high-pass filtered offline at 4 Hz (filter slope12 dB/octave). In the group mean waveforms, N1 was identified visually as a prominent negative peak within the first half of the time window. In order to aid in peak identification and

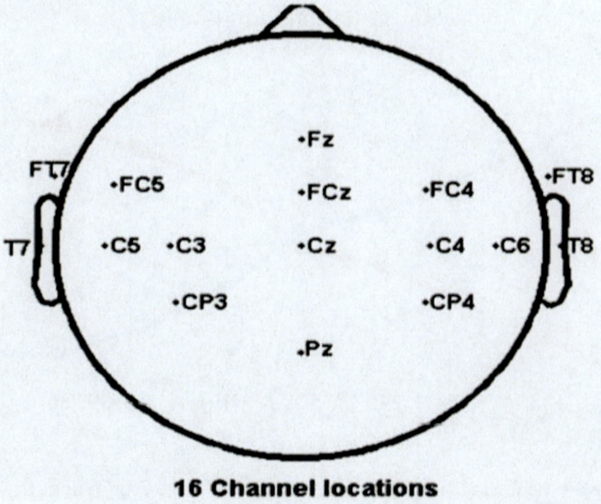

Fig. 3 Montage of electrodes on scalp

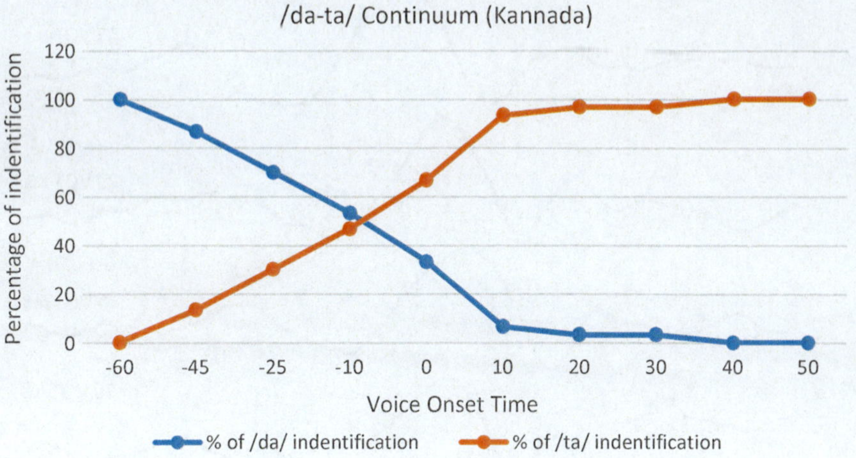

Fig. 4 Identification curve for /da-ta/ continuum for Kannada language

measurement in data from individual subjects, response windows of 150 ms were created around the peak in the group mean waveforms. Peak latencies and amplitude were detected based on the auto peak detection program from Scan 4 software. When the waveform contains a double peak within the defined time frame, the latency and amplitude were measured at the prominent peak visually. Group averaged waveforms were computed by averaging across the individual average waveforms. The 16 channels were grouped into 3 groups—average auditory potential across electrodes within each of three regions of interests (ROIs): Right channels, center channels, and left channels. The channels were grouped as follows—Right channels (C4, T8, C6,

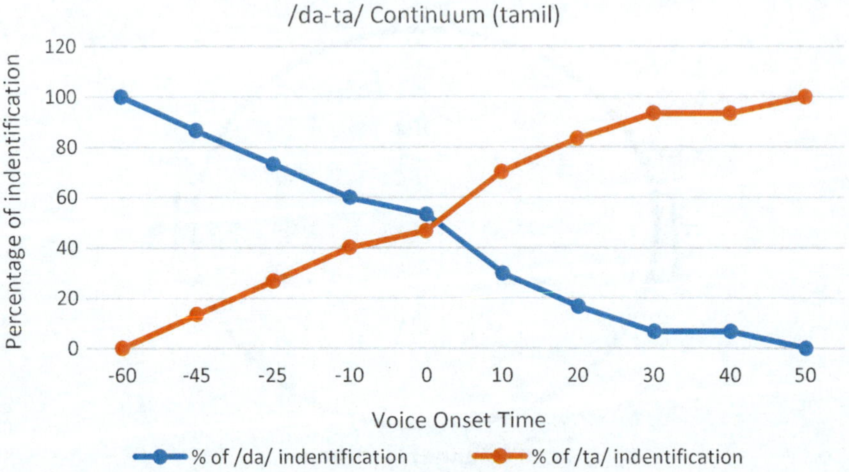

Fig. 5 Identification curve for /da-ta/ continuum for Tamil language

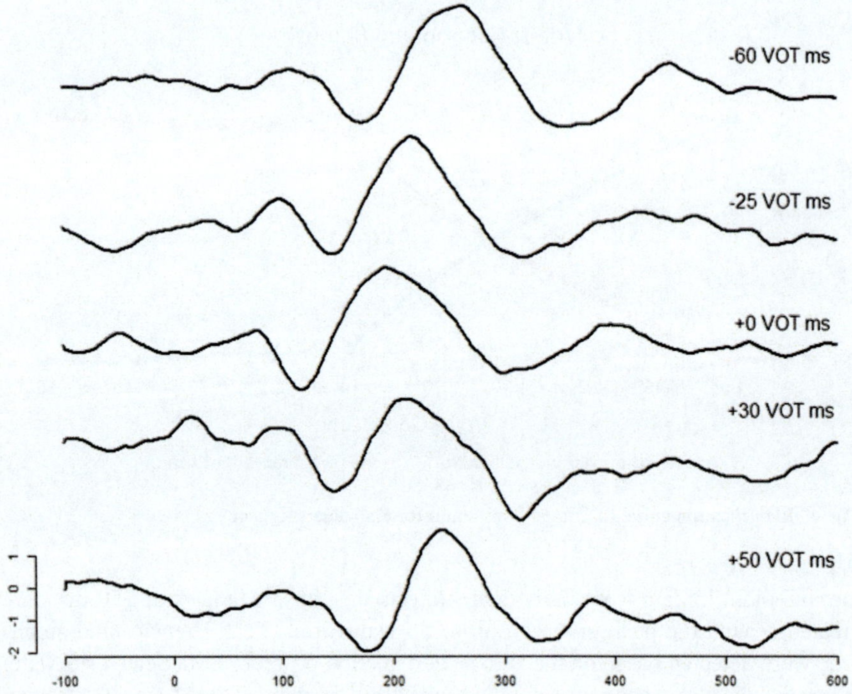

Fig. 6 N1 potential at Cz electrode for each token of the /da-ta/ continuum for Kannada

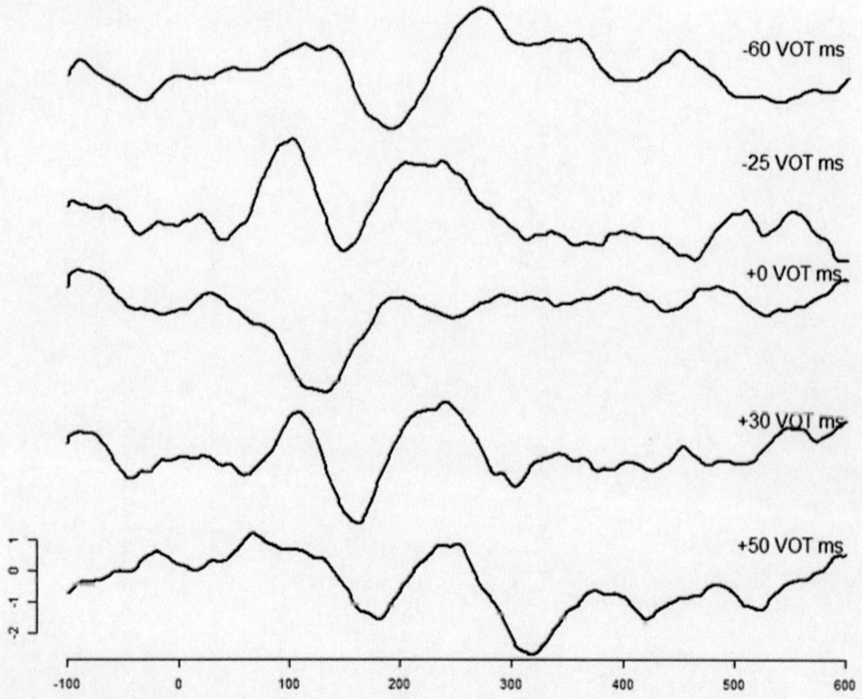

Fig. 7 N1 potential at Cz electrode for each token of the /da-ta/ continuum for Tamil

CP4, FC4, FT8), center channels (Fz, FCz, Cz, Pz), and left channels (C3, T7, C5, CP3, FC5, FT7). The rationale for including three separate ROIs was to check any difference exist between these three places which were already investigated by the Daliri and Max (2015a, b). Those researchers showed that the effect was larger at electrodes located over the central region and over the left hemisphere than over the right hemisphere (Daliri and Max 2015a). Hence, the present study also tried to find effect if any in N100 potential. It is important to note that there is no straightforward correlation between the neural generators of the EEG signal and the scalp location of surface electrodes where maximally strong signals (Michel et al., 2004).

Statistical analysis: For electrophysiological analysis, Shapiro–Wilk test was used to check the normality. Descriptive statistics (Mean, Median, and SD) was used to describe the basic data variance. Mann–Whitney U test was used to compare the N100 latency and amplitude across the two groups of participants for different tokens and this was performed separately for different regions of interests (ROIs), that is, right, center, and left scalp position. Friedman test was used to compare across region of interests (ROIs) for N100 latency and amplitude, and the same statistic test (Friedman) also was used to compare across the tokens for N100 latency and amplitude. The variables which showed significant difference in Friedman test have subjected to pairwise comparison using Wilcoxon signed-rank test. All the above

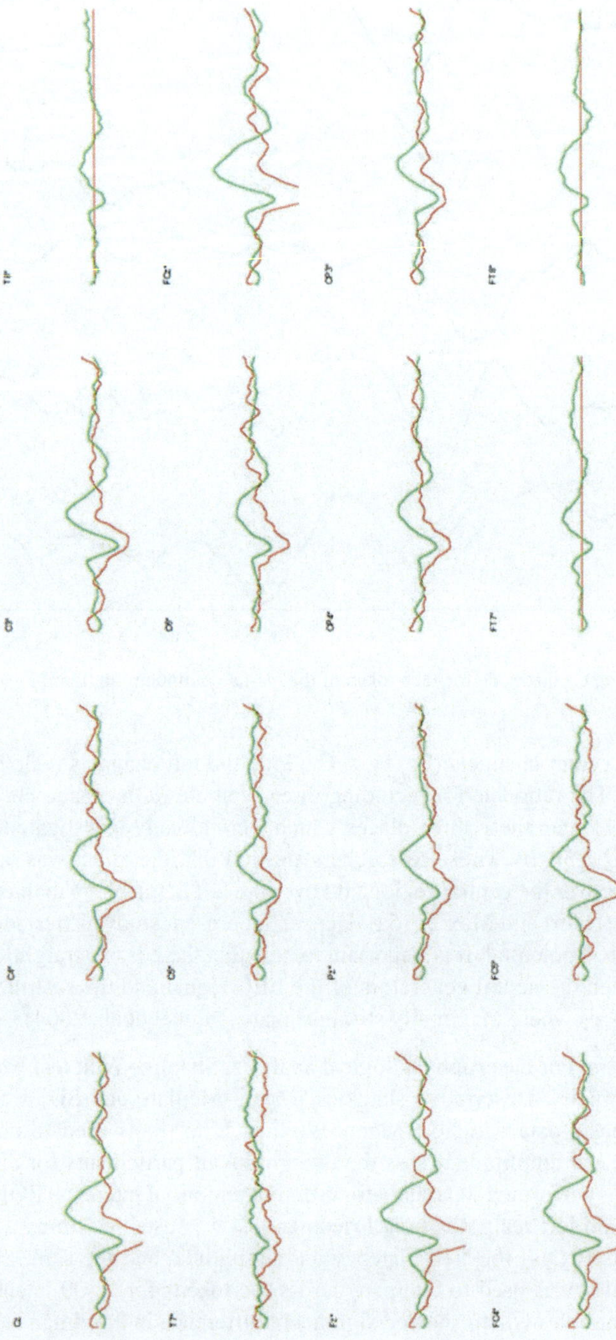

Fig. 8 N1 potential of /da-ta/ continuum for 0 VOT token of Kannada and Tamil (Green–Kannada and Tamil–Red)

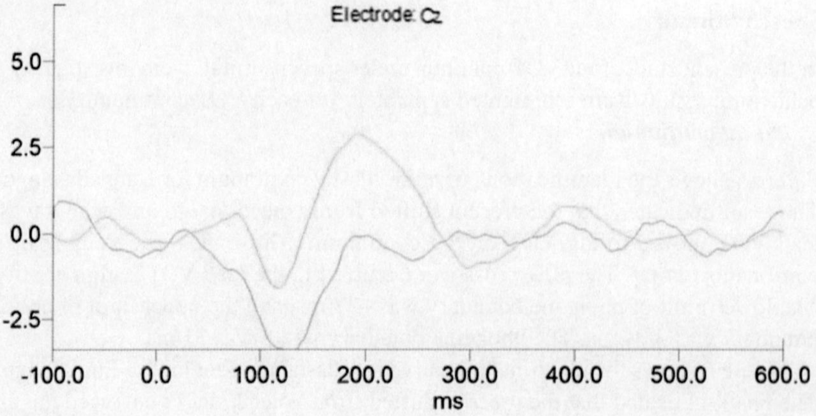

Fig. 9 N1 potential at Cz electrode of /da-ta/ continuum for 0 VOT token of Kannada and Tamil (Green–Kannada and Tamil–Red)

statistical tests were separately done for each speech continuum and non-speech continuum.

3 Results

The present study aimed to investigate the behavioral response changes in relation to voice onset time (VOT) and neuro-physiological changes in relation to voice onset time by measuring N1-evoked potential. Main objectives of the study were (i) behavioral response for VOT continuum in Kannada and Tamil speakers, (ii) neuro-physiological changes for VOT continuum in Kannada and Tamil speakers, (iii) the existence and the reliability of double peak (morphological change in N1) with the change in stimulus property (VOT), and (iv) the neuro-physiological changes for non-speech stimuli that mimic the speech stimuli. The results of the study were presented under the following titles:

Behavioral experiment
Electrophysiological experiment

Behavioral experiment

In behavioral experiment, VOT continua (/da-ta/) were used. A binary-forced choice identification task was used. Participants were instructed to listen to the tokens in the continuum and provide the response as which end of the continuum is heard. Each participant listens the same token three times randomly. The responses of the binary-forced choice identification task were averaged for each participant, and all the participants' responses were averaged to draw an identification curve for VOT continuum across the languages.

Speech Stimuli

In the present study, four VOT continua under speech stimuli were investigated. The behavioral results were represented separately for each VOT continuum.

/da-ta/ continuum

Figure 5 shows the identification curve for /da-ta/ continuum for Kannada language. The result indicated that the precept shifted from voiced /da/ to unvoiced /ta/ as the lead VOT moved to lag VOT on the continuum. Thus, VOT cue voicing of stop consonants /da-ta/. The 50% crossover occurred in the lead VOT region at –09 ms. The lower limit of phoneme boundary was –27 ms, and the upper limit of phoneme boundary was +04 ms. The phoneme boundary width was 31 ms.

Figure 6 shows the identification curve for /da-ta/ continuum for Tamil language. The result indicated that the precept shifted from voiced /da/ to unvoiced /ta/ as the lead VOT moved to lag VOT on the continuum. Thus, VOT cue voicing of stop consonants /da-ta/. The 50% crossover occurred in the lag VOT region at +03 ms. The lower limit of phoneme boundary was –25 ms, and the upper limit of phoneme boundary was +11 ms. The phoneme boundary width was 36 ms. Table 3 shows the comparison of /da-ta/ continuum response across Kannada and Tamil.

Electrophysiological experiment

The same VOT continuum used in the behavioral experiment was used in the electrophysiological experiment, that is, two VOT continua (/da-ta/, /da-t^ha) were used. However, the number of tokens used in the VOT continuum was reduced to five tokens for each continuum. These five tokens were covering the entire dynamic range of that particular VOT continuum.

The electrophysiological data were subjected to the normality test using Shapiro–Wilk test. The data was categorized for Kannada and Tamil for N100 latency and amplitude. Also, it was further sub-divided for five different tokens and three regions of interests (ROIs), that is, center, right, and left scalp positions. The similar type of normality check was done for each VOT continuum separately. Results showed that nearly half of the data falls under normal distribution (where $p > 0.05$) and other half falls under non-normal distribution (where $p < 0.05$). Since nearly half of data were non-normal distribution and the number of participants in the present study was ten individuals for each group (Kannada and Tamil), the non-parametric tests were selected for further statistical analysis. Similar to behavioral experiment, electrophysiological experiment results were also presented for each VOT continuum separately.

VOT continuum of /da-ta/

The descriptive quantitative result of N1 (N100) obtained from the electrophysiological experiment for /da-ta/VOT continuum was given in Tables 4 and 5 for Kannada and Tamil, respectively. The table shows the mean median and standard deviation values. Result of Tables 6 and 7 shows that N1 latency was different for all the tokens in the continuum. This variation was present in both the languages. However, there was no noticeable difference for N1 amplitude for all the tokens in the continuum.

Table 4 Mean, median, and standard deviation of /da-ta/ continuum for Kannada language

/da-ta/ continuum–Kannada

Stimulus		N1-latency			N1-amplitude		
		Mean	Median	SD	Mean	Median	SD
Central	Token 1	171.91	172.08	15.45	−2.54	−2.45	1.25
	Token 2	131.73	132.02	17.45	−1.61	−1.35	1.50
	Token 3	118.45	119.00	18.93	−2.32	−1.93	2.12
	Token 4	143.75	147.38	13.75	−2.43	−2.24	1.87
	Token 5	152.59	152.05	21.24	−2.69	−2.73	1.92
Left	Token 1	173.03	174.50	16.83	−2.33	−2.04	1.21
	Token 2	137.74	141.61	24.83	−1.69	−1.08	1.44
	Token 3	119.62	116.91	14.89	−2.26	−1.91	1.76
	Token 4	140.68	145.12	18.67	−1.84	−2.04	1.32
	Token 5	158.51	160.72	30.26	−2.37	−2.68	1.75
Right	Token 1	168.60	165.98	17.33	−2.15	−1.74	1.33
	Token 2	127.55	131.10	21.66	−1.17	−1.07	1.29
	Token 3	114.43	116.49	14.86	−2.25	−1.60	1.91
	Token 4	145.27	147.71	18.98	−2.21	−2.38	1.79
	Token 5	156.36	157.39	27.19	−2.26	−2.42	1.70

Table 5 Mean, median, and standard deviation of /da-ta/ continuum for Tamil language

/da-ta/ Continuum—Tamil

Stimulus		N1-latency			N1-amplitude		
		Mean	Median	SD	Mean	Median	SD
Central	Token 1	172.07	175.08	17.63	−2.68	−2.94	0.90
	Token 2	133.96	134.69	18.36	−1.81	−2.26	1.71
	Token 3	126.63	131.35	22.49	−2.48	−2.73	1.56
	Token 4	131.57	128.84	18.01	−2.49	−3.04	1.98
	Token 5	157.72	165.23	31.26	−1.99	−2.20	2.01
Left	Token 1	168.37	172.32	16.25	−2.25	−2.71	1.04
	Token 2	136.79	142.11	38.07	−1.96	−2.11	0.61
	Token 3	127.49	130.85	13.93	−2.61	−2.87	1.05
	Token 4	136.47	147.37	35.15	−2.38	−2.44	1.11
	Token 5	148.76	150.21	33.56	−1.92	−2.13	1.39
Right	Token 1	171.04	176.16	21.25	−2.64	−2.79	1.23
	Token 2	132.49	141.95	30.61	−1.33	−1.10	1.37
	Token 3	122.71	124.00	16.35	−2.38	−2.45	1.53
	Token 4	139.84	149.37	26.64	−1.88	−2.58	1.47
	Token 5	157.22	167.65	38.69	−1.81	−1.94	2.16

Table 6 Z- and P-values of Mann–Whitney U test for /da-ta/ continuum for the comparison of Kannada and Tamil

Mann–Whitney U test for /da-ta/ continuum–(Kannada versus Tamil)					
Stimulus		N1-latency		N1-amplitude	
		Z-value	p-value	Z-value	p-value
Central	Token 1	0.129	0.897	0.516	0.606
	Token 2	0.258	0.796	0.775	0.438
	Token 3	1.551	0.121	1.809	0.070
	Token 4	1.939	0.053	0.516	0.606
	Token 5	0.582	0.561	0.645	0.519
Left	Token 1	0.516	0.606	0.258	0.796
	Token 2	0.258	0.796	0.775	0.439
	Token 3	1.549	0.121	0.517	0.605
	Token 4	0.452	0.651	0.647	0.518
	Token 5	0.516	0.606	0.387	0.699
Right	Token 1	1.420	0.156	0.904	0.366
	Token 2	1.033	0.302	0.129	0.897
	Token 3	1.034	0.301	1.551	0.121
	Token 4	0.065	0.948	0.065	0.948
	Token 5	0.516	0.606	1.291	0.197

Table 7 Chi-square value for N100 latency and amplitude comparison across regions of interests for tokens of /da-ta/ continuum

Friedman test for /da-ta/ Continuum across channels (center, left, and right)				
Stimulus	N1-latency		N1-amplitude	
	Chi-square	p-value	Z-value	p-value
Token 1	0.143	0.931	1.857	0.395
Token 2	1.857	0.396	2.714	0.257
Token 3	4.000	0.135	4.571	0.109
Token 4	1.857	0.395	1.857	0.395
Token 5	2.286	0.319	4.714	0.093

But only token 2 shows the less amplitude compared to other tokens in both the languages. This trend was common among the two groups of participants.

Figure 7 shows the N1 potential for each token of the /da-ta/ continuum for Kannada language. Similarly, Fig. 8 shows the N1 potential for each token of the /da-ta/ continuum for Tamil language. From Fig. 7, it was observed that as the lead VOT reduces the N1 latency decreases, on the other hand, as the lag VOT increases, the N1 latency increases. Also when the VOT is zero, the N1 latency was minimum. The similar trend was observed in both the languages.

Mann–Whitney U test was used to compare the N100 latency and amplitude across the two groups of participants for different tokens and different regions of interests (ROIs), that is, right, center, and left scalp positions. Results showed no significant difference across group of participants. Table 6 shows Z-value of N100 latency and amplitude comparison between Kannada and Tamil across different tokens and regions of interest (ROIs). Since there was no difference between the groups (Kannada and Tamil), both the groups were combined for further analysis.

Friedman test was used to compare across regions of interests (ROIs), that is, center, left, and right scalp position for N100 latency and amplitude. Kannada and Tamil data were combined for Friedman test, since group did not show significant difference in Mann–Whitney U test. The results of the Friedman test showed that there is no significant difference in N100 latency and amplitude across the tokens of /da-ta/ Continuum. Table 7 shows N100 comparison across the regions of interests for N100 latency and amplitude for tokens of /da-ta/ continuum.

Again, Friedman test was used to compare across the tokens for N100 latency and amplitude. The groups and the region of interest (ROIs) data were companied for Friedman test because they did not show significant difference. The results of the Friedman test showed that there is significant difference ($p < 0.05$) in N100 latency and amplitude across the tokens for /da-ta/ continuum. Table 8 shows N100 latency and amplitude comparison across the tokens of /da-ta/ continuum.

Since there is significant difference in N100 latency and amplitude across the tokens of /da-ta/ continuum, pairwise comparison was done using Wilcoxon signed-rank test separately for latency and amplitude. The latency comparison showed that token 1 showed significant difference with all other tokens in /da-ta/ continuum. Token 3 showed significant difference with tokens 4 and 5. Also, token 2 showed significant difference with token 5. Table 9 shows the result of Wilcoxon signed-rank test for N100 latency across tokens for /da-ta/ continuum.

Pairwise comparison (Wilcoxon signed-rank test) was separately done for N100 amplitude. The amplitude comparison showed that token 1 showed significant difference with token 2. Also, token 2 showed significant difference with token 3 in /da-ta/ continuum. Table 10 shows the result of Wilcoxon signed-rank test for N100 amplitude across tokens for /da-ta/ continuum.

Table 8 Chi-square value for N100 latency and amplitude comparison across tokens of /da-ta/ continuum

Friedman test for /da-ta/ continuum across stimulus token		
	Chi-square	p-value
N1-latency	35.257	0.000[a]
N1-amplitude	10.686	0.030[b]

[a]$P<0.01$, [b]$P<0.05$

Table 9 Z-value and p-value of Wilcoxon signed-rank test for N100 latency across tokens for /da-ta/ continuum

Wilcoxon signed-rank test for /da-ta/ continuum across stimulus N1 latency		
	Z-value	p-value
Token 1 versus Token 2	3.296	**0.001**[a]
Token 1 versus Token 3	3.384	**0.001**[a]
Token 1 versus Token 4	3.286	**0.001**[a]
Token1 versus Token 5	2.198	**0.028**[b]
Token 2 versus Token 3	1.224	0.221
Token 2 versus Token 4	0.973	0.331
Token 2 versus Token 5	3.296	**0.001**[a]
Token 3 versus Token 4	2.040	**0.041**[b]
Token 3 versus Token 5	2.605	**0.009**[a]
Token 4 versus Token 5	1.726	0.084

[a]$P<0.01$, [b]$P<0.05$

Table 10 Z-value and p-value of Wilcoxon signed-rank test for N100 amplitude across tokens for /da-ta/ continuum

Wilcoxon signed-rank test for /da-ta/ Continuum across stimulus N1 amplitude		
	Z-value	p-value
Token 1 versus Token 2	2.794	**0.005**[a]
Token 1 versus Token 3	1.475	0.140
Token 1 versus Token 4	0.471	0.638
Token 1 versus Token 5	0.094	0.925
Token 2 versus Token 3	2.480	**0.013**[b]
Token 2 versus Token 4	1.005	0.315
Token 2 versus Token 5	1.350	0.177
Token 3 versus Token 4	1.475	0.140
Token 3 versus Token 5	0.847	0.397
Token 4 versus Token 5	0.596	0.551

[a]$P<0.01$, [b]$P<0.05$

4 Discussion

The aim of the present is to investigate the behavioral and neuro-physiological response for VOT continuum in native Kannada and Tamil speakers, also to check the existence and the reliability of double peak (morphological change in N1) in response to VOT continuum. Apart from those, the present study also investigated the neuro-physiological changes for non-speech stimuli that mimic the speech stimuli.

The behavioral response of the speech stimuli VOT continuum (/da-ta/) shows that there is noticeable difference in behavioral responses between Kannada and

Tamil languages. The 50% cross over for Tamil speaker is always lagged compared to Kannada, that is, Tamil speaker requires more length of VOT to discriminate the speech sounds from voice to unvoiced and voice to unvoiced aspirated. Secondly, the Kannada speaker had lesser boundary width compared to Tamil speaker except in voiced–unvoiced non-speech continuum. In Tamil language, voicing contrast is an allophone, that is, the meaning of the word will not change if the voicing feature of speech sound is replaced with unvoiced feature. Similarly, aspiration contrast is also an allophone in Tamil language. However, the voicing contrast and the aspiration contrast are being used in the spoken language but it is very negligible. Since Tamil language did not have good distinction between voiced and unvoiced and aspirated and unaspirated, Tamil speakers' discrimination of voiced to unvoiced continuum and voiced to unvoiced aspirated require more VOT clues to identify the change of voicing and aspiration features. Hence, there used longer lag VOT to perceive the invoicing and aspiration compared to Kannada speaker. For the same above reason, Tamil speaker had wider boundary width compared to Kannada speaker. This shows that Tamil speaker had longer duration of confusion to discriminate voicing and aspiration compared to Kannada speaker. The present finding is supported by Sreevidya and Savithri [11] studied perception of voicing and place contrasts in 90 adults. Participants were equally divided into three groups—Malayalam monolinguals, Tamil monolinguals, and bilinguals with Tamil as their native language and Malayalam as their second language. Result showed that voicing was well discriminated by Malayalam monolinguals and bilinguals with Tamil and Malayalam as first and second language, respectively. Voicing contrasts were poorly discriminated by Tamil monolinguals. Jayakumar et al. [12] investigated the perception of voicing in Tamil and Hindi speakers, using /ga-ka/ VOT continuum. Ten adult native Hindi speakers and ten adult Tamil speakers with normal hearing participated in this study. The subjects listened to the edited tokens and responded to a binary-forced choice response sheet. The results indicated that VOT was acute for the perception of voicing in both the languages. Also, Tamil speakers had wider boundary widths compared to Hindi speakers indicating perceptual inconsistency over a longer period of time. The 50% crossover of /ga-ka/ VOT continuum for Tamil speaker was –11 ms.

The results of electrophysiological test showed that there is no significant difference between Kannada and Tamil languages for N1 latency and amplitude for any of the speech stimuli and non-speech stimuli in the present study. Although these two groups of speaker belong to different native languages which had variation in voicing and aspiration perception of speech sounds, they did not show significant difference in electrophysiological test. However, they showed noticeable difference in the behavioral results of the same VOT continuums. This result strengths the viewpoint that N1 potential is obligatory evoked response which reflects sensory encoding of auditory stimulus attributes. The N1 component is often described as an onset response because it signals the neural encoding of sound onset [13]. The N1 component is majority affected by the stimulus properties rather than anything else like language or the linguistic exposer. This may be the reason why there was no significant difference between Kannada and Tamil speakers in terms of electrophysiological test. Probably, the potential which tabs the higher level processing like MMN and P300 would have been shown the stimulus discrimination ability more effectively than N1 potential. Sharma and Dorman [9] investigated native and nonnative phonetic categories using behavioral and electrophysiological responses from Hindi and English listeners. Bilabial CV stimulus that differed in VOT from –290 ms to 0 ms was used as stimulus. The pre-voicing (negative VOT) is not present in English language, which is present in Hindi language taken as stimulus to create a native and nonnative groups. The result showed that the change in N1 latency reflected the duration of pre-voicing across the stimulus continuum. This change in N1 latency was not significantly different for Hindi and English listeners. This study supports that language influence is not having the effect on the N1 component. This present result supports the neurophysiologic base for the perception of voicing which is not depending on language-specific [14]; Simos et al. 1997; [15–17], and this supports the notion that voicing distinction is determined by the capability to discern the sequential order of transient acoustic events.

In the present study, electrophysiological result of the N1 latency showed that as the lead VOT reduces, the N1 latency decreases, on the other hand, as the lag VOT increases the N1 latency increases. Also, when the VOT is zero, the N1 latency was minimum. This trend was seen for voice to unvoiced speech stimuli. This result suggests that the N1 potential is generated for the onset of the vowel from the VOT continuum or stop consonants. When the VOT continuum has a lead VOT (negative VOT or pre-voicing), there was no onset response (N1) for lead VOT, however, for the same stimulus from the time point of vowel onset, the N1 response generated. This may be the reason for the increment in the N1 latency as the lead VOT increases. This result was supported by [9]. They investigated native and nonnative phonetic categories using behavioral and electrophysiological responses from Hindi and English listeners. Bilabial CV stimuli (/ba/ to /pa/) that differed in VOT from -290 ms to 0 ms (pre-voicing which is not present in English language) were used as stimulus. The result showed that the increment in N1 latency reflected the duration of pre-voicing across the stimulus continuum. As the lead VOT increases in the duration, the latency of the N1 potential was increased. Horev et al. [10] studied the perception of a VOT continuum in Hebrew with the perception of an analogous non-speech continuum, using N1 electrophysiological measures. They used the VOT continuum which varies from –20 ms to + 40 ms on seven adults' participants. The results showed that VOT value had a significant effect on N1 latency.

The present study also showed that as the lag or positive VOT increases, the N1 latency increases. However, the lag VOT increment did not generate change in the morphology of the N1 waveform (N1& N1' or double peak) in the voiced to unvoiced continuum of speech stimulus and non- speech continuum. The possible reason can be vowel masks the neural response to the preceding, much softer acoustic event (i.e., lag or positive VOTs and the pre-voicing in negative VOTs) and also the backward masking of the long vowel on the N1 onset response) Horev et al. [10]. Previous research on longer positive VOT suggests that bifid N1 response. The first peak coincides with the release burst, and the second with the onset of voicing. Thus, N1 reflects the VOT value of the evoking stimulus in two ways: double N1 to long VOTs [8], Sharma et al. (2000) and/or an increase in N1 latency with an increase in VOT [9], Tremblay et al. (2003b). In the present study, the burst did not evoke the N1 responses for any of the voice to unvoiced VOT continuum. This result was supported by Tremblay et al. (2003b) investigated the synthetic speech tokens of /ba/-/pa/ voice-onset-time (VOT) continuum on adults with hearing loss and without hearing loss using N1 potential. The results of their study supported that as VOT increases, the latency of the N1 increase, and they did not find any morphological change in the N1 potential. King et al. [18] recorded cortical auditory evoked potentials (CAEPs) from six electrode sites in 10 participants. The stimulus used was a /da/-/ta/ continuum in which voice onset times (VOTs) varied from 0 to 60 ms. A forced choice identification task was performed. The point of 50% identification occurred at 32.5 ms. Cortical evoked responses elicited by stimuli with VOT values ranging from 0 to 60 ms (i.e., by stimuli perceived as /da/ and /ta/) were similar in morphology. There was no evidence of a "double-on" morphology for stimuli with long VOTs which is observed in adults. However, latency changes in the P1 and N2 components were observed as a function of VOT changes, i.e., shorter VOTs elicit a shorter peak latency than stimuli with longer VOTs.

The present study had one more group of VOT continuum, to explore more about the N1 morphological change or double peak of N1 (N1 and N1') was used, that is, the voice to unvoiced aspiration continuum. Voice to unvoiced aspirated speech stimuli and non-speech stimuli (/da-tha/, /ga-kha/and non-speech continuum) showed that N1 latency was decreased as the lead VOT reduces, on the other hand as the lag VOT along with aspiration increases, the N1 latency did not show noticeable change. But there was wading of N1 peak which was the morphological change observed. Also when the VOT is zero, the N1 latency was minimum. In the present VOT continuum (voice to unvoiced aspiration), as the lag VOT with aspiration increases, there was no noticeable change in N1 latency but there was wading of N1 peak that is different from the voice to unvoiced continuum. There were two reasons that can be postulated for this result. First one is that the aspiration in the present VOT is strong enough in the present VOT continuum and that the participant did not perceive the burst and vowel onset as two distinct events or in other words, the aspiration stimulated the auditory impulses as equal as vowel. Due to this reason, there was no change in N1 latency although log VOT was increased. The second reason may be that the aspirated lag VOT might have generated double N1 peak (N1 and N1') but when the averaging technique across the participants is employed for the purpose grand average of the waveform, the double peak would have got merged and generated the wider N1 peak. To investigate more details about the possibility of merged double peak, individual waveform has been subjected to the visual examination by three experienced persons in the field of auditory electrophysiology. The individual participant's visual examination N1 waveform showed that some (one-third participants) of the participants showed appearance of the double peak in N1 morphology; however, appearance was not consistent across participants. The double peak appears relatively frequent for non-speech continuum than speech continuum. The present double peak for the longer VOT and especially for the voice to unvoiced aspiration was supported by [8] and Sharma et al. (2000). Sharma and Dorman [8] examined neural correlates of VOT perception at the level of the auditory cortex in 16 adults in conjunction with behavioral testing using the /da/-/ta/continuum. The N1 response was measured for each stimulus from the /da/-/ta/. They observed that for stimuli with short VOTs (0–30 ms), a single negativity was apparent. However, for stimuli with long VOTs (50–80 ms), two distinct negative components (N1' and N1) were apparent. These data demonstrate a discontinuity in morphology of the AEP waveform in the region of the boundary between perceptual categories /da/ and /ta/.

To conclude, the present study showed noticeable behavioral difference across languages for the speech VOT continuum but N1 electrophysiological potential did not show the significant language differences, which suggests voicing distinction is determined by the sequential order of transient acoustic events rather than influence of language when we use N1 as potential measure. The presence of double peak for the longer VOT or change in the categorical perception was partially accepted with the present result; however, it was not consistent across participants and stimulus. N1 latency and morphology for the present study suggest no greater difference in the processing of speech.

5 Summary and Conclusion

Speech is transmitted in the form of acoustic energy. Speech perception is the process, wherein speech is decoded temporally and spectrally at the lower centers. The linguistic components are added only at the higher centers of the cortex. Speech perception requires tracking the acoustic features onto phonetic categories and ignoring the unwanted variability among signals that convey information about the same particular phoneme. Categorical perception means mode of speech perception in which participants can only discriminate between stimuli that they identify differently. Most common examples of categorical perception are between voiced and voiceless of initial stop consonants.

The aim of the present is to investigate the behavioral and neuro-physiological response for VOT continuum in native Kannada and Tamil speakers, and also to check the existence and the reliability of double peak (morphological change in N1) in response to VOT continuum. Apart from those, the present study also investigated the neuro-physiological changes for non-speech stimuli that mimic the speech stimuli. Present study had behavioral and electrophysiological experiment on ten adult native Kannada and Tamil speakers. Two types of stimuli were used in the present study. Type one was with speech VOT continuum (/da-ta/ continuum) these stimuli belong to voiced to unvoiced continuum and voice to unvoiced aspirated continuum. For behavioral experiment, nearly 9 tokens were used for each continuum, but for electrophysiological experiment only 5–3 tokens only used for each continuum.

The behavioral response of the speech stimuli VOT continuum (/da-ta/) shows that there is noticeable difference in behavioral responses between Kannada and Tamil languages. The 50% cross over for Tamil speaker always lags compared to Kannada, that is, Tamil speaker requires more length of VOT to discriminate the speech sounds from voice to unvoiced and voice to unvoiced aspirated. The same result was found for non-speech stimulus. The reason may be Tamil language did not have good distinction between voiced and unvoiced and aspirated and unaspirated, Tamil speakers discrimination of voiced to unvoiced continuum and voiced to unvoiced aspirated require more VOT clues to identify the change of voicing and aspiration features. Hence, there used longer lag VOT to perceive the invoicing compared to Kannada speaker.

The results of electrophysiological test showed that there is no significant difference between Kannada and Tamil languages for N1 latency and amplitude for any of the speech stimuli and non-speech stimuli in the present study. Although these two groups of speaker belong to different native languages which had variation in voicing and aspiration perception of speech sounds, they did not show significant difference in electrophysiological test. However, they showed noticeable difference in the behavioral results of the same VOT continuums. This result strengthens the viewpoint that N1 potential is obligatory evoked response which reflects sensory encoding of auditory stimulus attributes. The N1 component is often described as an "onset response because it signals the neural encoding of sound onset" [13]. The N1 component is majorly affected by the stimulus properties rather than anything else like language or the linguistic exposer. This may be the reason why there was no significant difference between Kannada and Tamil speakers in terms of electrophysiological test. Probably, the potential which tabs the higher level processing like MMN and P300 would have been shown the stimulus discrimination ability more effectively than N1 potential.

In the present study, electrophysiological result of the N1 latency showed that as the lead VOT reduces, the N1 latency decreases; on the other hand, as the lag VOT increases the N1 latency increases. This trend was seen for voice to unvoiced speech stimuli and non-speech stimuli (/da-ta/). This result suggests that the N1 potential is generated for the onset of the vowel from the VOT continuum or stop consonants. When the VOT continuum has a lead VOT (negative VOT or pre-voicing), there was no onset response (N1) for lead VOT; however, for the same stimulus from the time point of vowel onset, the N1 response generated. This may be the reason for the increment in the N1 latency as the lead VOT increases. The present study also showed that as the lag or positive VOT increases, the N1 latency increases. However, the lag VOT increment did not generate change in the morphology of the N1 waveform (N1& N1' or double peak) in the voiced to unvoiced continuum of speech stimulus and non-speech continuum. The possible reason can be vowel masks the neural response to the preceding, much softer acoustic event (i.e., lag or positive VOTs and the pre-voicing in negative VOTs) and also the backward masking of the long vowel on the N1 onset response Horev et al. [10].

The present study had one more group of VOT continuum, to explore more about the N1 morphological change or double peak of N1 (N1 and N1') was used, that is, the voice to unvoiced aspiration continuum. Voice to unvoiced speech stimuli showed that N1 latency was decreased as the lead VOT reduces; on the other hand, as the lag VOT along with aspiration increases, the N1 latency did not show noticeable change. But there was wading of N1 peak which was the morphological change observed. In the present VOT continuum (voice to unvoiced aspiration), as the lag

VOT with aspiration increases, there was no noticeable change in N1 latency but there was wading of N1 peak different from the voice to unvoiced continuum. There were two reasons to be postulated for this result. First one is that the aspiration in the present VOT is strong enough in the present VOT continuum and that the participant did not perceive the burst and vowel onset as two distinct events, or in other words, the aspiration stimulated the auditory impulses as equal as vowel. Due to this reason, there was no change in N1 latency although log VOT was increased. The second reason may be that the aspirated lag VOT might have generated double N1 peak (N1 and N1') but when the averaging technique across the participants is employed for the purpose grand average of the waveform, the double peak would have got merged and generated the wider N1 peak. To investigate more details about the possibility of merged double peak, individual waveform has been subjected to the visual examination by three experienced persons in the field of auditory electrophysiology. The individual participant's visual examination N1 waveform showed that some (one-third participants) of the participants showed appearance of the double peak in N1 morphology; however, appearance was not consistent across participants. The double peak appeared was relatively frequent for non-speech continuum than speech continuum.

To conclude, the present study showed noticeable behavioral difference across languages for the speech VOT continuum but N1 electrophysiological potential did not show the significant language differences, which suggests that voicing distinction is determined by the sequential order of transient acoustic events rather than influence of language when we use N1 as potential measure. The presence of double peak for the longer VOT or change in the categorical perception was partially accepted with the present result; however, it was not consistent across participants and stimulus. N1 latency and morphology for the present study suggest no greater difference in the processing of speech and non-speech signals.

References

1. Lisker L, Abramson A (1964) A cross language study of voicing in initial stops: Acoustical measurements. Word 20:384–422
2. Abramson, A., and Lisker, L. (1965). Voice onset time in stop consonants: Acoustic analysis and synthesis. P*roceedings of the 5th International Congress of Acoustics*. Liege: Imp. G. Thone
3. Studdert- Kennedy M (1976) Speech perception. In: Lass NJ (ed) Contemporary issues in experimental phonetics. Academic press, New York
4. Williams L (1977) The perception of stop consonant voicing bySpanish-English bilinguals. Percept Psychophys 19:285–289
5. Williams, L. (1980). Phonetic variation as a function of second language learning. In G. H
6. Savithri SR, Pushpavathi M, Sujatha S (1995) Development of speech production and perception. Some temporal aspects. Journal of Indian Speech & Hearing Association 14:1–10
7. Sathya, R.and Savithri, S., R. (1996). *Development of auditory perceptual processing in children.* Unpublished doctoral thesis submitted to University of Mysore, Mysore
8. Sharma A, Dorman MF (1999) Cortical auditory evokedpotential correlates of categorical perception of voice-onsettime. J Acoust Soc Am 106:1078–1083

9. Sharma A, Dorman M (2000) Neurophysiologic correlates ofcross-language phonetic perception. J Acoust Soc Am 107:2697–2703
10. Horev, N., Most, T., and Pratt, H. (2007). Categorical Perception of Speech (VOT) and Analogous Non-Speech (FOT) signals: Behavioral and electrophysiological correlates.*Ear Hearing,* 28(1), 111-28
11. Savithri, S. R. (1997). Development of speech perception in children. *Proceedings of International conference on computational Linguistics, Speech and Document Processing,* pp 368-384
12. Jayakumar T, Venugopal MB, Savithri SR (2006) Perception of voicing in Tamil and Hindi speakers. Proceeding of national acoustic society of India, Mysore
13. Martin, B. A., Tremblay, K. L., and Stapells, D. R. (2007). Principlesand applications of cortical auditory evoked potentials. In R. F.Burkard, J. J. Eggermont, & M. Don (Eds.), *Auditory evokedpotentials: basic principles and clinical application* (Chap. 23).Lippincott, Williams and Wilkins
14. Lie´geois-Chauvel, C., De Graaf, J. B., Laguitton, V., and Chauvel, P.(1999). Specialization of left auditory cortex for speech perceptionin man depends on temporal coding. *Cerebral Cortex,* 9,484–496
15. Steinschneider M, Schroeder CE, Arezzo JC, Vaughan HG (1994) Speech-evoked activity in primary auditory cortex: Effects of voice onset time. Electroencephalography and Clinical-Neurophysiology 92:30–43
16. Steinschneider M, Volkoc I, Noh M, Garell P, Howard MT (1999) Temporal encoding of the voice onset time phoneticparameter by field potentials recorded directly from humanauditory cortex. J Neurophysiol 82:2346–2357
17. Steinschneider M, Volkov IO, Fishman YI, Oya H, Arezzo JC, Howard MA (2005) Intracortical responses inhuman and monkey primary auditory cortex support a temporalprocessing mechanism for encoding of the voice onset timephonetic parameter. Cereb Cortex 5:170–186
18. King KA, Campbell J, Sharma A, Martin K, Dorman M, Langran J (2008) The representation of voice onset time in the cortical auditory evoked potentials of young children. Clin Neurophysiol 119(12):2855–2861
19. Dalebout SD, Stack JW (1999) Mismatch negativity to acoustic differences not differentiated behaviorally. Journal of American Academy of Audiology 10(7):388–399
20. Elangovan S, Stuart A (2011) A cross-linguistic examination of cortical auditory evoked potentials for a categorical voicing contrast. Neuroscience Lettan 490(2):140–144
21. Flege JE, Eefting W (1986) Linguistic and developmental effects on the production and perception of stop consonants. Phonetica 43:155–171
22. http://www.en.wikipedia.org/wiki/Kannada_language
23. http://www.en.wikipedia.org/wiki/Tamil_language
24. Kraus N, McGee T, Micco A, Sharma A, Carrell T, Nicol T (1993) Mismatch negativity in school-age children to speech stimuli that are just perceptibly different. Electroencephalogram Clinical Neurophysiology 88(2):123–130
25. Liberman AM, Harris KS, Hoffman HS, Griffith BC (1957) The discrimination of speech sounds within and across phoneme boundaries. J Exp Psychol 54:358–368
26. Liederman, J., Frye, R., Fisher, J. M., Greenwood, K., and Alexander, R. (2005). A temporally dynamic context effect that disrupts voice onset time discrimination of rapidly successive stimuli.*Psychology Bull Rev.*, 12(2), 380-6
27. Lisker L, Abramson A (1967) Voicing Contrast: Perceptual and productive voice onset time characteristics of adult. Journal of Acoustic society of America 56(3):981–994
28. Naatanen R, Picton TW (1987) The N1 wave of the humanelectric and magnetic response to sound: a review and ananalysis of the component structure. Psychophysiology 24:375–425
29. Pegg, J. E., and Werker, J. F. (1997) Adult and infant perception of two English phones.*Journal of Acoustic society of America,* 102(6), 3742–3753
30. Pegg JE, Werker JF (1997) Adult and infant perception of two English phones. Journal of Acoustic society of America 102(6):3742–3753

31. Sharma A, Marsh CM, Dorman MF (2000) Relationshipbetween N1 evoked potential morphology and the perception ofvoicing. J Acoust Soc Am 108:3030–3035
32. Simon C (1974) Some aspect of development of speech production and perception in children, *Speech communication seminar*, 7-15. Stockholm, John Wiley and sons, London
33. Simos PG, Molfese DL, Brenden RA (1997) Behavioraland electrophysiological indices of voicing-cue discrimination: Laterality patterns and development. Brain Lang 57:122–150
34. Sinex DG, McDonald LP, Mott JB (1991) Neural correlates of nonmonotonic temporal acuity for voice onset time. Journal of Acoustic society of America 90(5):2441–2449
35. Sreedevi N, Savithri SR (1991) Cross language study of stop perception. Journal of Indian Speech & Hearing Association 8:14–20
36. Tampas JW, Harkrider AW, Hedrick MS (2005) Neurophysiological indices of speech and nonspeech stimulus processing. Journal of Speech Language Hearing Research 48(5):1147–1164
37. Toscano, J. C., McMurray, B., Dennhardt, J., and Luck, S. J. (2010). Continuous perception and graded categorization: electrophysiological evidence for a linear relationship between the acoustic signal and perceptual encoding of speech.*Psychology Science*, 21(10), 1532-40
38. Tremblay KL, Piskosz M, Souza P (2003) Effects of age andage-related hearing loss on the neural representation of speechcues. Clin Neurophysiol 114:1332–1343
39. Werker, J. F, and Logan, J. S. (1985).Cross-language evidence for three factors in speech perception.PerceptionPsychophysic, 37(1), 35-44
40. Williams L (1977) The voicing contrast in Spanish. Journal of Phonetics 5:169–184
41. Williams L (1977) The perception of stop consonant voicing by Spanish-English bilinguals. Percept Psychophys 21:289–297
42. Yeni-Komshian, J. F. Kavanagh and C. A. Ferguson (Eds.), Child Phonology, vol.2: Perception, Academic press: NewYork, 1980
43. Zlatin M, Koenigsknecht R (1975) Development of the voicing contrast: Perception of stop consonants. J Speech Hear Res 18:541–553

Entropic Analysis of Garhwali Text

Manoj Kumar Riyal, Rajeev Kumar Upadhyay, and Sanjay Kumar

Abstract In the present study, a systematic statistical analysis has been performed by the use of words in continuous Garhwali speech corpus. The words of Garhwali in continuous speech corpus are taken from different sources of Garhwali, viz., Newspapers, storybooks, poems, lyrics of songs and magazines, and it showed that there is a quantitative relation between the role of content words in Garhwali and the Shannon information entropy [S] defined by the probability distribution. So far, very few researches have been conducted in Garhwali language. There is no previous knowledge about the syntactic structure of Garhwali language. We have taken finite continuous corpus of Garhwali language. The occurrences of words (frequency) are almost an inverse power law functions, i.e. Zipf's law and very close to 1.

Keywords Entropy · Garhwali · Rank · Frequency

1 Introduction

The randomness of the system is measured with the help of entropy. In our case, entropy measures the measure of randomness of different words present in our corpus. In many subjects viz. language, Science, Physics, Biology, Social Sciences and Computer Science entropy play an important role. Entropy has been measured on the basis of frequency of words or occurrences of words in the text. In scientific analysis of any language, words have more accurate and specific information about that language [1]. The entropy measure is a technique which fulfils a lot of natural

M. K. Riyal (✉)
Department of Physics, Government Post Graduate College, Kotdwara, Pauri Garhwal 246 149, Uttarakhand, India
e-mail: manoj.riyal@gmail.com

R. K. Upadhyay
Department of Physics, V. A. Government Degree College, Atrauly District Aligarh 202 280, India
e-mail: rku8@rediffmail.com

S. Kumar
Physics Lecturer, Government Senior Secondary Boys School, Phagwara, Punjab, India
e-mail: sanjaychadha54@gmail.com

M. Singh and Y. Rafat (eds.), *Recent Developments in Acoustics*, Lecture Notes in Mechanical Engineering, https://doi.org/10.1007/978-981-15-5776-7_3

language tasks [2]. This technique is widely used without previous knowledge as well as external knowledge. Therefore, this is the unique solution and it plays key role for text analysis of any language.

The most popular discipline in the domain of quantitative linguistics is word counting than latter or character. We have analysed number of occurrences of words and entropies of those words used in our corpus, which is very useful. This study is very helpful in usages and birth of new words as well as language teaching, analysing the pattern of the language, get information of most informative words, develop automatic speech recognizer, speech interface of Garhwali, grammatical studies and scientific analysis of the language. Garhwali dialect belongs to Indo-Aryan language family which is spoken in Garhwal region of the Uttarakhand state of India. Entropy analysis is the most powerful and informative tools in statistical physics and computer science during the many past decades, i.e. it describes the existence of randomness and order of the complex structure and behaviour characterized by the statistical systems. Entropy analysis is the critical phenomena for the investigation of such behaviour [3]. By the help of this type of study, we can develop language modelling through simulation. The Shannon entropy is a numerical representation of written text and the symbolic sequences generated are examined. It was found that entropy studies were sensitive to both language and text [3]. The dependencies of Shannon entropy reflect the dependencies of the text used in inverse law. Entropy may also give much more information about the quality of translation of the language.

One of the most powerful tools used for the comparison of literary text for translation to another language is Shannon entropy which provides the most accurate and useful information related to the quality of translation [1]. It is a tool to measure the average information (stored information) from the source in a single message [4]. The Shannon entropy gives the idea and information about language, i.e. information content of a particular word or letter in a language. The role of word length distribution in the natural language was examined on the basis of n-gram entropies and found that they were sensitive in the text language. The probability distribution for the single word length is changing crucially, and long words having length between five and ten words have the uniformity in probability distribution [5]. The n-gram entropy tells about word lengths, i.e. the role of distributions and their corresponding correlations of the length of the word in the specific language [5]. For finite-size text, the long-range correlation of pairs of letter and entropy were studied and found that entropy per letter and decay with a power law [6]. For ten European languages, the correlation of word length distribution and frequency distribution of words was studied and found the entropy of word length distribution of short words; the mean value of the Uralic (Finnish) corpus differs from others [7]. The rank–frequency distribution of occurrence of character of Garhwali language from continuous corpus having limited dictionary size has been analysed and found that the distribution of occurrences of consonants, vowels and all characters together followed Zipf–Mandelbrot law [8].

The objective of the study is to measure the Shannon entropy of document used in the study. This measures the randomness of the words used in the document. In this document, as the occurrences of different words increase, the entropy of the document is also increased. In the present study, we have measured entropy of different words of continuous corpus of Garhwali language. This study is very useful for the growth of new words, demise and origin of new words. This study is also useful as well as beneficial for the process governing the usages of new words, new language learning and teaching, time series analysis, frequency-domain analysis, grammatical studies, improved quality of translation, word length entropies and correlations between pairs of words of language. In the modern era, information is the most unpredictable in relation to the previous states of the important sources. Entropy also tells the uncertainty of appearing the words in the corpus [9].

1.1 Zipf's Law

Zipf's law tells about the relation between frequency of words (number of occurrence of words) and rank of the word, i.e. $f(n) \sim 1/n$, i.e. the frequency $f(n)$ of the nth commonest word in a database. The most basic statistical approach is by which some quantitative information about the use of words in a corpus of written language can be obtained by Zipf's law [10]. Basically, it consists of counting the number of occurrences of each different word or letter or symbol in the corpus, and then producing a list of these words sorted according to decreasing frequency. In more generalized form, if a large corpus is considered and ranks assigned to all words according to the order of decreasing frequencies of those words, then the frequency $f(n)$ of a word of rank n satisfies the relation given below:

$$f(n) = pn^q$$

where p and q are constants and q is the Zipf's exponent.

2 Methodology

The continuous speech corpus of Garhwali language was collected from different Garhwali sources, i.e. Garhwali newspaper, magazine, poems, lyrics, articles of the e-magazine by Bhishma Kukreti, songs, proverb, poems and stories [16]. All the corpus was collected from Internet. Then number of occurrences, i.e. frequency of each word has been calculated and the rank of corresponding word was given. A graph was drawn between frequency and entropy. For each word, it is possible to define a probability measure P_i over the partition as

$$P_i = \frac{f_i}{\sum_{j=1}^{p} f_j} \quad (1)$$

The quantity P_i stands for the probability of finding the word in part i, given that it is present in the corpus. The Shannon information entropy associated with the discrete probability distribution P_i reads

$$S = -\sum_{i=1}^{P} P_i \log P_i \tag{2}$$

The maximum entropy measure was computed as

$$E\text{max} = -\log\frac{1}{n} \tag{3}$$

Equation 3 represents the numerical value of maximum entropy present in our document.

Generally, the value of S is different for each word, and its value lies between 0 and $1 (0 \leq S \leq 1)$. As discussed below, the entropy of a given word provides a characterization of its distribution over the different partitions. All measurement and analysis and plots have been done with the help of MATLAB software.

3 Results and Discussion

A graph was plotted between number of occurrences and entropy of the words (Fig. 1), and the curve fitting it showed that the power law was the best fit, i.e. f(x) = a*x^b having coefficients (with 95% confidence bounds) a = 0.001564 (–0.001536,–0.001581), b = 0.8672 (0.865, 0.8793) and Goodness of fit is SSE: 0.0008221, R-square: 0.9762, adjusted R-square: 0.9766 and RMSE: 0.000649 (Fig. 1).

The graph between rank and frequency of occurrences was plotted on log–log scale [Rank–frequency graph (log–log scale)], and it followed inverse power law or Zipf's law, i.e. f(x) = a*x^b having coefficients (with 95% confidence bounds) a = 0.09842 (0.09701, 0.09813), b = –0.7898 (–0.7981, –0.7811) and for Goodness

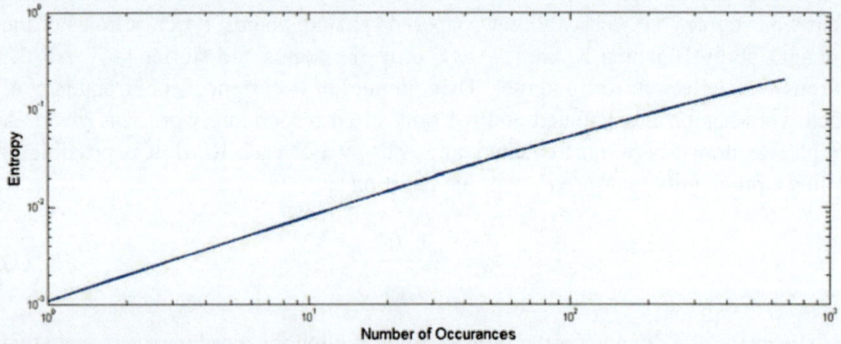

Fig. 1 Graph between number of occurrences and entropy of the words

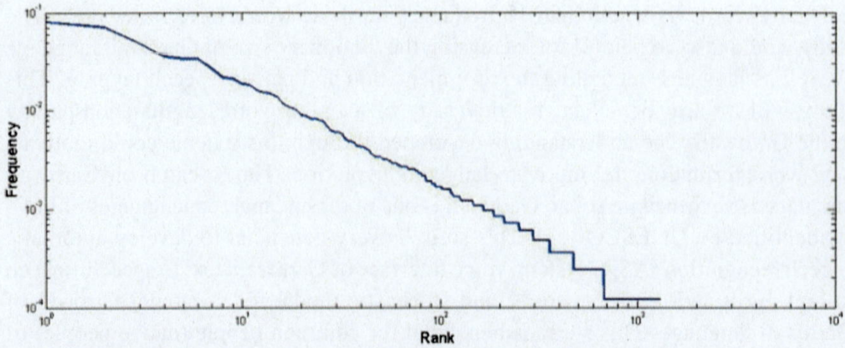

Fig. 2 Graph between rank and frequency of the words (log–log scale)

of fit is SSE: 0.001091, R-square: 0.9574, adjusted R-square: 0.9584 and RMSE: 0.0008145 (Fig. 2).

The R-square 0.9762 (Fig. 1) and R-square 0.9574 (Fig. 2) showed the good fitting in power law and inverse power law, and having the coefficient of determination was equally good. This showed that Garhwali language showed that the entropy of words followed the inverse square law. The graph between number of occurrences and entropy of the words (Fig. 1) was found to have the same behaviour as by graph of block entropy H(n) versus block size *n* [5]. The same linear behaviour is exhibited through the symbol sequences used in the text [3]. The graph of entropy versus the number of occurrences *n* for plays William Shakespeare also showed the same characteristics as in Fig. 1 [8]. In Fig. 2, the words (log–log scale) showed inverse power law, i.e. Zipf's law. This result is in agreement with other languages like Chinese, Japanese, English and Korean which also followed the power law with exponent close to 1 [11]. The power law is also followed for the population of lower tail cities in India [12]. From Figs. 1 and 2, curve fitting and value of R-square(coefficient of determination) showed that our results followed Zipf's law [13–15]. Therefore, randomness of the system, i.e. entropy, increases. In normal distribution, curve between frequency and entropy was right skewed or positive skewed, i.e. long tail is on right side. Therefore, Zipf's law gives the better fit for lower rank region [14].

4 Conclusion

We calculated the Shannon entropy and rank–frequency of continuous speech corpus downloaded from web and found that both exhibit the same pattern of dependencies on the text category and language. The Shannon entropy showed linear behaviour with number of occurrences, and inverse power law between rank and frequency. In Garhwali language, there are lot of single characters which are word in itself, i.e. many words have less than 10 frequency so those words have lower rank. The study would also be helpful for increasing the dictionary size of Garhwali language as well as language modelling, text segmentation and part-of-speech tagging. This study will be also beneficial for the study of average word lengths and sparsity in the Garhwali. The understanding developed through this scheme could motivate new word formation and motivate language evolution. The research on Garhwali language is very meagre so far. Garhwali is one of the endangered languages of India as identified by UNESCO [17]. This study is very beneficial to develop automatic speech recognition (ASR) system, voice interface of Garhwali, text to speech, speech to text, birth and death of words, and stochastic model for vocabulary growth of Garhwali language. This study is beneficial for common people (native people) of Uttarakhand (Garhwali speaking) as well as the farmers and rural natives of the area to get relevant information in their own language.

References

1. Lawnik M, Shannon's entropy in literary works and their translations (web: http://computer.scientific-journal.com/articles/1/23.pdf)
2. Nigam K, Lafferty J, McCallum A (1999) Using maximum entropy for text classification. In: IJCAI-99 workshop on machine learning for information filtering, vol 1
3. Papadimitriou C, Karamanos K, Diakonos FK, Constantoudis V, Papageorgiou H (2010) Entropy analysis of natural language written texts. Physica A 389(16):3260–3266
4. Shannon Claude E (1948) A mathematical theory of communications. Bell Syst Tech J 27:379–423
5. Kalimeri M, Constantoudis V, Papadimitriou C, Karamanos K, Diakonos FK, Papageorgiou H (2012) Entropy analysis of word-length series of natural language texts: Effects of text language and genre. Int J Bifurcat Chaos 22(09):1250223
6. Ebeling Werner, Pöschel Thorsten (1994) Entropy and long-range correlations in literary English. EPL (Europhysics Letters) 26(4):241
7. Kalimeri M, Constantoudis V, Papadimitriou C, Karamanos K, Diakonos FK, Papageorgiou H (2015) Word-length entropies and correlations of natural language written texts. J Quant Linguist 22(2):101–118
8. Montemurro MA, Zanette DH (2002) Entropic analysis of the role of words in literary texts. Adv Complex Syst 5(01):7–17
9. Ospanova R (2013) Calculating information entropy of language texts. World Appl Sci J 22(1):41–45
10. Zipf GK (1949) Human behaviour and the principal of least effort, reading. Addison-Wesley Publishing Co., MA
11. Cancho RFI, Sole´ RV (2002) Least effort and the origins of scaling in human language. Proceedings of the national academy of sciences of the United States of America, vol 100, pp 788–791
12. Devadoss S, Luckstead J, Danforth D, Akhundjanov S (2016) The power law distribution for lower tail cities in India. Physica A 15(442):193–196
13. Riyal MK, Rajput NK, Khanduri VP, Rawat L (2016) Rank-frequency analysis of characters in Garhwali text: emergence of Zipf's law. Curr Sci 110(3):429–434
14. Manin DY (2009) Mandelbrot's model for Zipf's law: can Mandelbrot's model explain Zipf's law for language? J Quant Linguist 16(3):274–285
15. Montemurro Marcelo A (2001) Beyond the Zipf-Mandelbrot law in quantitative linguistics. Physica A 300(3-4):567–578

16. http://e-agazineofuttarakhand.blogspot.in/2009/10/garhwali-kumaoni-himalayan-literature_6723.html
17. http://www.language-archives.org/language/gbm#language_descriptions

Effect of Age on Formant Frequencies in Kannada Speakers

V. Namitha, V. Namratha, and R. Rajasudhakar

Abstract Formants are the concentration of the acoustic energy around a frequency in the speech wave. Kaur and Narang (1) have noted the variation of pitch and formants in the different age groups in the English language. The current study aimed to find the relation between the aging and formant frequency (F1 and F2) in the Kannada language. The methodology of the study involved eight males and eight females (2 of 8–10 years of age, 2 of 18–20 years of age, 2 of 48–50 years of age, and 2 of above 60 years of age) who were native Kannada speakers. The participants were instructed to repeat the non-sense words V1CV2 in which V2 was /a/always. V1 consisted of three short vowels and three long vowels. The non-sense words used were aka, ika, uka, a: ka, i:ka, and u:ka. The subjects were asked to repeat thrice. In total, 288 tokens were collected for the study. The speech sample is recorded in the Praat software in an acoustically treated room and being the microphone placed 3 cm away from the mouth. The mean of the F1 and F2 of the steady part of the initial vowel (V1) excluding the onset and offset of the three trials is considered. Then, the F1 and F2 were compared across the age group and gender. The present study found that the F1 and F2 decrease as age increases in both males and females.

Keywords Formant frequency · Vowels · Kannada · Acoustic analysis · Aging effect

1 Introduction

Fant (1960) defines formant as the spectral peaks of the sound spectrum |P (f)|. It corresponds to resonant frequencies or pitch overtones of the vocal tract for articulating different types of voiced sounds, most notably vowels. The first peak in the vowel spectra is referred to as first formant frequency (F1) and the second peak in the spectra is referred to as second formant frequency. The specific formants F1 and F2 are typically evaluated for comparison of different vowels. F1 is the lowest

V. Namitha (✉) · V. Namratha · R. Rajasudhakar
Department of Speech-Language Sciences, All India Institute of Speech & Hearing, Mysuru, India
e-mail: namitha1202@gmail.com

M. Singh and Y. Rafat (eds.), *Recent Developments in Acoustics*, Lecture Notes in Mechanical Engineering, https://doi.org/10.1007/978-981-15-5776-7_4

formant and is inversely related to vowel height (or how close the tongue is to the roof of the mouth); the difference between F1 and F2 correlates roughly with tongue advancement.

Each human being goes through the process of aging. Our voice and speech patterns change from early childhood to old age. It is well known that acoustic and linguistic characteristics of children's speech are widely different from those of adult speech. Furthermore, characteristics of children's speech vary rapidly as a function of age due to the anatomical and physiological changes occurring during a child's growth and because children become more skilled in co-articulation with age.

Many researchers in the field of speech acoustics have studied the effects of age on the formants. Robb et al. (3) studied the developmental aspects of formant frequency and bandwidth in infants and toddlers. A cross-sectional study was done on 20 children who were in the age range of 4–25 months. The authors concluded that there was a slight rise in F1 at age of 18 months and average F1 and F2 values changed a little while the average bandwidths for F_1 (B_1) and F_2 (B_2) were found to significantly decrease as the age increases.

Sreedevi [6] studied formant frequency in three different age groups from children to adults in Kannada-speaking individuals. The sentences which had meaningful disyllabic test word (C1V1C2V2) with a carrier phrase were used for the study and results revealed that the formant frequencies (F1, F2, and F3) decreased from children to adolescents markedly and there was a further gradual decrease in adults. Further, the author noticed that there was a larger drop in F2 and F3 than in F1 from children to adults.

Mwangi et al. (2) studied the effect of vocal aging on the fundamental frequency and the formant in a longitudinal study wherein the speech of the Queen Elizabeth II's was considered over 50 years of age was considered. The results of this study revealed that vocal aging causes the decline of the fundamental frequency (F0) and first formant frequencies (F1) while second and third formants do not have an influence on aging.

Kaur and Narang [1] had studied the variation of pitch and formants in the different age groups in the English language. They analyzed the first three formants of vowels produced by the child, younger, and older of the same gender. This study showed that the child's formants are found to be higher than adults and it was more scattered. The older speakers had a significantly greater and larger numbers of inflections than the younger persons. The older women exhibit significantly higher minimum and significantly lower maximum intensity than the younger women. The voice of the aged female is characterized by a decrease in fundamental frequency.

Eichhorn et al. (4) aimed to study the effect of aging on vocal fundamental frequency (F0) and the vowel formants in men and women. Authors considered 96 participants (43 males and 53 females) who were in the age range of 20–90. The results indicated that as age increases there was a significant decrease in F0 in women. Significant differences in F1, F2, and F3 were noticed between vowels and gender. No significant differences in women were observed for the highest fourth formant (F4). The authors concluded that the women experience a significant decrease in F0, which is likely related to menopause.

From the above studies, it is clear that the effect of aging and development on formant frequencies in Indian languages are very few. Hence, the need of the present study arise.

2 Aim of the Study

The present study made an attempt to determine the effect of age on formant frequencies (F1 and F2) in Kannada-speaking individuals.

2.1 Method

Participants: The study involved eight males and eight females. They were divided into four age groups. Group I included four children of 8–10 years of age; Group II included the young adolescents of four in number whose age was between 18 and 20 years; Group III consisted of four subjects of 48–50 years of age; and Group IV consisted four subjects of above 60 years of age who were native Kannada speakers. Each group had two males and two females. Also, all the participants had Kannada as their mother tongue. Also, the participants did not report of any speech, language, and hearing difficulty at the time of study.

Material: The participants were instructed to repeat the non-sense words (V1CV2) in which V2 was always /a/. V1 consisted of three short vowels and three long vowels (aka, ika, uka, a: ka, i: ka, u:ka). V1 was succeeded always by the consonant /k/in the middle of the non-sense word. Hence, a total of six non-sense words were considered as material for the study.

Procedure: The objectives of the study were explained to the participants and informed written consent was obtained from them. All the participants were tested individually in a noise-free room. Participants were asked to repeat each non-sense word thrice in a normal rate. PRAAT software was used to record the samples by using a laptop (Lenovo G500). The microphone to mouth distance was maintained as 10 cm constantly for all the participants. A total of 288 tokens (16 participants $\times$ 6 vowels $\times$ 3 repetitions) were made for the study.

Analysis: The first (F1) and second (F2) formant frequencies were extracted by using PRAAT software. The steady portion of the vowel (V1) was selected to measure F1 and F2, excluding the onset and offset of the vowels. The average of the F1 and F2 was calculated by averaging the three repetitions for each participant and each of the vowels. Then the F1 and F2 were compared across the age groups and gender.

3 Results and Discussion

The results of the present study are discussed under the following three sub-headings:

(a) First formant frequency (F1) across different age groups,
(b) Second formant frequency (F2) across different age groups,
(c) Gender difference in F1 and F2, and

(a) F1 across different age groups.

Table 1 shows the average F1 values for both short and long vowels between males and females across different age groups. From Table 1, the average F1 values were higher for low vowel /a/when compared to vowels /i/and /u/(high vowels). Also, average F1 values tend to decrease from Group I to Group IV, that is, as age increases, the F1 value decreases.

The present study found that low vowel (/a/) had high F1 value compared to high vowels (/i/& /u/). This was observed in both short and long vowels and across both genders. The results of the present study are in consonance with the findings of Fant (1960) study who reported F1 is inversely proportional to tongue height. That is, when the height of the tongue is low, F1 would be higher and when the tongue height is high, the F1 would be lower. The results of present study are in consonance with the findings of Abolhasanizadeh, Karimabadi, Ayazi, and Moghadam's (2014) where they found that F1 decreases with age.

(b) F2 across different age groups

Table 2 shows the average F2 values for both males and females across different age groups. From Table 2, it can be observed that F2 values were lower for back vowel (/u/) when compared to mid and front vowels /i/. The front vowel (/i/) had the highest F2 value compared to mid /a/and back vowel (/a/). This was observed in both short and long vowels and in both genders. Also, average F2 values tend to

Table 1 Average F1 of short and long vowels across different age groups and genders

Groups	Gender	F1 value in Hz					
		Short vowels			Long vowels		
		/a/	/i/	/u/	/a:/	/i:/	/u:/
Group I	M	1004	405	475	1125	395	439
	F	1105	375	485	1173	326	526
Group II	M	683	336	748	1377	2129	876
	F	895	420	467	946	373	490
Group III	M	642	470	453	765	400	464
	F	950	464	501	1092	398	504
Group IV	M	664	324	440	820	332	383
	F	740	367	427	760	406	464

Table 2 Average F2 of short and long vowels across different age groups and genders

Groups	Gender	F2 value in Hz					
		Short vowels			Long vowels		
		/a/	/i/	/u/	/a:/	/i:/	/u:/
Group I	M	1703	1636	1329	1782	1853	1147
	F	1378	1651	876	1497	1729	1008
Group II	M	723	2289	384	1256	2253	1519
	F	1478	2679	1008	1387	2610	1005
Group III	M	1347	2274	963	1355	2452	807
	F	1574	2501	1096	1444	2180	1088
Group IV	M	1225	2386	1002	1214	2420	799
	F	1394	2266	941	1390	2348	950

decrease from group I to group IV. That is, as age increases, the F2 value decreases. The results of present study are in consonance with the findings of Abolhasanizadeh, Karimabadi, Ayazi, and Moghadam (2014), where they also reported reduced F2 values as a function of age (increase in age leads to decrease in F2 values).

The present study found that front vowel (/i/) had higher F2 value compared to back vowel (/u/). The F2 value of mid vowel (/a/) occurs between vowels /i/and /u/. This was observed in both short and long vowel counterparts. The results of the present study are in agreement with the findings of Fant (1960) where study reported that F2 is directly proportional to tongue advancement in the oral cavity. That is, more the tongue advanced/placed front as in case of high vowel (/i/), higher the F2 value.

The present study also found that average F1 and F2 values reduce (from group I) when age is increased further. The formant frequency depends on the size of the oral cavity. When the oral cavity size is small (in children), the resonance/formants would be higher. As the person grows further, the volume/size of the oral cavity becomes matured and increased, thereby reducing the resonance/formant frequency values.

(c) Gender difference

From Tables 1 and 2, the values of F1 and F2 were relatively higher for females compared to males. This was observed for both short and long vowels. The higher F1 and F2 values in females can be attributed to the source characteristics, that is, the fundamental frequency of voice is higher in females compared to males and further the resonance peaks would accordingly have amplified higher in them due to smaller vocal cavity (in females). The results of this study are in agreement with findings of Abolhasanizadeh et al. (5) and Sreedevi [6] which showed F2 values of vowels in females are more than that in males.

4 Summary and Conclusion

The present study aimed to analyze the change in formant frequencies across different age groups. The study considered four group of participants including children, adolescents, adults, and geriatric people as Group I, Group II, Group III, and Group IV, respectively. Subjects were asked to read six non-sense words. Using PRAAT, first formant frequency (F1) and second formant frequency (F2) were measured from the initial vowel of the non-sense word. The results showed that F1 and F2 were higher in females and as age increases F1 was reduced considerably than F2. As age increases, the anatomical structure of the vocal tract undergoes modifications and accordingly F1 and F2 values change. The results of the present study are not in consonance with results of previous study due to methodological differences.

5 Limitations

The present study considered less number of participants so to generalize the present findings will be difficult to whole population. Hence, more number of participants can be included in future study.

6 Future Direction

1. Study can be repeated with more number of subjects and different age groups.
2. Methodology can be adopted to study the formant changes in other Indian languages.

Acknowledgments We would like to thank Dr. Asha Yathiraj, Director, AIISH, Mysuru, for permitting us to carry out this study. Our heartful thanks to all participants in the study.

References

1. Kaur J, Narang V (2015) Variation of pitch and formants in different age group. Int J Multi Res Modern Edu 1:517–521
2. Mwangi S, Spiegl W, Honig F, Haderlein T, Maier A, Noth E (2009) Effects of vocal aging on fundamental frequency and formants. In: Marinus M. Boone (ed) Proceedings of the international conference on acoustics NAG/DEGA 2009. N.A.G, Rotterdam, pp 1761–1764
3. Robb MP, Chen Y, Gilbert HR (1997) Developmental aspects of formant frequency and bandwidth in infants and toddlers. FoliaPhoniatrica etlogopaedica 49:88–95
4. Eichhorn JT, Kent RD, Austin D, Vorperian HK (2017) Effects of aging on vocal fundamental frequency and vowel formants in Men and Women. J Voice. ISSN 0892-1997

5. Abolhasanizadeh V, Karimabadi H, Ayazi E, Sharifi Moghadam A (2014) The effect of age and sex on the acoustic characteristics of speech, language design 16:105–116
6. Sreedevi N (2000) Acoustic characteristics of vowel in Kannada. Submitted to the University of Mysore, Mysuru, Karnataka

Evaluation of Temporal Resolution Around the Tinnitus Frequency in Adults with Tonal Tinnitus

Prithivi Thanikaiarasu, Udhayakumar Ravirose, and Prashanth Prabhu

Abstract **Objective**: The perceptual characteristics of tinnitus are usually assessed by a matching procedure, where loudness and pitch of an external sound are matched to those of the tinnitus percept. For a complete assessment of tinnitus, central auditory processing abilities should be considered in addition to other routine evaluation. Temporal processing is one of the important auditory processing skills that is essential for complex higher level auditory processing. The gap detection test (GDT) and duration discrimination test are relatively simple psychoacoustic methods of measuring temporal resolution [1]. Hence, the present study is aimed at finding the duration discrimination threshold in tinnitus patients at the tinnitus frequency and half an octave above and below the frequency of tinnitus perception. **Method**: 15 participants with normal or minimal hearing loss with tinnitus in the age range of 18–40 years were enrolled in the present study. Duration discrimination test was administered on all the participants using MATLAB software (MLP toolbox) at the matched tinnitus frequency, half an octave below and above the frequency of tinnitus perception. **Results**: The results of the study revealed that there was a significant increase in duration discrimination thresholds at the tinnitus frequency compared to half an octave above and below the matched frequency. The result suggests that temporal resolution abilities are affected at the tinnitus frequency in individuals with tinnitus. **Conclusions**: The results of the study suggest that the temporal processing ability is affected in individuals with tinnitus especially at the frequency of tinnitus.

Keywords Tinnitus · Temporal resolution · Duration discrimination test

1 Introduction

Tinnitus is a perception of sound without an external source. It is a common disorder with prevalence estimates ranging from 7 to 20% among randomly selected populations [2]. Tinnitus is associated with a variety of disorders in the auditory system,

P. Thanikaiarasu (✉) · U. Ravirose · P. Prabhu
All India Institute of Speech and Hearing, Mysore, Karnataka, India
e-mail: prithiviarasu@gmail.com

M. Singh and Y. Rafat (eds.), *Recent Developments in Acoustics*, Lecture Notes in Mechanical Engineering, https://doi.org/10.1007/978-981-15-5776-7_5

whether generated peripherally or centrally, but may arise spontaneously, too. Studies have shown an association between the presence of tinnitus and neural activity disorders in the auditory central nervous system, from the peripheral level (dorsal cochlear nucleus) to the cortical level [3–6]. Evaluation of tinnitus may not be complete without assessing the effect of tinnitus on different auditory skills. For a complete assessment of tinnitus, central auditory processing abilities should be considered in addition to the routine psychological evaluation of tinnitus characteristics. Temporal resolution refers to the ability to detect changes in acoustic stimuli over time. It is important for resolving brief dips in the intensity of the interfering noise and, therefore, is critical for understanding speech in these situations [2, 7–9].

Measurement of duration discrimination threshold has traditionally been used for the identification of temporal discrimination deficits in a variety of clinical populations. It was hypothesized that neural activity in tinnitus patients might create deficits in their ability in temporal processing when compared to those non-tinnitus individuals. A research work on temporal resolution abilities in individuals with tinnitus using gap detection in noise and duration pattern test at 1 kHz frequency revealed some amount of temporal processing ability being affected in individuals with the perception of tinnitus [10]. However, previous studies have assessed temporal resolution in tinnitus patients only using GDT [10]. There are no studies which have used duration discrimination task to assess temporal processing in individuals with tinnitus. In addition, none of the studies have attempted to evaluate temporal processing around the perceived tinnitus pitch. The results of such a study would provide important information regarding the relationship between peripheral and central mechanisms of tinnitus generation that may affect auditory processing. Hence, the present study aimed at finding the duration discrimination threshold in tinnitus patients at the tinnitus frequency, half an octave above and below the frequency of tinnitus perception.

2 Method

A total number of 15 participants were considered for the study with the complaint of tinnitus. Individuals having continuous tinnitus, tonal perception, and in age range 18–45 years, with normal hearing status and/or minimal hearing loss, were included in the study. After obtaining informed consent and local ethics committee approval, all participants underwent the test procedures as follows.

2.1 Instrumentation

- A calibrated audiometer (Madsen Astra) coupled with an acoustically matched headphone (TDH-39) and a bone conductor (B-71) was utilized to estimate pure tone threshold, speech recognition threshold, and speech identification score.

- Tympanometry and acoustic reflex thresholds (ART) were obtained using a calibrated immittance meter (GSI-Tympstar).
- An otoacoustic emission was done using ILO software to check for the functioning of outer hair cell functioning, DPOAE was done.
- A Dell laptop with maximum likelihood procedure (MLP) toolbox implemented in MATLAB software version 7.10 was used to carry out duration discrimination test.
- A TDH-39 headphone was calibrated for the output of the computer at 60 dB SPL which was used to present the stimulus through the computer.

2.2 *Pure Tone Audiometry*

Participants were subjected to audiometric testing using Madsen Astra audiometer at the frequencies of 0.25, 0.5, 1.0, 2.0, 4.0, and 8.0 kHz for air conduction and between 0.25 and 4.0 kHz for bone conduction to confirm normal hearing status and/or with minimal hearing loss.

2.3 *Immittance Audiometry and Otoacoustic Emissions*

Immittance audiometry was done using GSI-Tympstar to carry out tympanometry and reflexometry to rule out any middle ear pathology and distortion product otoacoustic emissions were done using ILO software to check for outer hair cell functioning and to confirm the diagnosis of normal hearing sensitivity and/or minimal hearing loss.

2.4 *Tinnitus Psychoacoustic Assessment*

The psychoacoustic characteristics of the tinnitus were evaluated in all patients in order to define the pitch and loudness of the perceived tinnitus. Pitch matching attempts to quantify tinnitus in terms of its possible frequency, in that, pure tones were presented to the patient and the patients were asked to choose which one most closely matches the tinnitus that they hear. This matched frequency was taken as a reference signal for loudness matching.

In loudness matching, the intensity of the tone varied in 5 dB steps until the subject heard the sound equally loud as that of tinnitus perceived. This was helpful in recruiting participants with tonal perception of tinnitus and the matched tinnitus frequency was considered for carrying out duration discrimination test.

2.5 *Duration Discrimination Test*

The minimum difference in duration which a participant can discriminate was assessed. Duration discrimination thresholds (DDT) were measured for a pure tone of three frequencies which included at the matched tinnitus frequency, half octave and below the tinnitus frequency. The tone had raised cosine onset and offset gates of 10 ms. The minimum and maximum values of duration deviation used were 0.1 ms and 200.1 ms. On each trial of three blocks, two blocks had pure tones at a standard duration and other blocks selected at random contained a pure tone of variable duration, which was always longer than the standard duration. The participant's task was to identify the variable block. A 3AFC procedure was used by implementing MLP toolbox in the MATLAB software version 7.10 which was used to estimate the duration difference.

3 Statistical Analysis

Statistical analysis was done using SPSS software version 20.

4 Results

The mean and standard deviation of duration discrimination thresholds were determined and shown in Fig. 1. The figure shows that the duration discrimination thresholds were higher at tinnitus frequency compared to half octave above and below the tinnitus frequency.

Shapiro–Wilk test of normality was done to determine if the data was normally distributed. The result of test of normality shows that the data was not normally distributed ($p < 0.01$). Hence, non-parametric inferential statistics were done. Friedman's test was done to determine if there is any significant difference in DDT across the three conditions. The results of Freidman's test [χ^2 (2) = 28.09, $p < 0.01$] showed that there was significant difference across the conditions. Hence, Wilcoxon signed rank tests were done to separately compare two conditions. The results show that there was a significant increase ($p < 0.01$) in duration discrimination thresholds at the tinnitus frequency compared to half octave above [$Z = -4.04$, $p < 0.01$] and below the matched frequency [$Z = -4.04$, $p < 0.01$]. There was no significant difference [$Z = -1.08$, $p > 0.05$)] in DDT between half octave above and half octave below tinnitus frequency conditions. The result suggests that temporal resolution abilities are affected at the tinnitus frequency in individuals with tinnitus. The consequences of poor temporal resolution in individuals with tinnitus are discussed.

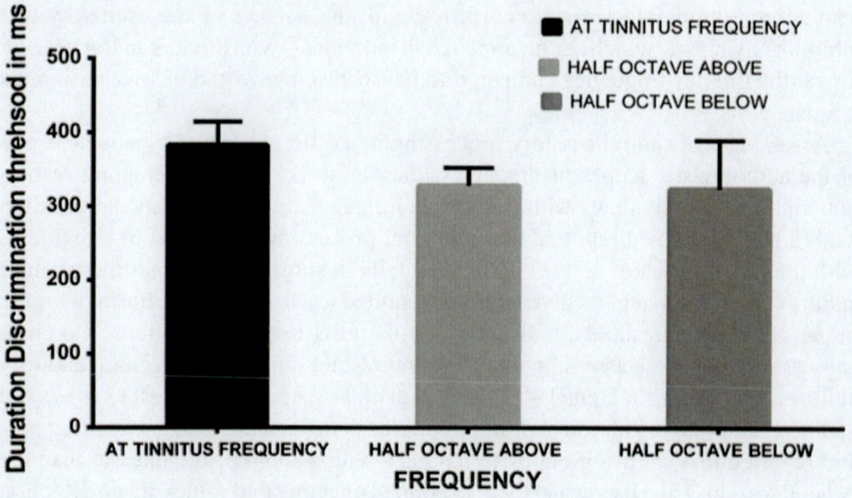

Fig. 1 Mean and SD of duration discrimination threshold at tinnitus frequency, half octave above and half octave blow the tinnitus frequency

5 Discussion

Tinnitus by definition is a phantom perception of sound without external source. Tinnitus is an otologic symptom and despite the great amount of research on this subject, the exact pathophysiological process of tinnitus remains unclear. Involvement of the whole auditory system, either peripheral or central, should be considered in the development of tinnitus. Some authors postulated that the presence of tinnitus has been associated with a disorder in the neural activity of the auditory system. A cochlear disorder, even when undiagnosed by pure tone audiometry, may initiate a series of processes in the nervous system that may result in tinnitus [11]. Temporal processing is one of the important auditory processing skills that are essential for complex higher level auditory processing. Temporal resolution is an auditory temporal processing skill that refers to the minimal time required to segregate or resolve acoustic events. The gap detection test (GDT) and duration discrimination test are relatively simple psychoacoustic methods of measuring temporal resolution [1].

Additionally, according to some published data, deficits in neural structures in the central auditory nervous system may result in the perception of tinnitus [12]. Bartels et al. [13] stated that an altered afferent input to the auditory pathway may be the initiator of a complex sequence of events, conclusively resulting in the generation of tinnitus at the central level of the auditory nervous system [13].

The present study investigated the ability of the duration discrimination test to differentiate between a group of patients with tinnitus who showed normal hearing sensitivity and/or minimal hearing loss. Good performance in auditory temporal resolution requires precise neuronal firing with the balance of the excitatory and inhibitory synapses, which can be impaired in individuals with tinnitus and/or hearing loss at the tinnitus frequency compared to half octave above and below the tinnitus frequency.

Assessment of central auditory processing in a different group of patients is one of the audiologist's scopes of practice. Gilani et al. [10] studied temporal resolution abilities in individuals with tinnitus using gap detection in noise and duration pattern test [10]. They reported that temporal processing is affected in individuals with tinnitus. Musiek et al. [14] performed GIN in subjects with confirmed central auditory nervous system involvement and reported a statistically significant increase in gap detection thresholds, indicating that the GIN test holds promise as a clinically useful tool in the assessment of temporal resolution, one of the central auditory abilities, in the clinical arena [14]. Sanches et al. [15] applied GIN test to assess the auditory temporal resolution skill in 18 tinnitus patients and 23 normal participants and reported that control group detected gaps with a shorter time interval than the patient's group [15]. Haas et al. [16] also pointed out threshold values of gap detection in tinnitus patients were longer in duration than non-tinnitus subjects and hypothesized that some changes in neural activity in tinnitus patients might prolong gap detection threshold (GDT) [16]. Fournier and Hebert [1] assessed gap detection in human with tinnitus and postulated that tinnitus group displayed a consistent deficit in gap processing at both low and high background noise frequencies, assuming that ongoing tinnitus masks the gap and results in their impaired gap detection [1].

In several studies, gap detection test and gap-in-noise test were done to assess temporal processing ability and whereas they have not attempted to assess using duration discrimination test and also at the frequency of tinnitus, half an octave below and above the tinnitus frequency. Hence, in the present study, reveals that the higher approximate threshold that the tinnitus patients required to detect tones of varying gap in order to discriminate tones detection, and is in accordance with the literature data, which revealed poorer temporal acuity abilities, especially at the tinnitus frequency.

6 Conclusions

In the present study, temporal processing difficulties were found in individuals with tinnitus having normal hearing and/or minimal hearing loss. The thresholds were poor especially at the tinnitus frequency when compared to that of the half octave frequency above and below of the tinnitus frequency. Thus, the implication of our study indicates the need for assessing the central auditory processing difficulties in individuals with tinnitus in addition to the routine testing of evaluating tinnitus. Further studies can use other tests of evaluating various auditory processing difficulties to incorporate a comprehensive test battery for more precise diagnosis for better remediation measures.

References

1. Fournier P, Hébert S (2013) Gap detection deficits in humans with tinnitus as assessed with the acoustic startle paradigm: does tinnitus fill in the gap? Hear Res 295:16–23
2. Hoffman HJ, Reed GW (2004) Epidemiology of tinnitus BT—tinnitus: theory and management. In: Tinnitus: theory and management, vol 3, pp 16–41
3. Eggermont JJ, Roberts LE (2004) The neuroscience of tinnitus. Trends Neurosci 27(11): 676–682
4. Weisz N et al (2013) The functional neuroanatomy of tinnitus: evidence for limbic system links and neural plasticity. Ear Hear 27(6):1479–1484
5. Lockwood AH, Salvi RJ, Coad ML, Towsley ML, Wack DS, Murphy BW (1998) The functional neuroanatomy of tinnitus: evidence for limbic system links and neural plasticity. Neurology 50(1):114–120
6. Weisz N, Muller S, Schlee W, Dohrmann K, Hartmann T, Elbert T (2007) The neural code of auditory phantom perception. J Neurosci 27(6):1479–1484
7. Dubno JR, Horwitz AR, Ahlstrom JB (2003) Recovery from prior stimulation: masking of speech by interrupted noise for younger and older adults with normal hearing. J Acoust Soc Am 113(4):2084–2094
8. Oxenham AJ, Bacon SP (2003) Cochlear compression: perceptual measures and implications for normal and impaired hearing. Ear Hear 24(5):352–366
9. Peters RW, Moore BCJ, Baer T (1998) Speech reception thresholds in noise with and without spectral and temporal dips for hearing-impaired and normally hearing people. J Acoust Soc Am 103(1):577–587
10. Gilani VM et al (2013) Temporal processing evaluation in tinnitus patients: results on analysis of gap in noise and duration pattern test. Iran J Otorhinolaryngol 25(73):221–225
11. Onishi ET, Fukuda Y, Suzuki FA (2004) Distortion product otoacoustic emissions in tinnitus patients. Int Tinnitus J 10:13–16
12. Bartels H, Staal MJ, Albers FWJ (2007) Tinnitus and neural plasticity of the brain. Otol Neurotol 28(2):178–184
13. Bartels H et al (2008) The additive effect of co-occurring anxiety and depression on health status, quality of life and coping strategies in help-seeking tinnitus sufferers. Ear Hear 29(6):947–956
14. Musiek FE, Shinn JB, Jirsa R, Bamiou D-E, Baran JA, Zaida E (2005) GIN (Gaps-In-Noise) test performance in subjects with confirmed central auditory nervous system involvement. Ear Hear 26(6):608–618
15. Sanches SGG, Sanchez TG, Carvallo RMM (2010) Influence of cochlear function on auditory temporal resolution in tinnitus patients. Audiol Neurotol 15(5):273–281
16. Haas M, Smurzynski R, Fagelson J (2012) The effect of tinnitus on gap detection. Am Tinnitus Assoc 10–11

Acoustic Analysis of Voice of Temple Priests

V. Priyadharshini, M. Vasupradaa, and K. Yeshoda

Abstract Professional voice users include singers, teachers, actors, professional speakers, and other entertainers. There are a group of non-elite voice users such as priests, vendors, bus conductors, etc., who are also dependent on their voice for daily living. Temple priests who chant Vedas and mantras the whole day without any form of amplification are at higher risk. Vedas are the foundational literature of Hinduism and it refers to infinitely large collection of mantras. The acoustical characteristics, in addition to the clear phonetic articulation of chants, have deep impact on the devotees. For the renditions to be clear and for it to have an impact, the priests unknowingly tend to put more effort into it. To assess the voice characteristics, perceptual and objective measures have been used extensively. Acoustic analysis is one of the gold standard measures to quantify the voice parameters. The literature mostly concentrates on the vocal load in elite professional voice users. Studies on priests are predominantly in terms of awareness of their voice problems through the use of questionnaires and have majority involved church priests. There are minimal studies done using quantitative acoustic analysis on voice among temple priests. Hence, this study aimed

- To investigate variations in acoustic measures across different tasks (reading, monologue, and chanting).

Method: 14 temple priests, in the age range of 35–55 years, who had undergone standard gurukula training, with minimum 5 years of experience were selected. All of them were native speakers of Tamil language. It was made sure that all the participants had only the occupation of being a temple priest, without any other sources of employment, in order to avoid the bias. The participants did the following tasks, namely, reading, chanting, and monologue in their habitual manner. The voice samples were recorded using a digital recorder (Olympus LS) in a quiet environment in the temple. Acoustic analysis was done using real-time pitch software from CSL4500. **Results**: Findings indicated that the mean frequency range and perturbations were noticeably higher compared to other groups of professional voice users. In general, there was no trend noticed.

V. Priyadharshini (✉) · M. Vasupradaa · K. Yeshoda
All India Institute of Speech and Hearing, Mysore-06, Mysore, Karnataka, India
e-mail: jayceechan7@gmail.com

M. Singh and Y. Rafat (eds.), *Recent Developments in Acoustics*, Lecture Notes in Mechanical Engineering, https://doi.org/10.1007/978-981-15-5776-7_6

Keywords Voice · Non-elite voice users · Vocal load · Temple priests · Acoustic analysis

1 Introduction

A normal voice production involves coordination between different subsystems, namely, the respiratory, articulatory, phonatory, and resonatory systems. Voice varies widely with respect to pitch, loudness, and quality. The size, length, tension, and mass of the vocal folds affect the loudness, quality, and pitch of the vocal output, which will be modified by other parts of the vocal tract [1]. Functional causes seem to be more prevalent due to today's lifestyle changes, multi-tasking in daily life, improper diet, lack of concern on voice usage, etc., which contribute to having an abnormal quality of voice. Both professional and non-professional voice users are prone to get affected quality of voice, due to any or combination of the causes.

Professional voice users include singers, teachers, actors, professional speakers, and other entertainers. By the nature of their professions, professional voice users are at greater risk for voice disorders. Professional voice users include not only the singers and actors, but also clergy, teachers, receptionists, sales personnel, physicians, and anyone else whose ability to earn a living is impacted negatively by loss of vocal quality and endurance [2]. There are a group of non-elite voice users such as priests, vendors, bus conductors, etc., who are also dependent on their voice for daily living. Temple priests who chant Vedas and mantras the whole day, without any form of amplification, are at higher risk.

Vedas are the foundational literature of Hinduism and it refers to infinitely large collection of mantras. The Vedas are chanted with many precise and accurate acoustical characteristics. It is for this reason that oral tradition has been one of the very efficient ways, in transmitting the Vedic chanting from master to disciple [3]. In 2003, UNESCO [4] proclaimed the notion that oral tradition of the Vedic chanting is an intangible cultural heritage of the world. The proclamation recognizes the fact that "*To ensure that the sound of each word remains unaltered, practitioners are taught from childhood complex recitation techniques that are based on tonal accents, a unique manner of pronouncing each letter and specific speech combinations*" [4]. Sir C.V. Raman in the year 1922 [5] opined that "*it would form a fascinating chapter of history to try and trace the gradual development of musical instruments and musical knowledge, from the rhythmic chanting of the Rig-Veda in the ancient home of Aryan race to the Indian music of the present day.*"

The sounds of the mantras chanted in worship places carry the listeners to spiritual experiences. The acoustical characteristics along with the clear phonetic articulation of chants have a deep impact on the devotees [6–8]. For the renditions to be clear and for it to have an impact, the priests unknowingly tend to put more effort into it.

Different kinds of analysis such as time-domain, frequency-domain, time-frequency domain, and non-linear dynamics are applied based on the parameters of interest. For the analysis of voice, different methods such as perceptual, aerodynamic, and acoustic measures have been proposed in the literature. Acoustic analysis seems to be more convenient measure of voice in terms of practicality, ease of procedure, and non-invasive nature.

The prevailing literature in professional voice users had mostly concentrated on the elite group of voice users. In line of non-elite professional voice users, limited studies have been noted and majority of them have focussed on the awareness of voice problems among the population of interest.

1.1 Studies on Voice Usage Among Priests

Hagelberg and Simberg [9] studied the prevalence of voice problems in priests and some risk factors contributing to them. They sent an electronic questionnaire to 2044 Lutheran priests in Finland and 44% of them had responded by filling in the questionnaires. Results indicated an increased prevalence of voice problems in priests with 26.7% of them stating that they had experienced frequently occurring vocal symptoms, 24.5% of them had conveyed that they had sought help for voice problems, 21% considered themselves as having voice problems, 18% diagnosed with voice disorder by a physician, and remaining 11.6% reported voice-related absenteeism.

In Indian context, Devadas et al. [10] studied the prevalence and risk factors of voice problems in priests in Kerala and had used a self-reported questionnaire for the same. 270 Marthoma priests with 1–35 years of professional experience were included in the study and the results revealed 17.8% had frequent voice problems during their career and it was also found that significantly higher number of priests with frequent voice problems missed their work.

In another study, done by Kasim and D'souza [11], Voice Handicap Index (VHI) and Voice-Related Quality of Life (VRQOL) scores were obtained among temple priests, in which 50 temple priests in the age range of 20–60 years were randomly selected and divided into four groups (20–30 years, 30–40 years, 40–50 years, and 50–60 years). The results revealed a high significant difference between the four groups for the VHI domains and no significant difference for VRQOL scores.

Studies in the domain of voice usage among priests have been done among church and temple priests, using standardized questionnaires to assess the awareness of voice problems that they face due to their professional voice demands at their workplaces. The need to do an acoustical analysis in terms of the perturbation measures is necessary to check the occupational vocal demands in that population. There had not been any quantitative acoustic analysis on voice among temple priests.

2 Aim of the Study

To study the vocal characteristics, in the voice of temple priests, using acoustic analysis.

Table 1 Participants details

Demographics	Mean	Range
Age (years)	40.57	35–55 (one subject was a 16 year old who was currently on gurukula training)
Number of years of experience	7.57	5–16
Average talking hours per day (excluding chanting)	10.92	8–14

3 Objective

- To investigate variations in acoustic measures across different tasks (reading, monologue, and chanting).

4 Method

4.1 Participants

- 14 temple priests, in the age range of 35–55 years (Table 1).
- All of them were native speakers of Tamil language.
- All the participants had only the occupation of being a temple priest, without any other sources of employment, in order to avoid the bias.
- All had undergone standard gurukula training and had with minimum 5 years of experience as temple priests.

Table 1 shows the participant details in terms of their age, experience, and their average voice usage per day.

These details were collected prior to orienting the participants toward the purpose and tasks of the study.

4.2 Tasks

The participants were given the following tasks, namely, reading, chanting, and monologue to be done in their habitual manner. A common reading passage was given to all the 14 participants and the passage was taken from the Bhagavat Gita, the Hindu scripture in Tamil which is a part of the epic *Mahabharata* (Chaps. 23–40 of the 6th book of Mahabharata) for the reading task. The participants were asked to chant the lines of "Ganesh sloka" (i.e., "Vakrathunda Mahakaya") for the chanting task. They were asked to speak about themselves for monologue recording.

4.3 Procedure

The information about name of the priest, the address of the temple in which they are working, and the number of years completed in this profession were noted. The purpose of the study, assurance of confidentiality was clearly explained to the participants and the authors made sure that their participation in the study was voluntary. The participants were recruited for the study only after informed consent. The recordings were carried out individually in a quiet environment in the temple. The Participants were asked to be seated comfortably and the authors recorded their voice samples using a digital recorder (Olympus LS 100). The microphone to mouth distance was maintained at 10 cm, while audio recording of the data.

4.4 Analysis

The recorded samples were transferred to the external module of CSL 4500 (Pentax, USA) for the acoustic analysis using Real-Time pitch software. Reading, chanting, and the monologue were analyzed using Real-time pitch. The following acoustic parameters were extracted for each of the three tasks:

- Mean frequency,
- Minimum frequency,
- Range,
- Maximum frequency,
- Standard deviation in F_0 (SD of F_0), and
- Variation in F_0 (vF_o).

The 42 samples of reading, chanting, and monologue data (14 participants X 3 tasks = 42 recordings) were analyzed using Statistical Package for the Social Sciences (SPSS) software version 17. Descriptive statistics were extracted. The details are provided in the results section and the same are discussed.

5 Results and Discussion

The mean, minimum, and maximum frequencies; standard deviation F_0 (SDF_0); variation in F_0 (Vf_0); and the range were obtained as shown in Table 2. These measures were compared among three tasks, namely, chanting, reading, and monologue.

From Table 2, it is observed that the Mean (M) was highest for chanting, followed by monologue and reading. In minimum frequency, M was lowest for monologue and highest for reading. For the maximum frequency, M was highest for monologue and lowest for reading. This indicates that the range was highest for monologue,

Table 2 Mean (M) and standard deviation (SD) for the different tasks

Parameters	Chanting		Reading		Monologue	
	M	SD	M	SD	M	SD
Mean frequency	153.11	24.59	139.91	23.75	145.94	25.49
Minimum frequency	116.02	23.18	120.25	19.04	90.63	11.26
Range	151.45	84.79	81.59	73.79	234.27	44.50
Maximum frequency	267.48	77.12	201.85	72.32	325.05	43.04
SDF_0	33.67	24.17	23.55	18.73	34.22	9.36
vF_0	0.20	0.11	0.22	0.03	0.41	0.11

followed by chanting and reading. The M for standard deviation F_0 was highest for monologue and lowest for reading. In terms of vF_0, M was highest for monologue.

Shapiro–Wilk test for normality reveals that the obtained data follows normal distribution, i.e., $p > 0.05$. Hence, repeated measure ANOVA was carried out to see the significant difference between the three tasks. Further, post hoc analysis, that is, Bonferrine's pairwise comparison test was done to check for the pairwise significant difference between the tasks, in each parameter.

Pairwise comparison revealed that there was a significant difference, i.e., $(p > 0.05)$ in mean frequency between chanting and reading. Among the other parameters, there was a significant difference noted between monologue and reading. Pairwise comparison was done for all except the standard deviation (SDF_0) and variation in f_0 (vf_0), as there was no significant differences observed for those two parameters across the tasks.

The aim of the study was to investigate variations in acoustic measures across different tasks (reading, monologue, and chanting). In general, the results revealed significant difference in mean frequency. In comparing between the tasks across the parameters, it was found that the mean frequency varied significantly between two tasks, namely, the chanting and reading tasks.All the other parameters such as minimum and maximum frequencies, variation in F_0, standard deviation of F_0, and range showed significant difference between reading and monologue tasks.

This indicates that the temple priests were expressive in their monologue followed by chanting as noticed by increased range of F_0. The variation in F_0 was maximum for monologue and least for reading.

Increased mean MF_0 in chanting indicates that the tension of the vocal folds is increased compared to other two tasks. This may result in additional strain on the vocal folds. Considering this result, it may be opined that chanting is more vocally loading compared to monologue and reading. The range is maximum in monologue and least is reading. The range of chanting is in between the two tasks. It may be speculated that maximum variations in F_0 are always good characteristic contributing to expressiveness in speech.

6 Conclusion

It can be concluded from the results that there is a significant difference between mean frequencies in chanting and reading. According to Jiang and Titze [12], impact stress for vocal fold tissue increases as a function of F_0, vocal intensity, and adduction forces resulting in increased vocal load. The increase in mean frequency during chanting indicates that the priests tend to exert more effort in order to deliver the articulatory and acoustic precisions while rendering it. This effort tends to increase the tension at the level of vocal folds, which in turn causes the fundamental frequency to rise. More studies can be done in this line to check for the significant difference between different tasks and using same or different stimuli across the tasks to note the perturbation changes in voice. This can also serve as a baseline for assessment and treatment.

Acknowledgements The authors would like to thank all the participants who participated in the study. We also like to thank Director, All India Institute of Speech and Hearing—Mysuru, for providing us with sophisticated instruments to carry out the data collection and analysis part.

Ethical standards Participants of the study were included only after informed consent. The study protocol has been approved by the All India Institute of Speech and Hearing research committee on human research, Mysore, Karnataka, India.

References

1. Titze IR (1995) The utility of acoustic measures of voice quality. In: Proceedings of the workshop on acoustic voice analysis, University of Iowa, Iowa city
2. Sataloff RT (2000) Professional voice users: the evaluation of voice disorders. Occupational Medicine (Philadelphia, Pa.) 4(16): 633–647
3. Prasad MG (2013) Oral tradition in Indian civilization. In Chapter 6, Perspectives on the origin of Indian civilization, New Delhi: Center for Indic Studies and D K Printworld
4. Prasad MG, Rajavel B (2013) Acoustics of chants, conch-shells, bells and gongs in hindu worship spaces, M.S. Narayanan's memorial lecture, New Delhi, India
5. Raman CV (1922) The acoustical knowledge of the ancient Hindus. Ashutosha Mukherji Silver Jubilee 2:179–185
6. Mukhopadhyaya S (1999) Importance of sound in the tradition of vedic chanting. In: SC Malik (ed) Dhvani: nature and culture of sound, Indira Gandhi National Centre for the Arts, New Delhi
7. Prasad MG (2004) Mantras to music: acoustics in Hinduism. In: 147th Acoustical society of America Meeting, New York, USA, 2004. www.acoustics.org/press/147th/prasad.htm
8. Prasad MG (2010) On the role of acoustics in the Vedic Hindu tradition and philosophy. J Acoust Soc Am 3(1983):127
9. Hagelberg AM, Simberg S (2015) Prevalence of voice problems in priests and some risk factors contributing to them. J Voice 29(3):389.e11–389.e18
10. Devadas U, Jose N, Gunjawate D (2016) Prevalence and influencing risk factors of voice problems in priests in Kerala. J Voice 30(6):771.e27–771.e32
11. Huda Kasim N, D'souza JM VHI &VRQOL in temple Priests. Lang India 16(4)
12. Jiang JJ, Titze IR (1994) Measurement of vocal fold intraglottal pressure and impact stress. J Voice 8(2):132–144

Perceptual Judgments of Resonance, Speech Understandability, and Speech Acceptability in Children with Repaired Cleft Palate Across Words and Sentences

M. Pushpavathi, Ajish K. Abraham, S. R. Mahadeva Prasanna, and K. S. Girish

Abstract ***Introduction***: In children, even with early repair of the cleft, speech errors such as hypernasality, atypical consonant production, and abnormal nasal airflow persist which affects overall speech intelligibility. Perceptual evaluation is considered as the gold standard in the speech assessment of individuals with cleft lip and palate. Speech intelligibility is an important and essential measure of disordered speech; furthermore, it is a major goal of therapeutic intervention. Various factors such as resonance, speech understandability, and speech acceptability vary across stimuli. ***Aim and Objectives***: The present study investigated resonance, speech understandability, and speech acceptability across stimuli (words versus sentences) through perceptual judgment and also correlation among the same was analyzed. ***Method***: The evaluation included 20 native speakers of Kannada language with repaired cleft palate in the age range between 6 and 12 years. The stimuli considered were ten meaningful Kannada words and ten oral Kannada sentences, both loaded with pressure consonants. Responses were collected and were perceptually rated using Henningsson's rating scale by three experienced speech-language pathologists. ***Results***: The results revealed no significant difference between words and sentences across resonance, speech understandability, and speech acceptability in the paired t-test. Cronbach's alpha was computed for words and sentences which indicated higher inter-judge reliability among the three judges ($\alpha > 0.70$). There was a significant correlation ($p < 0.05$) among resonance, speech understandability, and speech acceptability in both words and in sentences. ***Conclusion***: The present study concludes that resonance, speech understandability, and speech acceptability do not vary across stimuli in children with cleft lip and palate.

M. Pushpavathi (✉)
All India Institute of Speech and Hearing, Mysuru, Karnataka 570006, India
e-mail: pushpa19@yahoo.co.in

A. K. Abraham · K. S. Girish
Department of Electronics, All India Institute of Speech and Hearing, Mysuru, Karnataka 570006, India

S. R. Mahadeva Prasanna
Department of Electrical & Electronics Engineering, IIT, Guwahati, Guwahati, Assam, India

M. Singh and Y. Rafat (eds.), *Recent Developments in Acoustics*, Lecture Notes in Mechanical Engineering, https://doi.org/10.1007/978-981-15-5776-7_7

Keywords Repaired cleft lip and palate · Resonance · Speech intelligibility · Speech understandability · Speech acceptability · Henningsson's scale

List of Abbreviations

CLP	Cleft lip and palate
RCLP	Repaired cleft of lip and palate
SLP	Speech-language pathologist
SD	Standard deviation

1 Introduction

Cleft lip and palate (CLP) is one of the most common congenital birth defects whose speech is primarily characterized by abnormalities in oral resonance, nasal emission, and errors in articulation which affect the speech intelligibility [1, 10, 13]. Children with CLP require surgical intervention to establish appropriate oral motor skills that are adequate for normal speech production. However, even with early surgical repair of a cleft palate, speech errors persist as a result of associated velopharyngeal dysfunction. In very few children, these surgical interventions alone are sufficient to develop normal speech and language, yet a huge percentage of children with repaired cleft palate (RCLP) require speech therapy for better speech and language abilities. This calls for the multi-disciplinary approach in the management of children with CLP and RCLP.

It is noticeable that there are certain speech errors that are commonly observed in children with RCLP, like hypernasality, atypical consonant production, compensatory articulation, and abnormal airflow persist which affects overall speech intelligibility. These speech characteristics related to articulation and resonance problems in children with cleft palate requires a detailed assessment by speech-language pathologists (SLPs).

Although many tests procedures and protocols are used to evaluate speech production errors and nature of the disorders in CLP population, each individual must be carefully examined because a myriad of factors can contribute to the error patterns and individuals with CLP form a heterogeneous group. The perceptual assessment was considered as the gold standard [7] in the analysis of speech. Resonance and speech intelligibility are crucial measures of disordered speech and an important goal of therapeutic intervention for children with RCLP. Thus, carrying out a perceptual judgment of resonance and intelligibility is important in order to measure the speech outcomes and to determine appropriate treatment plans. Hence, Henningsson et al. [4] developed a framework of universal parameters for detailed documentation of speech outcomes in persons with cleft palate with a standardized four-point rating scale for assessing the severity of hypernasality and speech intelligibility.

Several studies have explored the impact of age, gender, stimuli, and the type of cleft on speech intelligibility in children with cleft palate. Van Lierde et al. [14] evaluated the impact of the type of cleft on speech intelligibility across 19 children with unilateral cleft lip and palate and 18 children with bilateral cleft lip and palate. Spontaneous speech and repetition of sentences were used as the stimuli and it was subjected to perceptual evaluation by three SLPs. They concluded that the speech intelligibility of children with bilateral cleft lip and palate was comparatively poorer than children with unilateral cleft lip and palate, but the difference was not significant.

Whitehill and Chau [16] investigated word intelligibility across the type of cleft, age, and gender in Cantonese-speaking population with cleft lip and palate. They considered 15 individuals with repaired cleft lip and palate in the age range of 5–44 years. 60 Chinese words loaded with pressure consonants were selected and it was made to repeat. Eight Cantonese-speaking individuals were selected as listeners and the responses recorded were given to the listeners for identification. They concluded that there was no effect of the type of cleft, age, and gender on speech intelligibility.

In the Indian context, Gnanavel and Pushpavathi [3] evaluated the impact type of cleft and stimuli on speech intelligibility between sentences and words in cleft lip and palate population. They considered a total of 20 participants in different cleft types (unrepaired cleft palate, repaired cleft lip and palate, repaired cleft palate, and repaired submucous cleft palate) with an equal number of participants in each category in the age range of 10–25 years. They were made to repeat the words and sentences loaded with pressure consonants in the Kannada language. Responses were recorded and presented to four SLPs for perceptual evaluation of speech intelligibility on a five-point rating scale given by Ray [12]. The study revealed that there was no stimuli effect on speech intelligibility. There was high inter-judge reliability among the four SLPs which was statistically significant (<0.05). Hence, they concluded from their study that speech intelligibility can be enhanced with the help of speech therapy soon after the surgical management.

Speech intelligibility is majorly affected by hypernasality or nasal air emission which causes the individual with CLP or RCLP to learn compensatory articulation errors (Myers [11]. Also, Lee et al. [9] stated that hypernasality is associated with various other speech problems and it eventually affects the speech intelligibility in children with cleft lip and palate. Very few studies have been carried out in correlating hypernasality and speech intelligibility. Kang et al. [6] assessed the correlation between speech intelligibility and speaking factors like language, articulation, nasalance score, and reading ability in 11 children with cleft palate and an equal number of typically developing children. They concluded that there was no statistically significant correlation between nasalance score and speech intelligibility.

However, there are drawbacks and disagreement in rating speech intelligibility using traditional rating scales [15]. So speech understandability and speech acceptability were proposed as global speech parameters in assessing cleft lip and palate population, which are closely related to the notion of intelligibility. "Speech understandability is the degree to which the speaker's message can be comprehended by the listener" and "Speech acceptability is the degree to which speaker's message is acceptable in the speaker's language or dialect" [4, 15]. For example, an individual with CLP may use glottal stops and the speech can be understood by the listener, but still the speech is not acceptable for that respective language.

Few investigations have been conducted on the effect of stimuli on speech intelligibility, especially in children. However, there is a dearth of studies in both Indian and Western context on perceptual assessment of resonance, speech understandability, and speech acceptability in children with RCLP using a standardized assessment protocol. Hence, there is a need to investigate the same across stimuli in children with RCLP. Thus, the present study aimed to investigate the resonance, speech understandability, and speech acceptability across stimuli (words versus sentences) through perceptual judgment in children with a repaired cleft palate. It also aimed to find the correlation among resonance, speech understandability, and speech acceptability in both words and sentences separately.

2 Method

2.1 Participants

Twenty non-syndromic children with RCLP in the age range of 6–12 years with an equal number of males and females were considered for the present study. All the children considered for the study had Kannada as their native language. The language abilities of all the children with RCLP were age adequate. Children with other associated problems like hearing loss, intellectual disability, and nasal pathologies were excluded from this study.

2.2 Materials and Procedure

The speech stimuli comprised of ten meaningful words and ten meaningful oral sentences, both loaded with pressure consonants which were developed by Jayakumar and Pushpavathi [5]. Participants were made to repeat the sentences and words after the tester. The responses were recorded using Brüel & Kjær Sound Level Meter (Type 2250-s handheld analyzer) in a sound-treated room. The recorded speech samples were presented to three speech-language pathologists (SLPs) for perceptual evaluation. The order of speech samples which were presented to SLPs was randomized. A standardized four-point rating scale was used for perceptual evaluation of resonance, speech understandability, and speech acceptability [4], where '0' indicates "within normal limits", '1' indicates "mildly affected", '2' indicates "moderately affected", and '3' indicates "severely affected". An individual's total score varied from 0 to 30 in both words and in sentences. Instructions for perceptual evaluation were provided to the SLPs both orally and in written form. The SLPs rated the samples in a quiet room situation.

2.3 Statistical Analysis

Obtained data were subjected to statistical analysis using IBM (International Business Machines Corporation) Statistical Package Social Sciences software (version 20). Average mean values and standard deviation of resonance, speech understandability, and speech acceptability were calculated separately in both words and sentences. Shapiro–Wilk test of normality was applied to check the normality, where the test revealed the normal distribution of the data ($p > 0.05$). So, the paired t-test was carried out to check if there is any difference in resonance, speech understandability, and speech acceptability across words and sentences. Cronbach's alpha was calculated to find the inter-judge reliability across three SLPs. Pearson's correlation coefficient was calculated to check the correlation among resonance, speech understandability, and speech acceptability in both words and sentences.

3 Results and Discussion

The present study is one of the preliminary attempts in the Indian context to investigate the effect of stimuli on the variables such as resonance, speech understandability, and speech acceptability. The study also attempts to highlight the correlation among the variables mentioned above within the stimuli in children with RCLP. The obtained results are discussed under following sub-sections.

3.1 Comparison of Resonance, Speech Understandability, and Speech Acceptability Across Words and Sentences in Children with RCLP

The total mean score and standard deviation of resonance, speech understandability, and speech acceptability were calculated in both words and sentences and it is shown in Table 1.

The table depicts the mean and the standard deviation score for resonance, speech understandability, and speech acceptability across words and sentences. The total mean scores of resonance, speech understandability, and speech acceptability rated by the three SLPs are comparatively greater for words than for sentences. The mean scores of resonance, speech understandability, and speech acceptability in percentage were calculated in both words and sentences and it is represented in Fig. 1.

Table 1 Mean and standard deviation of raw scores across variables in words and sentences

	Mean		SD		p-value
	Words	Sentences	Words	Sentences	
Resonance	13.75	12.75	5.118	4.621	0.279
Speech understandability	14.15	13.50	5.896	5.042	0.408
Speech acceptability	15.90	14.90	5.350	6.017	0.089

Note SD—Standard Deviation; *p ≤ 0.05

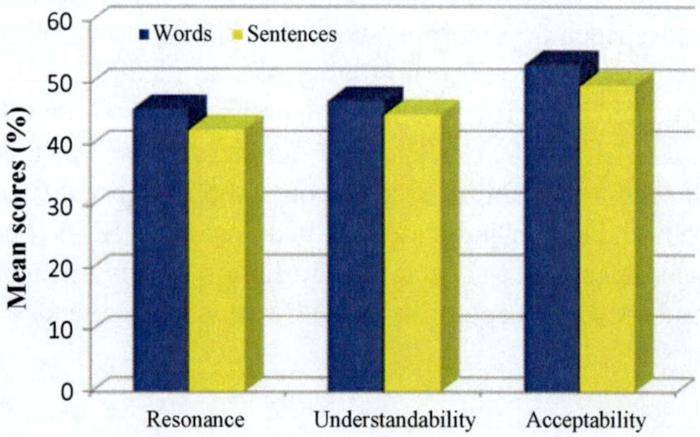

Fig. 1 Mean scores in percentage were calculated in both words and sentences

A paired-sample t-test was conducted to compare resonance, speech understandability, and speech acceptability across words and sentences. It was observed that among all the three variables there was no significant difference in the scores, i.e., $t(19) = 1.114$, $p > 0.05$ for resonance, $t(19) = 0.846$, $p > 0.05$ for speech understandability, and $t(19) = 1.795$, $p > 0.05$ for speech acceptability across words and sentences.

This result is in consensus to the findings of Gnanavel and Pushpavathi [3] where there was no difference in the intelligibility scores across words and sentences. There are many factors which contribute to this result where number of participants considered is one of them. Present study considered only 20 participants, a significant difference could have been obtained with the inclusion of larger sample size. Among the factors that can influence the perceptual judgment of hypernasality, speech understandability, and speech acceptability, type of speech stimuli stands out as a major one.

Few authors stated that hypernasality is distinguished only during spontaneous speech and it is considered to be severe in this type when compared with words and sentences [7, 8]. The present study compared variables across words and sentences, significant differences could have been obtained with the comparison of spontaneous speech with sentences or words. Also, the rating scale used in the present study is a four-point rating scale, where "very mild", "mild to moderate", and "moderate to severe" categories are used. During the perceptual rating of the variables, a difference in the severity of the rating could have been obtained if a larger scale was used; [2] used a seven-point rating scale to assess hypernasality in persons with dysarthria, where '1' represented "normal resonance" and '7' represented "severely hypernasality".

3.2 Correlation of Resonance, Speech Understandability, and Speech Acceptability Within Words and Sentences in Children with RCLP

The correlation among the mean scores of resonance, speech understandability, and speech acceptability were checked within words as well as within sentences. Pearson correlation coefficient was administered and the results are represented on a scatter plot as given in Fig. 2.

The results of the present study indicated that, in words, there was a positive correlation between resonance and speech understandability ($r = 0.563$, $n = 20$, $p = 0.010$), resonance and speech acceptability ($r = 0.714$, $n = 20$, $p = 0$), speech understandability and speech acceptability ($r = 0.960$, $n = 20$, $p = 0$). Even in sentences, there was a positive correlation between resonance and speech understandability ($r = 0.701$, $n = 20$, $p = 0.001$), resonance and speech acceptability ($r = 0.813$, $n = 20$, $p = 0$), and speech understandability and speech acceptability ($r = 0.946$, $n = 20$, $p = 0$).

Overall, there was a strong positive correlation among resonance, speech understandability, and speech acceptability in both words and sentences as depicted in the above figure. This indicates that there is a linear association among resonance, speech understandability, and speech acceptability within words and sentences, i.e., as the severity of hypernasality increases, speech understandability

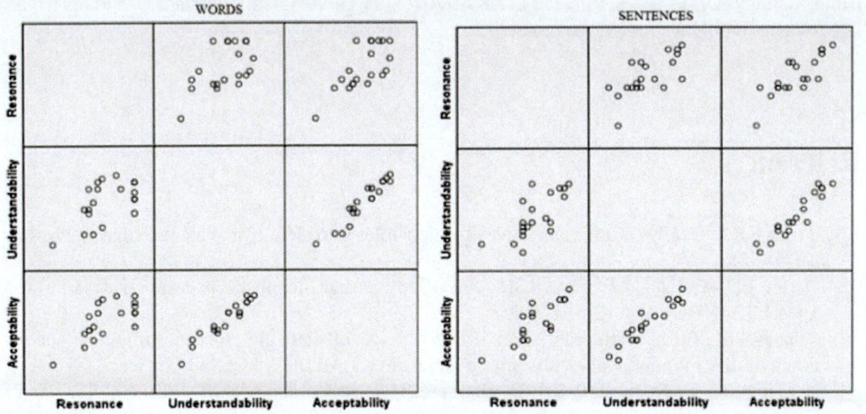

Fig. 2 Relationship among resonance, speech understandability, and speech acceptability in words and sentences

and speech acceptability decrease and vice versa. This result is in consonance to the findings of Myers [11], who also reported a positive correlation between speech intelligibility and resonance. It also agrees with the findings of Lee et al. [9] who stated that hypernasality eventually affects the speech intelligibility in children with cleft lip and palate. However, the findings of the present study were not in agreement with the results of Kang et al. [6] who found no correlation between nasalance score and speech intelligibility in children with cleft palate. This difference might be because of the variables such as stimuli used, language, sample size, and the linguistic environment.

4 Conclusions

The present study investigated resonance, speech understandability, and speech acceptability across words and sentences through perceptual judgment using Henningsson's rating scale and the correlation among the same was checked. The results revealed no significant difference between words and sentences across resonance, speech understandability, and speech acceptability and there was a higher inter-judge reliability among the three judges. There was a significant correlation among resonance, speech understandability, and speech acceptability in both words and in sentences. Hence, the current study concludes that there is a minimal effect of stimuli on resonance, speech understandability, and speech acceptability in children with repaired cleft lip and palate. However, a similar study can be done using a large number of samples. These three parameters are essential for deciding the quality of speech in persons with cleft palate.

Acknowledgements This is a part of ongoing research on "NASOSPEECH: Development of Diagnostic System for Severity Assessment of the Disordered Speech", funded by the Department of Biotechnology (DBT - No. SH/PL/DBT (AKA)/2016-17), Government of India. The authors would like to thank DBT for funding the project.

References

1. Bzoch KR (ed) (1997) Communicative disorders related to cleft lip and palate, 4th edn. Pro-Ed, Austin, TX
2. Darley FL, Aronson AE, Brown JR (1969) Differential diagnostic patterns of dysarthria. J Speech Lang Hear Res 12(2):246–269
3. Gnanavel K, Pushpavathi M (2012) Effect of type of cleft and speech stimuli on speech intelligibility in individuals with cleft lip and palate. J All India Inst Speech Hear 31
4. Henningsson G, Kuehn DP, Sell D, Sweeney T, Trost-Cardamone JE, Whitehill TL (2008) Universal parameters for reporting speech outcomes in individuals with cleft palate. Cleft Palate-Craniofacial J 45(1):1–17
5. Jayakumar T, Pushpavathi M (2005) Normative score for nasometer in Kannada. Student research at AIISH (Articles based on dissertation done at AIISH) 7:44–53

6. Kang J, Park B, Paek S, Kwon S, Lee J, Choi Y (2009) the relationship between speaker factors and speech intelligibility of children with cleft palate. Commun Sci Disord 14(3):338–348
7. Kuehn DP, Moller KT (2000) Speech and language issues in the cleft palate population: the state of the art. Cleft Palate-Craniofacial J 37(4):348–348
8. Kummer AW (2014) Speech evaluation for patients with cleft palate. Clin Plast Surg 41(2):241–251
9. Lee GS, Wang CP, Yang CC, Kuo TB (2006) Voice low tone to high tone ratio: a potential quantitative index for vowel [a:] and its nasalization. IEEE Trans Biomed Eng 53(7):1437–1439
10. McWilliams BJ, Morris HL, Shelton RS (1990) Cleft palate speech, 2nd edn. BC Decker, Philadelphia
11. Myers EN (2008) Operative otolaryngology: head and neck surgery E-Book: 2-Volume Set (Vol. 2). Elsevier Health Sciences, 782–783
12. Ray J (2002) Treating phonological disorders in a multilingual child: A case study. Am J Speech-Language Pathol 11(3):305–315
13. Stengelhofen J (1989) The nature and causes of communication problems in cleft palate. In: Stengelhofen J (ed) Ceft palate. Churchill Livingstone, Edinburgh
14. Van Lierde KM, De Bodt M, Van Borsel J, Wuyts FL, Van Cauwenberge P (2002) Effect of cleft type on overall speech intelligibility and resonance. Folia Phoniatrica et Logopaedica 54(3):158–168
15. Whitehill TL (2002) Assessing intelligibility in speakers with cleft palate: a critical review of the literature. Cleft Palate-Craniofacial J 39(1):50–58
16. Whitehill TL, Chau CHF (2004) Single-word intelligibility in speakers with repaired cleft palate. Clin Linguist Phon 18(4–5):341–355

Continuous Speech Recognition Technologies—A Review

Shobha Bhatt, Anurag Jain, and Amita Dev

Abstract Speech recognition is the most emerging field of research, as speech is the natural way of communication. This paper presents the different technologies used for continuous speech recognition. The structure of speech recognition system with different stages is described. Different feature extraction techniques for developing speech recognition system have been studied with merits and demerits. Due to the vital role of language modeling in speech recognition, various aspects of language modeling in speech recognition were presented. Widely used classification techniques for developing speech recognition system were discussed. Importance of speech corpus during the speech recognition process was described. Speech recognition tools for analysis and development purpose were explored. Parameters of speech recognition system testing were discussed. Finally, a comparative study was listed for different technological aspects of speech recognition.

Keywords Speech recognition · Feature extraction · Continuous speech · Classification · Language model · HMM

1 Introduction

Speech recognition is the method of transforming spoken utterances into text. It has been researched for decades. Speech recognition can be applied for voice interfaces, voice-driven machines, speech-enabled browsers, and Internet-based services. For speech recognition, the speech signal is first transformed into parametric form, and acoustic models are generated from the extracted features. At the recognition

S. Bhatt (✉) · A. Jain
U.S.I.C.T, Guru Gobind Singh Indraprastha University, Sector-16, Dwarka, Delhi, India
e-mail: bhattsho@gmail.com

A. Jain
e-mail: anurag@ipu.ac.in

A. Dev
Indira Gandhi Delhi Technical University for Women, New Delhi, India
e-mail: amita_dev@hotmail.com

M. Singh and Y. Rafat (eds.), *Recent Developments in Acoustics*, Lecture Notes in Mechanical Engineering, https://doi.org/10.1007/978-981-15-5776-7_8

step, parametric observations are attained, which have higher probabilities matching corresponding to input speech. Automatic Speech Recognition (ASR) system recognizes the utterances from spoken words [1]. Speech recognition systems can be characterized into isolated word, connected word, continuous speech, spontaneous speech, and speaker adapted speech. The words are spoken in isolation for isolated word recognition, and words are spoken with some pauses in connected word recognition. Continuous speech recognition deals with the continuously spoken word. Spontaneous speech is a natural way; the humans speak.

Speaker-adaptive systems, the speech recognition system is developed independently, and then this model is used for distinctive characteristics of target speaker [2, 3]. Researchers have experimented with improving recognition with advancements in signal processing, acoustic modeling, language modeling, and search algorithms. State of the art has been shifted from isolated word to speaker adaptation, context-dependent modeling, and discriminative training [4] during the years. Progress in speech recognition was made possible due to better and accurate speech modeling techniques, different features, extraction methods, and efficient search algorithms [5, 6]. Speech recognition systems using Hidden Markov Model (HMMs), Artificial Neural Networks (ANNs), and hybrid of HMMs and ANNs have been experimented [7, 8].

Extracting useful information from speech signal is very significant phase toward speech recognition. It is required to transform the signal into low-dimensional space, which is called feature extraction [9]. Further, extracted features should be resilient to environmental changes. For taking out these features from the speech signal, perception- and production-based methods are widely used. Perception-based methods work on the principle of the human hearing system while production-based methods work on how speech is produced. Different perception-based scales such as Mel and Bark are used for processing due to the nonlinear perception of the human hearing system. Further, the selection of the windowing method and pre-emphasis are also important criteria for pre-processing of the speech [10–12]. Various methods were experimented using cepstral mean normalization, energy thresholding, Linear Discriminant Analysis (LDA), and application of filter banks to improve the basic features for speaker independence and speech recognition in different noisy conditions. Other improvements were carried out using the transformation of basic features by performing differentiation and concatenation on these static features. For this, first difference and second difference of static features are widely used [13, 14]. It is also a crucial decision to decide the sub-word unit for speech recognition. Different sub-word models, such as syllable based and phoneme based, are used [15]. Scholars in speech recognition face problems due to variability in speech [16, 17]. Classifiers are used in the speech recognition process during recognition. The classification methods such as HMMs, ANNs, and discriminative training are presented in literature [18–22]. The complexity of the ASR system is increased when it advances from isolated to spontaneous speech and speaker-dependent to the speaker-independent mode. Speaker independent system are more difficult to develop than speaker dependent system [25]. Another challenge is coarticulation effect in phone segments. The phone segments are affected by its neighboring phones [26]. Further, problems that need addressing are lack of speech corpus, pronunciation dictionaries, and transcription files for under-resourced languages. Design of speech corpus is also crucial for the development of a speech recognition system [36, 37].

The purpose of writing this paper is to present a review of different technologies and trends in speech recognition. Readers will be able to know how the speech recognition system works, its challenges, and trends in speech recognition after going through this paper. It is an effort to highlight different aspects of speech recognition. The paper describes speech recognition issues such as feature extraction, classification methods, language modeling, speech corpus, widely used tools for the development of the spoken systems, and speech recognition parameters.

The paper is structured as follows. Section 2 describes continuous speech cognition and the structure of a speech recognition system. Section 3 explains various feature extraction methods. Section 4 describes speech corpus and widely used classification techniques. Section 5 deals with the role of language modeling in speech recognition. Section 6 describes different widely used speech recognition tools. Section 7 illustrates parameters for speech recognition testing. Finally, comparison of different classification and feature extraction techniques is presented.

2 Speech Recognition Structure

The continuously spoken words are converted into text during continuous speech recognition. Speech recognition system can be represented in Fig. 1 at a basic level. Feature extraction block converts speech signal into a suitable parametric feature. These features are used for generating acoustic models of speech utterances. The acoustic model is prepared from speech parameters. Language model block contains all the issues related to language modeling in speech recognition. The output from both the block is fed into a speech recognition engine. Speech recognition engine outputs the recognized word based on inputs from both the blocks.

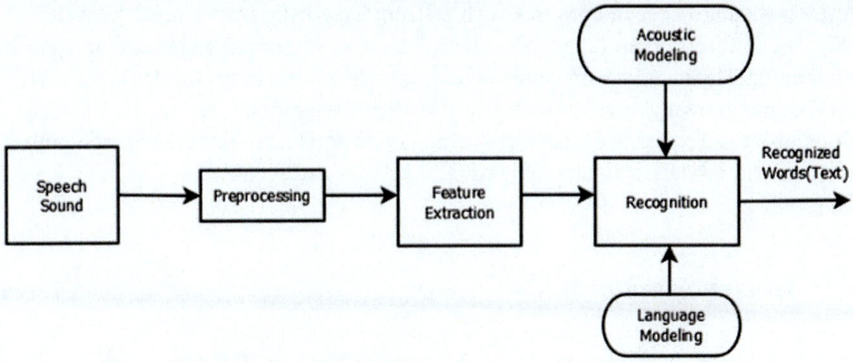

Fig. 1 Speech recognition process

3 Feature Extraction Techniques

Feature extraction is a crucial block in the speech recognition structure. Feature extraction is the process of calculating a set of meaningful properties from speech wave, which are significant for further processing. These properties are termed as features of the speech signal. Feature extraction approaches are divided into production based and perception based. Production-based features are calculated on the principle of how speech is produced, and perception-based features are based on how we perceive the sound. Linear Predictive Coding (LPC) is implemented using the concept of voice production mechanism while Mel Frequency Cepstral Coefficients (MFCC) and Perceptual Linear Prediction (PLP) perceptually motivated coefficients are used for feature extraction. Extracted features are further modified using different dynamic coefficients. Feature extraction methods are also categorized based on temporal analysis and spectral analysis [4, 38, 39].

For feature extraction, working of the vocal tract model and auditory model is captured. The main steps are pre-emphasis and windowing. The short-term spectrum is calculated using windowing methods such as Hamming. The spectrum can be represented as the signal's Fourier coefficients or as the set of power values at the outputs from a bank of filters. Further, the cepstrum is also calculated, which is the inverse Fourier transform of the logarithm of the spectrum. Different nonlinear logarithmic scales are used for the features based on the behavior of the auditory model [14, 36].

4 Speech Corpus and Classification Methods of Speech Recognition

Development of speech is essential corpus significant and crucial steps toward speech recognition are essential. For developing a speech recognition system, enough samples of speech are required. Different efforts have been made by researchers to develop standard speech corpuses such as from Linguistic Data Consortium (LDC) [27]. For low-resourced languages, classification is important to process in speech recognition. Here we have explained widely known models based on HMMS and artificial neural network-based model. The classification methods are used for the recognition process. Experiments have also been conducted to explore speech recognition using hybrid models with a combination of HMMs and ANNs.

4.1 Hidden Markov Model

HMMs represent the temporal nature of a speech signal. In HMMs, only observation can be seen, and the states behind observations cannot be directly observed.

Any HMM is represented by three features ($\prod$, A, B),

where $\prod$ is the initial state distribution vector,

A = [aij] is the state transition matrix which shows the probability of transition from state ai to state aj at a particular time, and

B = [bjk] is the matrix of observing symbols vk at current state qj.

The HMMs are used for observation sequence probability estimation, classification of utterances, and model parameter estimation by using forward–backward from Baum and Viterbi algorithm [5, 12, 28]. The most important phases using HMM-based speech recognition are training and testing. The advantages of using HMMs are that time and spectral variability can be modeled parallelly. Other advantages are HMMs can be initialized by using an initial estimate of the HMM parameters without labeling the data [4, 27–29, 35].

4.2 Artificial Neural Networks (ANNs)

In speech recognition, the artificial neural network has mostly used the method after HMM. These can be used independently or with the combination of HMMs. ANNs are very effective classifier. These classifiers learn and organize the data based on input data during the training phase. ANNs are capable of adapting when the data is unknown. Most popular ANN architectures are multilayer perceptron, self-organizing maps, and radial basis function [8, 31, 32].

5 Language Model

Knowledge of spoken language is production based for speech recognition. This includes knowledge of linguistic sounds, pronunciation variation, interpretation of the meaning of words, grammatical structure, and the possible meaning of sentences. By using language models, the search space for the recognition engine is reduced. Natural language processing is used for the implementation of these features in automatic speech recognition. Example of some of the steps is designing pronunciation dictionary and defining a grammar for the utterances [5, 7]. Nowadays, statistical language models are used that give probabilities of a sequence of words based on the previous set of words. Most widely used language models for speech recognition are N-gram language models, which predict the probabilities of words based on occurrences of its previous N-1 words [27].

6 Speech Recognition Tool Kit

There are different types of open-source speech recognition tools available. These systems have been developed using different programming languages like Java, C++, and C. These tool kits are based on HMMs, ANNs, and combination of both. The following section describes each in detail.

6.1 *Htk*

The HTK is an HMM-based tool for developing HMMs, especially for building the speech recognition systems. It was initially developed at Cambridge University by Speech Vision and Robotics Group. Main modules for developing the speech recognition systems are data preparation, training, and testing. The module HCopy is used for feature extraction with other tools. The module HCompv is used for flat start initialization. The module HeRest is used for re-estimation, and HVite is used for recognition. Speech recognition results can also be analyzed with tool HResult. The well-documented material is available as HTK book for developing different types of speech recognition system [3, 30].

6.2 *Sphinx*

Large vocabulary ASR systems were developed using Sphinx in the late 80s. Nowadays, versions of Sphinx which are available are 2, 3, 3.5, and 4 decoder versions and one common training tool which is named as Sphinx train. Sphinx can also be used to train context-dependent HMMs. It supports MFCC and PLP speech features with dynamic coefficients [5, 31, 33, 34].

6.3 *Kaldi*

Kaldi is a speech recognition tool kit available under Apache License. It is written in C ++. It was developed by Daniel Povey and others. Stable release was made in October 2013. Kaldi requires basic concepts of speech recognition and basic processing of speech signal. For better and efficient use of Kaldi, basic knowledge of scripting language like bash, Perl, and Python are also required. Kaldi has support for the deep neural network also [2].

7 Speech Recognition Evaluation

ASR systems are evaluated based on different parameters such as recognition accuracy, out of the vocabulary word rejection, response time, and error recovery [3]. For calculating speech recognition, accuracy reference string and recognized strings are used. There are three types of errors, namely, substitution, insertion, and deletion error [3]. [18, 36] is for speech recognition. Parameters are defined below:

$$\%\,\text{Correct} = (\text{N} - \text{D} - \text{S}) / \text{N} \times 100 \tag{1}$$

$$\%\text{Accuracy} = (\text{N} - \text{D} - \text{S} - \text{I}) / \text{N} \times 100 \tag{2}$$

where D denotes deletion error, I shows error due to insertion, S denotes the error caused by substitution, and N shows the entire number of labels in reference transcription. Another metric Word Error Rate (WER) is also used for describing the performance of the speech recognition system. It is specified as given below [35, 36]:

$$\%\text{WER} = (\text{D} + \text{S} + \text{I}) / \text{N} \times 100 \tag{3}$$

8 Comparative Analysis of Classification and Feature Extraction Techniques

Every classification and feature extraction technique has its own merits and demerits. Table 1 [7, 33] shows a comparative analysis of HMM- and ANN-based classification techniques. HMM-based model best describes the temporal nature of the speech signal and ANN-based system is better in learning and needs fewer steps in comparison to other speech recognition techniques.

Table 2 [7, 33] shows a comparative analysis of different feature extraction techniques. LPC-based coefficients are easy to compute. The perception-based coefficients such as MFCCs and PLPs provide better recognition due to the use of different nonlinearity functions to mimic the auditory model.

9 Conclusion

This paper highlights technical details which are required during speech recognition. An attempt was made to review the speech recognition area during the past years. The focus was to review the different trends and techniques used in speech recognition. In the past year, several technical advancements occurred. Research findings show that

Table 1 Classification techniques [7, 33]

Classification technique	Advantages	Disadvantages
Hidden Markov model	It best describes the temporal nature of speech recognition	It is based on the probabilistic model. Present state probability depends on the previous state
	It is easy to model	
	It can work in both discrete and continuous modes	
Artificial neural network	It is self-learning and adaptive model	A large amount of training data is needed
	Many steps, like traditional speech recognition, are not required in ANNs	
	ANNs are adaptive to a new environment	
	Low-level features can also be used for speech recognition	

Table 2 Feature extraction techniques [7, 33]

Linear Predictive Coefficients (LPC)	Mel Frequency Cepstral Coefficients (MFCCs)	Perceptual Linear Prediction (PLP)
The speech signal model based on the human voice production system is well represented by LPCs Further, it provides linear characteristics and reasonable source vocal tract separation. LPC approaches are easy to understand and implement	MFCCs are perceptually motivated coefficients and offer good discrimination and a small correlation Further features of the gradually varying parts are focussed in the low cepstral coefficients MFCCs can be manipulated to generate different variants Individual features of MFCC appear just inadequately correlated, which serve for the development of a statistical-model-based system	PLP coefficients provide a better approximation of the speaker-independent system. PLP coefficients are generated in the short-term spectrum of the speech signal using perceptual scales PLP-dominant frequencies are quite insensitive to vocal tract length

significant work has been done toward different feature extraction, classification, and development of resources for research in speech recognition. Several standard speech corpora were designed to meet the requirements for the developments of the speech recognition system. It was also observed that different open-source tools based on HMM and ANN are also available for research work. This work also presented a comparative analysis of their specific features and classification techniques with their merits and demerits. Language models limit the search space in speech recognition, so the importance of language models has been explained. Finally, evaluation parameters for speech recognition have been discussed. However, despite a lot of research work, there is a need for the development of robust speech recognition system, especially for under-resourced language to bridge the gap of the digital divide. Surely this research paper will help the research community to go deeper in speech recognition research.

Acknowledgments The authors would like to acknowledge the Ministry of Electronics and Information Technology (MeitY), Government of India, for providing financial assistance for this research work through "Visvesvaraya Ph.D. Scheme for Electronics and IT".

References

1. Sarma BD, Mahadeva Prasanna SR (2017) Acoustic–phonetic analysis for speech recognition: a review. IETE Tech Rev 1–23
2. kaldi-asr.org/doc/kaldi_for_dummies.html
3. Furui S (2007) Speech and speaker recognition evaluation. In: Dybkjær L, Hemsen H, Minker W (eds) Evaluation of text and speech systems. Text, speech and language technology, vol 37. Springer, Dordrecht
4. Saon George, Chien Jen-Tzung (2012) Large-vocabulary continuous speech recognition systems: a look at some recent advances. IEEE Signal Process Mag 29(6):18–33
5. Kacur J, Rozinaj G (2008) Practical issues of building robust HMM models using HTK and SPHINX systems, speech recognition, France Mihelic and Janez Zibert (ed), InTech. https://doi.org/10.5772/6376
6. Bahl LR et al (1999) Context dependent modeling of phones in continuous speech using decision trees. HLT
7. Cutajar M et al (2013) Comparative study of automatic speech recognition techniques. IET Signal Process 7(1):25–46
8. Lippmann Richard P (1989) Review of neural networks for speech recognition. Neural Comput 1(1):1–38
9. Vimala C, Radha V (2015) Isolated speech recognition system for Tamil language using statistical pattern matching and machine learning techniques. J Eng Sci Technol (JESTEC) 10(5):617–632
10. Picone Joseph W (1993) Signal modeling techniques in speech recognition. Proc IEEE 81(9):1215–1247
11. Fook CY et al (2013) Comparison of speech parameterization techniques for the classification of speech disfluencies. Turkish J Electric Eng Comput Sci 21(1):983–1994
12. Scharenborg OE, Bouwman AGG, Boves LWJ (2000) Connected digit recognition with class specific word models
13. Nieuwoudt C, Botha EC (1999) Connected digit recognition in Afrikaans using hidden Markov models
14. Bhiksha R, Singh R (2011) Design and implementation of speech recognition systems. Carniege Mellon School of Computer Science
15. Davel M, Martirosian O (2009) Pronunciation dictionary development in resource-scarce environments
16. Wu T (2009) Feature selection in speech and speaker recognition. Katholieke Universiteit Leuven
17. Kumar K, Kim C, Stern RM (2011) Delta-spectral cepstral coefficients for robust speech recognition. In: 2011 IEEE International conference on acoustics, speech and signal processing (ICASSP). IEEE
18. Aggarwal RK, Dave M (2012) Integration of multiple acoustic and language models for improved Hindi speech recognition system. Int J Speech Technol 15(2):165–180

19. Bush M, Kopec G (1987) Network-based connected digit recognition. IEEE Trans Acoust Speech Signal Process 35(10):1401–1413
20. Singhal S, Dubey RK (2015) Automatic speech recognition for connected words using DTW/HMM for English/Hindi languages. In: 2015 Communication, control and intelligent systems (CCIS). IEEE
21. He ZG, Liu ZM (2012) Chinese connected word speech recognition based on derivative dynamic time warping. In: Advanced materials research, vol 542. Trans Tech Publications
22. Bernardis G, Bourlard H (1998) Improving posterior based confidence measures in hybrid HMM/ANN speech recognition systems. In: Fifth international conference on spoken language processing
23. Bourlard H, Morgan N (1998) Hybrid HMM/ANN systems for speech recognition: overview and new research directions. In: Adaptive processing of sequences and data structures. Springer, Berlin, pp 389–417
24. Livescu Karen, Fosler-Lussier Eric, Metze Florian (2012) Subword modeling for automatic speech recognition: past, present, and emerging approaches. IEEE Signal Process Mag 29(6):44–57
25. Renals S, McKelvie D, McInnes F (1991) A comparative study of continuous speech recognition using neural networks and hidden Markov models. In: 1991 International Conference on Acoustics, Speech, and Signal Processing. ICASSP-91. IEEE
26. Saini P, Kaur P, Dua M (2013) Hindi automatic speech recognition using htk. Int J Eng Trends Technol (IJETT), 4(6), 2223–2229 versité de Aix-en-Provence, 1998
27. Makhoul John, Schwartz Richard (1995) State of the art in continuous speech recognition. Proc Natl Acad Sci 92(22):9956–9963
28. Klatt Dennis H (1977) Review of the ARPA speech understanding project. J Acoust Soc Am 62(6):1345–1366
29. Jelinek Frederick (1976) Continuous speech recognition by statistical methods. Proc IEEE 64(4):532–556
30. Levinson SE, Rabiner LR, Sondhi MM (1983) An introduction to the application of the theory of probabilistic functions of a Markov process to automatic speech recognition. Bell Syst Tech J 62(4): 1035–1074
31. htk.eng.cam.ac.uk
32. Dev Amita S, Agrawal S, Roy Choudhury D (2003) Categorization of Hindi phonemes by neural networks. AI & SOCIETY 17(3–4):375–382
33. Anusuya MA, Katti SK (2011) Front end analysis of speech recognition: a review. Int J Speech Technol 14(2):99–145
34. https://cmusphinx.github.io/
35. Bhatt S, Dev A, Jain A Hindi speech vowel recognition using hidden Markov model. In: Proceedings of The 6th International Workshop on Spoken Language Technologies for Under-Resourced Languages, pp 196–199
36. Bhatt, Shobha, Dev, Amita Jain, Anurag. Hidden Markov Model Based Speech Recognition-A Review. In: 12 th INDIACom 2018, 5th International conference on "computing for sustainable global development, 1st to 3rd March, 2018. http://bvicam.ac.in/news/INDIACom%202018%20Proceedings/Main/papers/712.pdf
37. Bhatt S, Jain A, Dev A (2017) Hindi Speech recognition: issues and challenges. In: 11th INDIACom 4rd International conference on computing for sustainable global Development. 1st to 3rd March, 2017. http://bvicam.ac.in/news/INDIACom%202017%20Proceedings/Main/papers/936.pdf
38. Agrawal SS, Prakash N, Jain A (2010) Transformation of emotion based on acoustic features of intonation patterns for Hindi speech. Afr J Math Comput Sci Res 3(10): 255–266
39. Madan A, Gupta D (2014) Speech feature extraction and classification: a comparative review. Int J Comput Appl 90(9)

Effect of Number of LPCs and Formant Deletion Methods of Inverse Filtering on Acoustic Parameters of Voice

Vimala J. Kasturi, Rashika Sharma, and Santosh Maruthy

Abstract Acoustic analysis is one of the efficient, non-invasive, and quantitative methods of voice assessment. The estimation of glottal flow parameters using acoustic analysis is achieved through the method of Inverse Filtering (IF). This study aims at describing effects of the two available methods of inverse filtering—Formant-Based Inverse Filtering (FBIF) and Linear Prediction-Based Inverse Filtering (LPBIF) on acoustic parameters. The effects of formant deletion and number of LPCs on the various vocal parameters—Fundamental frequency based, intensity based, perturbation based, and noise-based measures—were studied. Phonation samples of /a/ at a comfortable pitch and loudness by 30 healthy participants (15 males and 15 females) were recorded on to a PC in a noise-free environment. In the LPBIF-based method, the recorded voice sample was analyzed under five different conditions, i.e., by varying the number of LPCs. The standard value of LPC used in the Vaghmi software is 18. However, in the current study, the number of LPCs was set to 14, 16, 18, 20, and 22, respectively in each of the analysis conditions with other settings remaining the same. In the FBIF-based method, the recorded voice sample was analyzed under four different conditions, i.e., by deleting different formants from the sample. Formant deletion was accomplished using manual methods. The four conditions used were deletion of F1, deletion of F2, deletion of F3, and deletion of F1, F2, and F3 formants. Acoustic parameters of F0, F0 min, F0 max, I0, I0 min, I0 max, JF0, JT0, RAP3, RAP5, SHdB, APQ5, APQ11, HNR, and GNE were compared in all the different conditions mentioned above. A non-parametric Friedman's test of differences among repeated measures was conducted. The results revealed no change in the value measured across all conditions in both FBIF and

V. J. Kasturi (✉)
Department of Speech Language Pathology, All India Institute of Speech and Hearing, Mysuru, India
e-mail: k.vimalajayakrishna@gmail.com

R. Sharma
Speech Language Pathologist, Jawahar Lal Nehru Medical College and Hospital, Ajmer, Rajasthan, India

S. Maruthy
Department of Speech-Language Sciences, All India Institute of Speech and Hearing, Mysuru, India

M. Singh and Y. Rafat (eds.), *Recent Developments in Acoustics*, Lecture Notes in Mechanical Engineering, https://doi.org/10.1007/978-981-15-5776-7_9

LPBIF methods for the parameters of F0, F0 min, F0 max, I0, I0 min, and I0 max. Significant differences across the four formant conditions under FBIF method were found on all perturbation and noise-related measures, i.e., JF0, JT0, RAP3, RAP5, SHdB, APQ5, APQ11, HNR, and GNE. Significant differences across the five LPC conditions were found only on the parameters SHdB, APQ5, APQ11, HNR and GNE. FBIF could be assumed to be more sensitive than LPBIF as the differences across conditions in FBIF were significant in all measured parameters unlike LPBIF where the differences were significant in only five of the parameters. The pros and cons of using each of the two methods for acoustic analysis of voice are discussed in the current study. Further research needs to be done to investigate the effect of varied parameters of inverse filtering in disordered population.

Keywords Acoustic analysis · Inverse filtering · Vaghmi · Voice · LPC · Formant-based inverse filtering

1 Introduction

There are three common objective approaches for clinically evaluating voice and its disorders—acoustic assessment, aerodynamic analysis, and endoscopic imaging. Of these methods, acoustic analysis is the most commonly used instrumental measure. Very many computer-assisted techniques and software have been identified to extract the vocal parameters through acoustic analysis like the Multi-Dimensional Voice Program (MDVP) and Praat which are commonly used in clinical and research settings [1]. Yet another computer-based program which has provision for acoustic analysis of voice is 'Vaghmi', developed by Voice and Speech Systems, Bangalore, India. The Vaghmi system has three modules, namely, Diagnostic, Therapy, and Utility. The therapy and diagnostic modules are further divided into voice and speech modules. The utility module provides a platform for signal recording, playback, display, and generation of reports. The current study makes use of the diagnostic module of Vaghmi system for acoustic analysis of voice.

Voice-diagnostic module consists of algorithms to perform inverse filtering of voice signal and compute frequency and intensity measurements, jitter and shimmer measurements, noise-related measures, Long-Term Average Spectrum (LTAS), and Maximum Phonation Duration (MPD). Vaghmi uses two methods of inverse filtering—Formant-Based Inverse Filtering (FBIF) and Linear Prediction-Based Inverse Filtering (LPBIF). Miller in 1959, proposed the concept of inverse filtering [2]. "The Source-Filter Theory of Speech Production leads naturally to the principle of inverse filtering as a means for extracting the acoustic features of the speech wave" [3]. Whereas, the vocal tract filters the source signal, inverse filtering performs an inverse of vocal tract filter, in effect, removing the filtering characteristics of the vocal tract. The accuracy of glottal estimation in inverse filtering depends on the accuracy of the glottal filters. Quantification of the accuracy of acoustic measurements, thus, lies in the filter settings in a given method of inverse filtering.

The vocal tract filter has a set of resonances called formants labeled as F1, F2, etc. In inverse filtering, one anti-resonant filter replaces each formant filter. Anti-resonant filter must be tuned to the formant to be canceled. Both the formant frequency and bandwidth are to be tuned. In Formant-Based Inverse Filtering (FBIF) an initial estimate of formant frequencies and bandwidths are displayed in a table. However, the user can edit this information to tune the formant data to greater accuracy and thereby arrive at accurate estimates of vocal parameters during acoustic analysis. In most methods of analysis, the validity and reliability of acoustic parameters are inherently limited to analyzing a sustained vowel. In the diagnostic module of Vaghmi, FBIF method of analysis is provided as default setting when a phonation sample is processed. But it also offers a second method of analysis—Linear Prediction-Based Inverse Filtering (LPBIF) which is more commonly used for processing speech samples. Predictive coding is a remarkably simple concept, where prediction is used to achieve efficient coding of signals [4]. Early work exploring the application in speech coding includes the work of Atal and Schroeder [5] and that of Itakura and Saito [6]. Linear Predictive Coding (LPC) analyses the speech signal by estimating the formants, removing their effects from the speech signal, and estimating the intensity and frequency of the remaining buzz. The process of removing the formants is called inverse filtering, and the remaining signal after the subtraction of the filtered modeled signal is called the residue. By minimizing the ‘energy’ of the difference between the actual speech samples and the predicted ones, a unique set of predictor coefficients can be determined. These coefficients provide an estimate of the filter function, and can therefore be used to determine the inverse filter. By default the number of LPCs is the sampling frequency in 2 kHz + 2. In the current study, the sampling frequency was set to 16 kHz as a default setting in Vaghmi and hence, the prescribed number of LPCs was set to 18. Vaghmi offers provision to vary these given settings to analyze the voice samples. Thus, IF needs to be done interactively or manually if FBIF method of analysis is selected, whereas, an automatic procedure of analysis is selected when the LPBIF method is selected. Validity and reliability of acoustic analysis performed with different tools were previously shown to be affected by many factors. These include, for example, microphone type, noise levels, data acquisition system, sampling rate, and software used for analysis, hence a studying effects of manipulation of such factors, in the present study software used may aid in better understanding of acoustic outcomes [7, 8]. In the current study, an attempt was made to study the effects of formant deletion and number of LPCs (filter order) on the various vocal parameters (fundamental frequency based, intensity based, perturbation based, and noise-based measures).

2 Method

2.1 Participants

A total of 30 healthy participants (15 males and 15 females in the age range of 18–24 years with a mean age of 23.1 years and SD of 1.37) were recruited for the current study. These individuals had no obvious voice pathology or any other significant health-related problems as confirmed after clinical review.

2.2 Instrumentation

A commercial PC-based digital signal processing software—Vaghmi developed by Voice and Speech Systems, Bangalore, was used in the present study. This software–hardware unit is developed for assessment and management of voice disorders. The Voice-Diagnostic module in Vaghmi has algorithms to perform inverse filtering of the speech signal, frequency and intensity measurement, jitter and shimmer measurements, harmonic-to-noise ratio, long-term average spectrum (LTAS), and Maximum Phonation Duration (MPD). In the diagnostic module of Vaghmi, FBIF method of analysis is provided as default setting when a phonation sample is processed. But it also offers a second method of analysis—Linear Prediction-Based Inverse Filtering (LPBIF) which is more commonly used for processing speech samples.

2.3 Procedure

The study was initiated after an approval from AIISH Ethical Committee. An informed consent was obtained from all the participants after explaining the due procedures of the study. The participants were made to sit comfortably and were asked to phonate the vowel /a/ at a comfortable pitch and loudness with the microphone being placed at a distance of 15 cm from the mouth of the speaker. For recording of the phonation samples, a Shure SM48 cardioid dynamic microphone with a frequency response of 55–14000 Hz was used. The phonation sample thus obtained was directly recorded on to a PC through the Voice-Diagnostic module of Vaghmi in a sound-free room. Each of the recorded phonation samples was analyzed in two methods available in the diagnostic module of Vaghmi software—Linear Prediction-based inverse filtering (LPBIF) and Formant-based inverse filtering (FBIF). In the LPBIF-based method, the recorded voice sample was analyzed under five different conditions, i.e., by varying the number of LPCs. The standard value of LPC used in the Vaghmi software is 18. However, in the current study, the number of LPCs was set to 14, 16, 18, 20, and 22, respectively, in each of the analysis conditions with other settings remaining the same. In the FBIF-based method, the recorded voice sample was analyzed under four different conditions, i.e., by deleting different formants from the sample. Formant deletion was accomplished using manual methods and formants were edited if the software estimated formants were found to be incorrect. The four conditions used were deletion of F1, deletion of F2, deletion of F3, and deletion of F1, F2, and F3. Acoustic parameters of Frequency measures: Fundamental Frequency (F0), Minimum Fundamental Frequency (F0 min), Maximum Fundamental Frequency (F0 max); Intensity measures: Intensity (I0), Minimum Intensity (I0 min), Maximum Intensity (I0 max); Jitter measures: Jitter in Fundamental Frequency (JF0), Jitter in Pitch Period (JT0), Three-period Relative average perturbation (RAP3), Five-period Relative average perturbation (RAP5); Shimmer measures: Absolute shimmer given in (SHdB), Five-point amplitude perturbation quotient (APQ5), Eleven-point amplitude perturbation quotient (APQ11); and Noise measures: Harmonic-to-Noise Ratio (HNR), Glottal-to-Noise excitation ratio (GNE) were measured in all the different conditions mentioned above. A non-parametric Friedman's test of differences among repeated measures was conducted for the Perturbation measures and Noise measures under each of the analysis methods.

3 Results

The mean values of acoustic parameters of Frequency and Intensity measures; Perturbation measures and Noise measures are depicted in Tables 1, 2, 3, 4, and 5, respectively. For F0 & I0 parameters the mean and SD values did not show any difference across conditions in both methods of analysis and hence, no further statistical analysis was employed. When the non-parametric Friedman's test of differences among repeated measures was conducted for the Perturbation measures and Noise measures under each of the analysis methods (Tables 6 and 7), in the FBIF method, all the Perturbation and Noise measures were found to be statistically significant ($p < 0.05$) across the four formant conditions. However, in the LPBIF method, only three of the perturbation measures (ShdB, APQ5, and APQ11) and both the noise measures (HNR and GNE) were found to be significantly different at $p < 0.05$. Among the perturbation and noise-related measures, except those of APQ 11 and APQ 5, it was found that values of acoustic measures were higher on usage of FBIF when compared to LPBIF.

4 Discussion

In the current study, two methods of analysis—LPBIF and FBIF were studied. The mean values obtained for F0 and I0 related measures did not differ across conditions and across methods. "This was because both the methods essentially employed LP analysis. LPC analysis has been particularly powerful and convenient because it generates numeric data for formant frequencies and bandwidths" [9]. In the FBIF

Table 1 Frequency-related acoustic parameters under both methods of IF

Table.	LPC-based analysis					Formant-based analysis				
F0-based parameters	Condition (LPC)	Males		Females		Condition	Males		Females	
		Mean	S.D	Mean	S.D		Mean	S.D	Mean	S.D
F0	14	126.67	16.06	229.21	18.40	F1 DEL	126.67	16.06	229.21	18.40
	16	126.67	16.06	229.21	18.40	F2 DEL	126.67	16.06	229.21	18.40
	18	126.67	16.06	229.21	18.40	F3 DEL	126.67	16.06	229.21	18.40
	20	126.67	16.06	229.21	18.40	FT DEL	126.67	16.06	229.21	18.40
	22	126.67	16.06	229.21	18.40	–	–	–		–
F0 MIN	14	125.03	16.63	224.72	17.44	F1 DEL	125.03	16.63	224.72	17.44
	16	125.03	16.63	224.72	17.44	F2 DEL	125.03	16.63	224.72	17.44
	18	125.03	16.63	224.72	17.44	F3 DEL	125.03	16.63	224.72	17.44
	20	125.03	16.63	224.72	17.44	FT DEL	125.03	16.63	224.72	17.44
	22	125.03	16.63	224.72	17.44	–	–	–	–	–
F0 MAX	14	130.62	18.35	232.53	18.54	F1 DEL	130.62	18.35	232.53	18.54
	16	130.62	18.35	232.53	18.54	F2 DEL	130.62	18.35	232.53	18.54
	18	130.62	18.35	232.53	18.54	F3 DEL	130.62	18.35	232.53	18.54
	20	130.62	18.35	232.53	18.54	FT DEL	130.62	18.35	232.53	18.54
	22	130.62	18.35	232.53	18.54	–	–	–	–	–

Table 2 Intensity-related acoustic parameters under both methods of IF

	LPC-based analysis					Formant-based analysis				
I0-based parameters	Condition (LPC)	Males		Females		Condition	Males		Females	
		Mean	S.D	Mean	S.D		Mean	S.D	Mean	S.D
I0	14	106.83	3.26	108.55	2.47	F1 DEL	106.83	3.26	108.55	2.47
	16	106.83	3.26	108.55	2.47	F2 DEL	106.83	3.26	108.55	2.47
	18	106.83	3.26	108.55	2.47	F3 DEL	106.83	3.26	108.55	2.47
	20	106.83	3.26	108.55	2.47	FT DEL	106.83	3.26	108.55	2.47
	22	106.83	3.26	108.55	2.47	–	–	–	–	–
I0 MIN	14	104.69	3.83	106.75	2.62	F1 DEL	104.69	3.83	106.75	2.62
	16	104.69	3.83	106.75	2.62	F2 DEL	104.69	3.83	106.75	2.62
	18	104.69	3.83	106.75	2.62	F3 DEL	104.69	3.83	106.75	2.62
	20	104.69	3.83	106.75	2.62	FT DEL	104.69	3.83	106.75	2.62
	22	104.69	3.83	106.75	2.62	–	–	–	–	–
I0 MAX	14	108.41	3.15	110.01	2.83	F1 DEL	108.41	3.15	110.01	2.83
	16	108.41	3.15	110.01	2.83	F2 DEL	108.41	3.15	110.01	2.83
	18	108.41	3.15	110.01	2.83	F3 DEL	108.41	3.15	110.01	2.83
	20	108.41	3.15	110.01	2.83	FT DEL	108.41	3.15	110.01	2.83
	22	108.41	3.15	110.01	2.83	–	–	–	–	–

Table 3 Frequency-related perturbation parameters under both methods of IF

Perturbation-based parameters	LPC-based analysis					Formant-based analysis				
	Condition	Males		Females		Condition	Males		Females	
		Mean	S.D	Mean	S.D		Mean	S.D	Mean	S.D
JF0	14 LPC	1.03	0.74	1.28	0.33	F1 DEL	3.47	9.96	6.43	9.07
	16 LPC	1.29	1.24	90.61	346.18	F2 DEL	1.03	0.52	1.38	0.56
	18 LPC	1.08	0.78	1.23	0.25	F3 DEL	0.87	0.62	1.18	0.28
	20 LPC	1.19	1.12	1.25	0.29	FT DEL	15.17	15.57	4.71	4.69
	22 LPC	1.35	1.69	1.23	0.27	–	–	–	–	–
JT0	14 LPC	0.94	0.73	1.28	0.32	F1 DEL	3.47	9.98	6.73	9.56
	16 LPC	1.25	1.20	1.23	0.27	F2 DEL	1.02	0.52	1.37	0.56
	18 LPC	0.99	0.57	1.22	0.25	F3 DEL	0.86	0.61	1.17	0.28
	20 LPC	1.16	1.07	1.25	0.29	FT DEL	15.80	16.12	5.11	5.44
	22 LPC	1.27	1.66	1.23	0.27	–	–	–	–	–
RAP 3	14 LPC	0.62	0.45	0.81	0.21	F1 DEL	2.27	6.60	4.38	6.25
	16 LPC	22.68	84.74	0.78	0.18	F2 DEL	0.64	0.34	0.87	0.37
	18 LPC	0.65	0.47	0.77	0.17	F3 DEL	0.52	0.39	7.87	27.72
	20 LPC	0.72	0.68	0.79	0.20	FT DEL	10.30	10.60	3.17	3.38
	22 LPC	0.81	1.06	0.78	0.18	–	–	–	–	–
RAP 5	14 LPC	0.58	0.33	0.78	0.18	F1 DEL	1.61	4.08	4.23	6.30
	16 LPC	0.72	0.61	0.77	0.17	F2 DEL	0.59	0.21	0.83	0.28
	18 LPC	24.31	91.75	0.76	0.16	F3 DEL	0.50	0.26	7.75	27.29
	20 LPC	0.65	0.49	0.77	0.18	FT DEL	8.50	8.46	3.44	4.00
	22 LPC	0.73	0.77	0.76	0.17	–	–	–	–	

Table 4 Intensity-related perturbation parameters under both methods of IF

Noise-based parameters	LPC-based analysis					Formant-based analysis				
	Condition	Males		Females		Condition	Males		Females	
		Mean	S.D	Mean	S.D		Mean	S.D	Mean	S.D
SHdB	14 LPC	0.54	0.27	0.25	0.05	F1 DEL	0.33	0.90	0.40	0.55
	16 LPC	0.64	0.32	0.23	0.04	F2 DEL	0.13	0.10	0.07	0.02
	18 LPC	0.59	0.24	0.23	0.04	F3 DEL	0.13	0.09	0.06	0.02
	20 LPC	0.54	0.22	0.22	0.04	FT DEL	1.05	0.98	0.50	0.39
	22 LPC	0.57	0.26	0.22	0.04	–	–	–	–	–
APQ 11	14 LPC	5.19	1.93	3.78	0.84	F1 DEL	2.44	5.66	3.60	4.98
	16 LPC	6.91	3.43	3.39	0.68	F2 DEL	1.18	0.84	0.62	0.20
	18 LPC	5.92	2.08	3.31	0.58	F3 DEL	1.22	0.62	0.58	0.16
	20 LPC	5.72	2.29	3.26	0.60	FT DEL	7.56	6.27	4.33	3.54
	22 LPC	5.96	2.73	3.24	0.59	–	–	–	–	–
APQ 5	14 LPC	4.01	1.54	2.48	0.55	F1 DEL	1.87	4.30	3.29	5.03
	16 LPC	5.16	3.26	2.24	0.51	F2 DEL	0.82	0.65	0.50	0.18
	18 LPC	4.53	1.65	2.15	0.44	F3 DEL	0.90	0.53	0.45	0.13
	20 LPC	4.31	2.01	2.14	0.44	FT DEL	6.62	5.35	3.90	3.45
	22 LPC	4.61	2.38	2.12	0.43	–	–	–	–	–

Table 5 Noise-related acoustic parameters under both methods of IF

Noise-based parameters	LPC-based analysis					Formant-based analysis				
	Condition	Males		Females		Condition	Males		Females	
		Mean	S.D	Mean	S.D		Mean	S.D	Mean	S.D
HNR	14 LPC	23.78	2.29	27.45	1.73	F1 DEL	26.30	2.23	29.41	1.24
	16 LPC	24.40	1.67	28.00	1.18	F2 DEL	26.25	1.75	29.44	0.98
	18 LPC	24.34	1.69	27.67	1.68	F3 DEL	25.54	1.59	29.20	1.10
	20 LPC	24.37	1.67	27.92	1.18	FT DEL	23.62	2.13	27.52	0.86
	22 LPC	24.38	1.69	27.89	1.17	–	–	–	–	–
GNE	14 LPC	−1.11	0.64	−1.18	0.73	F1 DEL	−21.17	1.40	−2.52	1.82
	16 LPC	−1.00	0.52	−1.09	0.70	F2 DEL	−21.32	0.89	−1.58	1.04
	18 LPC	−0.86	0.45	−1.08	0.77	F3 DEL	−21.37	0.68	−21.43	0.76
	20 LPC	−0.83	0.44	−1.14	0.79	FT DEL	−4.55	2.60	−24.86	2.46
	22 LPC	−0.83	0.46	−1.17	0.80	–	–	–	–	–

method, all the Perturbation and Noise measures were found to be statistically significant ($p < 0.05$) across the four formant conditions. However, in the LPBIF method, only three of the perturbation measures (ShdB, APQ5, and APQ11) and both the noise measures (HNR and GNE) were found to be significantly different at $p < 0.05$. One

Table 6 Friedman's test of differences for FBIF

Acoustic Parameter	χ^2	df	Sig.
JF0	44.64	3	0.000**
JT0	46.36	3	0.000**
RAP 3	38.68	3	0.000**
RAP 5	36.88	3	0.000**
SHdB	38.99	3	0.000**
APQ 11	39.32	3	0.000**
APQ 5	44.32	3	0.000**
HNR	40.40	3	0.000**
GNE	31.24	3	0.000**

*$p < 0.05$, and **$p < 0.01$

Table 7 Friedman's test of differences for LPBIF

Acoustic parameter	χ^2	df	Sig.
JF0	6.58	4	0.159
JT0	4.35	4	0.360
RAP 3	8.27	4	0.082
RAP 5	4.25	4	0.372
SHdB	10.11	4	0.038*
APQ 11	16.48	4	0.002*
APQ 5	12.68	4	0.013*
HNR	17.36	4	0.002*
GNE	23.46	4	0.000**

*$p < 0.05$, and **$p < 0.01$

can assume that FBIF is a more sensitive method of analysis, given that the formant details are edited accurately, since, values of all acoustic measures differed across all the conditions, whereas, LBPIF did not present variations across all parameters. FBIF gives a smoother IF signal. LPBIF gives a noisy IF signal. Jitter and shimmer values are higher when LPBIF is used. However, the same was not found in the results of the present study. Except for values of APQ 11 and APQ 5, it was found that values of acoustic measures were higher on usage of FBIF when compared to LPBIF. This may be because, only formants were manipulated whereas, the number of LPCs was kept constant at 18 for all the conditions—deletion of F1, deletion of F2, deletion of F3, and deletion of F1, F2, and F3. The developers of Vaghmi recommend that appropriate manipulations be made in terms of formant measurements as well as setting the filter order for further processing, where formant estimation is inaccurate [10]. Appropriate adjustments for filter order, taking into consideration the characteristics of both the speaker and the speech sample to be analyzed, thus, becomes an important aspect of analysis [9]. There are two common guidelines to set the filter order: (a) set it to the number of formants expected plus two, or (b) set it to sampling frequency in kHz. "In 2004 Vallabha et al. a heuristic that can be used to determine the optimal filter order for either a corpus of vowels or a single vowel" [11].

5 Conclusions

When using PC-based software applications for acoustic analysis, users should take care to understand the settings and requirements for any analysis system they use. It is particularly important to take special care with respect to variables such as sampling rate of input signals, adjustment of LPC filter order, and smoothing of formant tracks. Analysis software, may, at times, give erroneous results; hence, clinical users of these systems should carefully follow instructions provided in manuals and take special care to examine data for flaws or inconsistencies. Therefore, it is recommended that researchers note and report analysis settings used in research projects, including downsampling, LPC filter order, and special adjustments made for speaker characteristics [9]. Such inputs will guide the further research in investigating voices in disordered population.

References

1. Amir O, Wolf M, Amir N (2009) A clinical comparison between two acoustic analysis softwares: MDVP and Praat. Biomed Signal Process Control 4(3):202–5
2. Miller RL (1959) Nature of the vocal cord wave. J Acoust Soc Am 31:667–679
3. Fant G Acoustic theory of' speech production, s' Gravenhage, 2nd
4. Atal BS (2006) The history of linear prediction. IEEE Signal Process Mag 23(2):154–161
5. Atal BS, Schroeder MR (1979) Predictive coding of speech signals and subjective error criteria. IEEE Trans Acoust Speech Signal Process 27:247–254
6. Itakura F, Saito S (1970) A statistical method for estimation of speech spectral density and formant frequencies. Trans IECE Jpn 53-A:36–43
7. Deliyski DD, Shaw HS, Evans MK (2005) Influence of sampling rate on accuracy and reliability of acoustic voice analysis. Logop Phoniatr Vocology 30(2):55–62
8. Deliyski DD, Shaw HS, Evans MK, Vesselinov R (2006) Regression tree approach to studying factors influencing acoustic voice analysis. Folia Phoniatrica et Logopaedica 58(4):274–88
9. Burris C, Vorperian HK, Fourakis M, Kent RD, Bolt DM (2014) Quantitative and descriptive comparison of four acoustic analysis systems: vowel measurements. J Speech Lang Hear Res 57(1):26–45
10. Ananthapadmanabha TV (2010) Voice analysis using Vaghmi diagnostics software: case studies inverse filtering, voice and speech system, Bangalore
11. Vallabha G, Tuller B, Slifka J, Manuel S, Matthies M (2004) Choice of filter order in LPC analysis of vowels. From Sound Sense 50:B148–63

Impact of Timing of Surgery on the Development of Speech in Toddlers with Repaired Cleft Lip and Palate

M. Pushpavathi, V. Kavya, and V. Akshatha

Abstract ***Introduction***: Cleft lip and palate (CLP) is one of the most prevalent congenital deformities seen in Indian children. This condition hinders effective communication in early childhood due to speech and language difficulties. The development of speech and language is mainly based on the type and severity of CLP and the age at which surgical intervention followed by speech therapy was initiated. Early intervention plays an important role in enhancing the communication skills in children with CLP. ***Objectives***: The present study endeavours to compare and profile the speech characteristics of toddlers with repaired cleft of lip and palate (RCLP) with respect to the frequency, type and pattern of speech sound inventory of those who had undergone surgical intervention at an early age versus those who had delayed surgery. ***Participants***: Six Kannada speaking toddlers in the age range of 2–3 years, who were diagnosed to have Expressive Language Delay secondary to repaired cleft of lip and/or palate were considered for the Early Intervention Programme. The six participants were placed into two groups based on the age of surgical intervention. Thus there were three participants in the *early intervention group* (EIG—who underwent surgery before 1.6 years of age) and three in the delayed intervention group (DIG—who underwent surgery after 1.6 years of age). ***Method***: The baseline was established and measures such as type and pattern of the vowels and consonant inventory were analysed for toddlers in both the groups. Speech and language therapy was given for 20 sessions. Focused stimulation technique was used to enhance speech and language parameters. Post-therapy measurement was done after 20 sessions. ***Main outcome measures***: Descriptive statistics was used to compare the differences in speech measures for the pre-therapy and post-therapy conditions between early and delayed intervention groups. ***Results***: The results revealed certain points of interest.

M. Pushpavathi (✉) · V. Kavya · V. Akshatha
Department of Speech-Language Pathology, All India Institute of Speech and Hearing, Mysore, India
e-mail: pushpa19@yahoo.co.in

V. Kavya
e-mail: kavya.vijayan@gmail.com

V. Akshatha
e-mail: akshathavshwnth@gmail.com

M. Singh and Y. Rafat (eds.), *Recent Developments in Acoustics*, Lecture Notes in Mechanical Engineering, https://doi.org/10.1007/978-981-15-5776-7_10

Although both groups showed an improvement in the quality and quantity of phonetic inventory between pre- and post-therapy conditions, it was interesting to note that the mean values in the EI group were consistently higher across phonemes when compared to the DI group. The results clearly highlight the positive effects of early surgical and speech intervention. The speech patterns of each group and the consequent implications have been discussed in detail. ***Conclusions***: The present study is one of the few attempts to investigate the impact of timing of surgical correction on the development of phonetic inventory in children with CLP and also highlights the clinical implications of early surgical and subsequent speech and language intervention.

Keywords Cleft palate · Early intervention · Delayed intervention · Vowel and consonant inventory · Surgical correction

1 Introduction

A cleft of lip or palate is a birth defect which occurs when there is a failure in the fusion of tissues or palatal shelves during early stages of pregnancy between the eighth and twelfth weeks of foetal development. Cleft of lip and palate (CLP) is one of the most prevalent congenital deformities seen in children around the world. With a high incidence of 1 in 650 live births being born with cleft and related orofacial deformities, it is the most common congenital deformity in the Indian subcontinent [1]. Also a rise in the number of cases has been reported wherein approximately 40,000 new patients are being registered every year [2]. CLP occurs due to multifactorial condition and it can result in a variety of associated problems affecting both speech and language such as limited sound inventory, hypernasality, nasal air emission, reduced speech intelligibility, compensatory articulation, limited vocabulary, reduced communication attempts and expressive language delay [3, 4].

This condition hinders effective communication in early childhood due to speech and language difficulties. Thus children with CLP require a multidimensional care by a team of professionals as well as non-professional members such as parents, family members and caregivers. However, with early intervention in terms of surgical correction and speech therapy the children with CLP can obtain an expectable quality of life. The verbal output is mainly based on the variability and severity of the cleft. The age at which surgical intervention was done followed by intensive speech therapy are also crucial factors as early intervention plays an important role in enhancing the communication skills in children with CLP.

Surgical reconstruction of cleft lip (cheiloplasty) and cleft palate (palatoplasty) ought to be done at an optimal age. This will restore a normal facial appearance, aid in proper development of speech and language and also prevent the recurrence of middle ear infections [5]. The age at which children with CLP undergo surgical repair of lip or palate varies and is dependent on the type and severity of the cleft and the health condition of the child. In the present scenario, some debates occur regarding the optimal age for primary palatoplasty with some claiming that early surgery aids in better speech development versus others who state that delaying the repair of the hard palate helps in better facial development [6]. However, researchers over the years have come to a common consensus that when conditions are favourable, lip closure should be done between 3 and 6 months and the primary palatal closure should be carried out between 9 and 12 months [7–9]. The emphasis on an early surgery is thus laid in order to aid in better speech and language development in the initial stages of the child and also to prevent the maladaptive articulatory behaviours from developing.

Studies conducted over the last two decades have investigated the effects of the timing of primary palatal and lip closure on the development of speech and language. Most of the studies highlight the positive effects of early surgery whereas some have also concluded that surgery at an early age does not result in significant improvement in speech. Dorf and Curtin could be credited as the pioneers in investigating the effect of early palatal closure on speech adequacy. The authors compared the speech repertoire of toddlers who had undergone surgery early versus late surgery in the pre-operative and post-operative conditions. The study highlighted a clear trend depicting better speech adequacy for toddlers who had undergone early surgery. The authors established that the babies who had undergone surgery within the first year of life had a better chance of developing good speech skills and a reduced potential for acquiring compensatory articulatory behaviours. The authors further established that the 'articulation age' of toddlers should act as the determinant for surgical readiness of a child [10].

A study was carried out to assess the speech skills in pre- and post-operative conditions of adolescents in Sri Lanka who had undergone late palatal surgery. The children underwent intensive articulatory training for a period of 8 months post-surgery in order to change the deviant articulatory patterns. The findings of the study clearly showed that surgical correction alone did not lead to spontaneous improvement in speech skills. When surgery was followed by regular and intensive speech therapy there were significant improvements in the articulatory abilities, nasality and velopharyngeal closure abilities [11].

The phonological development in children who had undergone surgical correction at various ages was examined in a study. The study investigated the prevalence in percentage of velopharyngeal inadequacy and compensatory articulatory errors (CAE) and clearly indicated the positive effects of early surgery. The phonological development of toddlers was followed till the age of 3–4 years and a retrospective analysis was carried out. It was found that infants who underwent palatal surgery at the age of 6 months had around 5% CAE whereas those who underwent surgery at 12 months had 35% CAE. As the age of surgery increased the percentage of CAE drastically increased. Surgery age of 2 years had a CAE of 55% and those who underwent at 4 years had a massive CAE score of 75% [12].

A recent study examined the phonological skills of two groups of toddlers aged 3 years, wherein one group had undergone primary soft palate (SP) repair at 4 months and hard palate (HP) correction at 12 months [13]. The second group was yet to undergo repair of HP or had an HP which was unoperated but the SP was operated.

The author used a picture-naming test to elicit the target phonemes from both the groups. The study found that the toddlers in both groups had a limited speech sound repertoire compared to the typically developing children, but the second group with unoperated HP had a greater number of speech sound errors compared to the group which had undergone surgery. There have also been studies which have contradicted the beneficial effects of early surgical correction. A recent study has also shown that primary palatal surgery done at or around 12 months of age did not result in significant changes in speech skills by the age of 3 years [14].

In the Indian context, there have been limited studies exploring the relation between timing of palatal surgery and its effect on speech and language. A retrospective study was conducted wherein the speech skills such as articulation, resonance and speech intelligibility were examined in individuals who had undergone primary palatal surgery after the age of 10 years [15]. The findings of the study depicted that in the post-operative condition, the children who underwent late palatal surgery showed a nominal improvement in speech skills including intelligibility. This improvement was spontaneous in nature as none of the children underwent regular speech therapy. This study highlights the importance of timing of intervention in terms of surgical correction and intensive speech therapy in the early stages of a child's life.

The efficacy of early speech and language intervention in children with CLP has been documented in the past few years and most of the studies opine that early intervention is of utmost benefit when carried out between 1.6–5 years. Various studies have reported the positive effects of early speech therapy for children with repaired cleft of lip and palate (RCLP) [16–18]. The children with RCLP showed an improvement in communication skills, linguistic abilities and articulatory abilities when provided with early language intervention. In order to implement and schedule an early linguistic intervention programme, the structural orofacial deformities need to be surgically repaired at the optimum age based on various factors of the child. Thus there is widespread research across the globe related to this topic in order to provide the most advantageous treatment for each child with CLP so as to better their future.

Although there have been many studies highlighting the pre-operative speech characteristics in unoperated CLP patients and post-operative speech abilities after late palatal surgery, there have been limited studies especially in the Indian context which have investigated the effects of early surgery on speech characteristics. Therefore, the present study endeavours to compare and profile the speech characteristics of toddlers with RCLP with respect to the frequency, type and pattern of speech sound inventory of those who had undergone surgical intervention at an early age versus those who had delayed surgery. The objectives of the present study are as follows:

1. To investigate the frequency of occurrence of vowels in the early versus delayed surgery groups before and after speech and language intervention and
2. To determine the frequency of occurrence of consonants in the early versus delayed surgery groups before and after speech and language intervention.

2 Method

2.1 *Participants*

Six Kannada speaking toddlers in the age range of 2–3 years, who were diagnosed to have Expressive Language Delay secondary to repaired cleft of lip and/or palate served as the participants for the study. The six participants were placed into two groups based on the age of surgical intervention. Thus there were three participants in the *early intervention group* (EIG—who underwent surgery before 1.6 years of age) and three in the delayed intervention group (DIG—who underwent surgery after 1.6 years of age). A detailed language evaluation was carried out by a qualified Speech Language Pathologist using Receptive Expressive Emergent Language Scale [19]. All the participants were enrolled for the Early Language Intervention Programme (ELIP) at the Unit for Structural and Orofacial Anomalies (U-Sofa). An informed consent (approved by the Ethical Committee) was obtained from the parents prior to their participation in the study. The detailed description of the subjects is shown in table 1.

2.2 *Data Collection and Processing*

A detailed pre-therapeutic evaluation was carried out to establish the baseline of the children. The audio–video recordings were conducted in a quiet room during unstructured play sessions between the Speech Language Pathologist and the child, with the mother also being involved in the sessions. Focused stimulation technique was used to enhance speech and language parameters. Recordings were done using a Handycam recorder (Sony DCR-SR88). The recorder was placed on a tripod stand at a distance of approximately 8–10 ft from the child. The spontaneous utterances recorded after the baseline audio–video recording, a post-therapy audio–video recording was done on 20th session. Thus a total of 2 video recordings and analyses for each participant were carried. Thus the frequency of occurrence of vowels and consonants in the child's speech inventory was calculated.

2.3 *Speech and Language Therapy*

Speech language therapy was conducted in the clinical setup for sessions lasting forty-five minutes each. Speech therapy sessions were carried out by a qualified Speech Language Pathologist. This was done by preparing a master lesson plan by taking up specific goals and activities. Play way method was used to improve vowel and consonant inventory, functional communication skill and to increase the frequency of meaningful utterances.

Table 1 Detailed description of participants of the study

Participants	Type of cleft	Age of surgery	Age of enrollment for therapy	Language age	
				RLA	ELA
Early Intervention Group (EIG)					
1	Complete cleft of primary palate	1 year (Palate)	1.6 years	16-18 months	9-10 months
2	Unilateral complete cleft of lip and palate	5 months (lip) & 1 year (Palate)	3 years	33-36 months	20-22 months
3	Unilateral complete cleft of lip and palate	7 months (lip) & 1.1 years- (Palate)	2 years	22-24 months	18-20 months
Delayed Intervention Group (DIG)					
1	Bilateral complete cleft of hard and soft palate	1.8 years (palate)	2.4 years	27-30 months	12-14 months
2	Cleft of soft palate	1.6 years (palate)	2.6 years	27-30 months	16-18 months
3	Unilateral complete cleft of lip and palate	7 months (lip) & 3 years (palate)	3.9 years	36-40 months	18-20 months

Note REELS—Receptive Expressive Emergent Language Scale (Bzoch and League [19])
RLA—Receptive language age
ELA—Expressive language age

Focused stimulation approach was demonstrated to the mother during therapeutic sessions using low-cost materials and commonly available toys. A multisensory approach was used to increase the oromotor movements. Visual feedback using mirror work and tactile cues with interesting oromotor activities was provided appropriately wherever required. A corpus of vocabulary was prepared which contained the most commonly used functional words by toddlers. This was done with the aim of increasing the child's functional vocabulary. The positive behaviours and behaviours which were attempted correctly by the child were reinforced with token and tangible reinforcements to improve their cooperation and confidence in doing the activity. The mothers were also told to carry out a similar training programme at home in the absence of direct supervision of the SLP. A total number of 20 therapy sessions were conducted.

2.4 Data Analysis

Each of the video and audio-taped session of mother–child interaction during unstructured play was transcribed using International Phonetic Alphabet (IPA) symbols. The frequency of occurrence of vowels and consonants were calculated. Vowels were classified as high (/i/), mid (/e/,/æ/) and low (/a/) based on the tongue height and their frequencies were calculated. The frequency of occurrence of different consonants was calculated based on place of articulation (bilabials, glottals, labiodentals and semi-vowels). The tabulated data was entered in a Microsoft excel sheet and the difference in the percentage of vowels and consonants was calculated. Descriptive statistics was used to compare the differences in speech measures for the pre-therapy and post-therapy conditions between early and delayed intervention groups.

3 Results

The present study is one of the preliminary efforts to investigate the effect of the timing of surgical correction on the speech-related skills of toddlers with RCLP. The study also attempts to highlight the changes brought about in the phonetic inventory of toddlers in both early and late surgery groups with therapy, thereby stressing the importance of an early intervention programme. The pre-test and post-test measures for the phonetic inventory of toddlers have been compared in both EI and DI groups and the findings obtained have been depicted below.

3.1 Frequency of Occurrence of Vowels Before and After Early Intervention Across Groups

The effect of early versus late surgery on speech-related measures of toddlers were analysed by calculating the frequency of occurrence of vowels based on their tongue height in the baseline and post-therapy conditions.

The bar graph given above (Fig. 1) shows the frequency of occurrence of vowels with respect to tongue height in the pre-and post-therapy conditions in both the experimental groups. The graph clearly indicates a trend wherein the EI group consistently exhibited a greater frequency across all the three vowels when compared to the DI group. Among the vowels, /a/ was frequently occurring followed by /i/and /u/. This was seen in both early and delayed intervention groups. However, children who underwent early surgery exhibited a greater frequency across all vowels compared to the delayed intervention group.

With respect to the post-therapy condition, all the vowels showed an increase in the 20th session compared to the baseline in both the groups. Statistical analysis could not be carried out due to the small sample size. The results thus indicated better

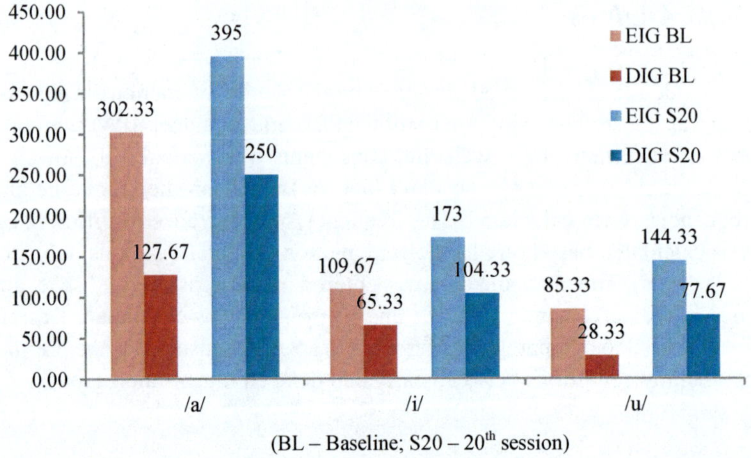

Fig. 1 Frequency of occurrence of vowels in the EI versus DI groups in the pre- and post-therapy conditions

Table 2 Frequency of vowels in EI and DI groups comparing pre- and post-test conditions

Vowels	BL		S1	
	EIG	DIG	EIG	DIG
/a/	302.33	127.66	395	250
/i/	109.67	65.33	173	104.33
/u/	85.33	28.33	144.33	77.66

(BL—Baseline; S20–20th session)

scores for the EI group in terms of number of vowels which is depicted in the table (Table 2).

3.2 *Frequency of Occurrence of Consonants Before and After Early Intervention Across Groups*

The graph below (Fig. 2) shows the frequency of occurrence of consonants in the pre-and post-therapy conditions in both the groups. Among the consonants, EIG exhibited more number of consonants compared to the DIG group. This was seen in the baseline. After 20 sessions of intervention, there was improvement seen in the occurrence of consonants in both the groups. A comparable trend was seen across all consonants wherein the EI group consistently had a greater frequency when compared to the DI group.

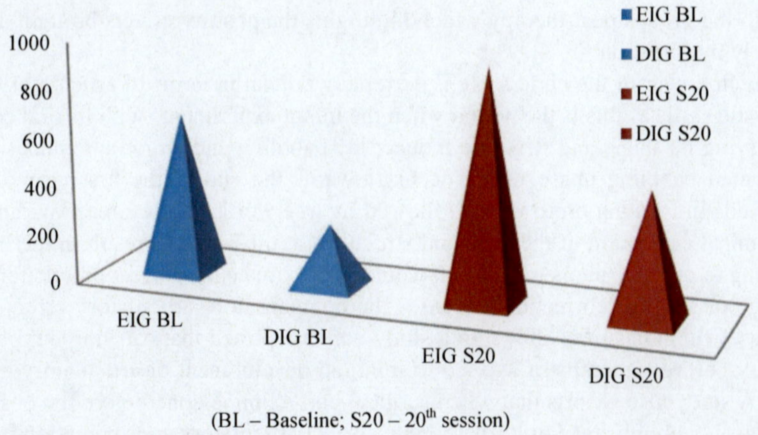

Fig. 2 Comparison of pre-therapy scores versus post-therapy scores for the consonants in the EI and DI groups

The pyramid graph above (Fig. 2) depicts the total number of consonants in the toddler group's repertoire and clearly indicates that the EI group had a greater number of consonants in their repertoire when compared to the DI group. It was also observed that the variety of consonants did not differ between the two groups. Both the groups had a variety of consonants such as bilabials, dentals, alveolars, retroflexes, velars, affricates, fricatives, glottals and semi-vowels, although clearly the early intervention group had a greater frequency when compared to the delayed intervention group.

4 Discussion

The results of the present study clearly indicate that the toddlers who had undergone early surgery had better phonetic abilities in terms of the frequency of both vowels and consonants when compared to the delayed surgery group. The DIG group had a lesser corpus of phonemes in their repertoire. This supports the findings of previous studies which reported that toddlers who had undergone surgery early or within the first year of life had a better speech skills and a reduced potential for acquiring compensatory articulatory behaviours [10, 13]. They also highlighted that when compared to toddlers who had not undergone surgery the children who had undergone early surgery showed fewer articulatory errors.

The present study, therefore, highlights the importance of an early intervention programme both in terms of surgical intervention followed by intensive speech therapy. The results have made it apparent that toddlers who underwent lip and palatal repair in the first year of life followed by speech therapy had a better phonemic inventory when compared to toddlers who underwent surgery after the age of 1.6 years. It should also be noted that there was a significant increase in the phonemic repertoire of both the groups post-therapy which highlights the positive effects brought about by early intervention.

The first year of the child's life is extremely crucial in terms of articulatory and linguistic skills as this is the period when the infant experiments with its oral cavity by varying its shape and structure induced by babbling and jargon utterances. The variegated babbling phase which occurs towards the end of the first year of the child aids in forming proto words followed by true words. Therefore, providing an anatomical correction of the orofacial structures at this age will result in the infant learning to produce phonemes which would approximate the correct production. As the age of surgical correction increases the compensatory articulatory errors may increase which has been shown in a study which reported that carrying out palatal surgery before 6 months of age led to minimal development of articulatory errors [12]. A study also reports that when surgical correction is done before the onset of canonical babbling stage, it leads to more correct articulatory approximation by the infant. The importance of surgical readiness with respect to the 'articulatory age' has also been supported previously [10]. The results of the present study have also been contradicted by a previous study which reported that primary palatal surgery done in the first year of life did not result in significant changes in the speech skills of children by the age of 3 years [14].

With respect to the type and variety of phonemes it was seen that the variety of consonants did not differ in the two groups. Both the groups had a variety of consonants such as bilabials, dentals, alveolars, retroflexes, velars, affricates, fricatives, glottals and semi-vowels, although clearly the early intervention group had a greater frequency in their repertoire. The children preferred the usage of these phonemes in simple syllabic structures such as CV, VCV and CVCV. These syllabic structures were incorporated in the true and proto words acquired by them. This suggests that the palatal correction and speech therapy at the prelinguistic phase would aid in greater articulatory efforts as the oromotor abilities of the toddlers would adapt in a better way. A recent study has reported that the articulatory and resonatory characteristics of toddlers who underwent surgery before and after 6 months within the first year are equivalent. This denotes that palatal correction in the first year or before 12 months would benefit the speech and linguistic skills of the children with CLP [20].

The post-therapy analyses of both the early and delayed groups showed a trend which was comparable wherein both the groups showed significant improvements in the number of phonemes acquired. However, the DIG group showed a marginally lesser improvement after the early intervention programme, whereas the EI group showed a greater change in the phonetic repertoire post-therapy. Although the type of consonants acquired remained the same further studies conducted over a longer period with intensive speech and language intervention may yield greater results in terms of the type of phonemes and vocabulary development. These results support the findings of various studies which have highlighted that an early parent-implemented intervention programme led to an increase in the number and accuracy of speech sound inventory and also aided in articulatory development [16–18].

The effectiveness of a systematic early language intervention programme has been investigated previously in toddlers with expressive language delay secondary to repaired cleft of palate [18, 21]. The results of the present study corroborate the findings of recent studies which have revealed that intensive speech and language stimulation leads to significant changes in the phoneme repertoire and vocabulary of toddlers with RCLP [18, 21–23]. In the present study also a significant increase was observed for the frequency of speech sounds, for both vowels and consonants which illustrate the positive effects of a structured and parent-implemented intervention program for children with CLP especially during the formative years. Therefore, the need of the hour is to create awareness in the community regarding the importance of regular speech and language training in this population.

5 Conclusions

The timing of the surgery of lip or palate thereby plays a crucial role in shaping the future communicative abilities of the child as it may affect the subsequent speech and language intervention. As linguistic intervention cannot begin without the surgical repair this will mean that the more the surgery gets delayed the speech and linguistic skills of the child may suffer. Similar studies carried out in future exploring the effect of timing of surgery on the linguistic skills of children with respect to the vocabulary development, mean length of utterance, frequency of different words, articulatory pattern development and speech intelligibility will give a fair idea regarding the changes if any, brought about by the timing of surgery on the language skills.

Acknowledgments This is a part of ongoing research on 'Efficacy of Early Language Intervention Program for Children with Repaired Cleft lip and Palate', funded by the Department of Science and Technology (DST—No. SB/SO/HS/02/2014), Government of India. The authors would like to thank DST for funding the project. The authors would also like to thank the Director, All India Institute of Speech and Hearing, Mysuru, for providing the infrastructure to carry out the study.

References

1. Raju S (2000) In search of a smile-study of children born with cleft lip and palate in India. Tata Institute of Social Sciences: Mumbai. www.smiletrain.org. Accessed 15 Aug 2009
2. Shrivatsav S (2013) Parents still don't bring cleft lip, palate kids for treatment. Times of India, Nagpur, 18 January 2013
3. Sunitha R, Jacob M, Jacob MS, Nagarajan R (2004) Providing intervention services for communication deficits associated with cleft lip and/or palate- A retrospective analysis. Asia Pac Disabil Rehabil J 15:78–85
4. D'Antonio LL, Scherer NJ (2008) Communication disorders associated with cleft palate. In: Losee J, Kirschner R (eds) Comprehensive cleft care, pp 64–78. McGraw Hill
5. Patel PK (2016) Cleft palate repair. http://emedicine.medscape.com/article/1279283-overview
6. Rohrich RJ, Love EJ, Byrd HS, Johns DF (2000) Optimal timing of cleft palate closure. Plast Reconstr Surg 106(2):413–422
7. Rosenstein SW, Dado DV (2005) Early bone grafting with the functional cleft lip repair. Semin Plast Surg 19(4):302–312

8. Guerrero CA (2012) Cleft lip and palate surgery: 30 years follow-up. Ann Maxillofac Surg 2(2):153–157
9. Kobus K, Kobus-Zalesna K (2014) Timing of cleft lip and palate repair. Dev Period Med 18(1):78–83
10. Dorf DS, Curtin JW (1982) Early cleft palate repair and speech outcome. Plast Reconstr Surg 70(1):74–79
11. Sell DA, Grunwell P (1990) Speech results following late palatal surgery in previously unoperated Sri Lankan adolescents with cleft palate. Cleft Palate-Craniofac J 27(2):162–168
12. Ysunza A (2000) Phonological Development related with Cleft Palate. http://www.michiganspeechhearing.org/docs/Ysunza%20PHONOLOGICAL%20DEVELOPMENT%20IN%20CLP%20-%20EARLY%20REPAIR.pdf
13. Willadsen E (2012) Influence of timing of hard palate repair in a two-stage procedure on early language development in Danish children with cleft palate. Cleft Palate-Craniofac J 49(5):574–595
14. Klinto K, Svensson H, Elander A, Lohmander A (2013) Speech and phonology in Swedish-speaking 3-year-olds with unilateral complete cleft lip and palate following different methods for primary palatal surgery. Cleft Palate-Craniofac J 51(3):274–282
15. Murthy J, Sendhilnathan S, Hussain SA (2009) Speech outcome following late primary palate repair. Cleft Palate-Craniofac J 47(2):156–161
16. Scherer NJ, Kaiser A (2007) Early intervention in children with cleft palate. Infants Young Child 20:355–366
17. Meinusch M, Romonath R (2011) Early language intervention for children with cleft lip and/or palate: a systematic review. Evid Based Commun Assess Interv 5(4):197–215
18. Pushpavathi M, Kavya V, Akshatha V (2017) Efficacy of Focused Stimulation in Early Language Intervention Program for Toddlers with Repaired Cleft Palate. Glob J Otolaryngol 9(1):1–8
19. Bzoch KR, League R (1971) Assessing language skills in infancy. University Park Press, Baltimore
20. Luyten A, Bettens K, D'haeseleer E, De Ley S, Hodges A, Galiwango G, Bonte K, Vermeersch H, Van Lierde K (2014) The impact of palatal repair before and after 6 months of age on speech characteristics. Int J Pediatr Otorhinolaryngol 78:787–798
21. Pushpavathi M, Kavya V, Akshatha V (2017) Efficacy of early intervention program for children with cleft lip and palate: a case study. JCLAD 5(1):31–42
22. Chapman KL, Hardin MA (1991) Language input of mothers interacting with their young children with cleft lip and palate. Cleft Palate Craniofac J 28(1):78–85
23. Scherer NJ, D' Antonio L, McGahey H (2008) Early intervention for speech impairments in children with cleft palate. Cleft Palate Craniofac J 45:18–31

Acoustics Materials

Sonic Crystals for Highway Noise Reduction

Debasish Panda and A. R. Mohanty

Abstract Sonic crystals are noise barriers which have come to picture since the last two decades for their sound attenuation properties. They consist of sound scatterers arranged periodically inside a host material. The scatterers have high impedance and are put in a fluid of low impedance. Sound attenuation takes place due to multiple scattering of sound waves by the rigid sound scatterers, within particular frequency bands known as Band Gaps. In this paper, a finite element study has been performed on a 2-D sonic crystal having circular scatterers arranged in square pattern. The scatterers are assumed to be sound hard, which imposes that the normal velocity and normal acceleration at their boundary are zero and the arrangement is periodic which is because of the cyclic symmetry of the structure. Relevant boundary conditions have been incorporated into the design which aims in determining the Band Gaps and the corresponding transmission losses through the sonic crystal. Results of eigenfrequency and frequency response analysis of the scatterers are done using a commercial finite element software, COMSOL Multiphysics are presented in this paper.

Keywords Sonic crystals · Sound attenuation · Band gaps · Transmission loss · Eigenfrequency

1 Introduction

Noise is an unpleasant sound. Humans tend to get irritated by noise which affects their mental health adversely. Prolonged exposure to noise can cause permanent hearing damage in human beings and other animals. Today, traffic and environmental noise pollution have become one of the major environmental concerns worldwide. This has led to new developments in the field of noise control in terms of noise barriers.

D. Panda (✉) · A. R. Mohanty
Mechanical Engineering Department, IIT Kharagpur, Kharagpur, India
e-mail: debasishlitu@yahoo.co.in

A. R. Mohanty
e-mail: amohanty@mech.iitkgp.ac.in

M. Singh and Y. Rafat (eds.), *Recent Developments in Acoustics*, Lecture Notes in Mechanical Engineering, https://doi.org/10.1007/978-981-15-5776-7_11

When installed properly noise barriers can reduce vehicular noise level in highways significantly.

Sonic crystals are noise barriers with periodic arrangement of sound scatterers arranged in different lattice configurations and embedded in a matrix or host material. The scatterers have high acoustical impedance compared to the matrix which is generally a fluid medium, e.g. air, water. Sonic crystals are analogous to photonic crystals [1] used for forbidding the propagation of light in certain frequency bands in a way that sonic crystals attenuate sound propagation at selective frequency ranges known as band gaps as a consequence of multiple scattering. Band gaps appear in the regions of wavelengths that are comparable with the periodicity of the lattice. Sonic crystals are also optically transparent which adds to the aesthetic aspect of the urban landscape. They can be used in highways, townships, and workshops for noise control. The earliest realization of sonic crystals was made by Martinez et al. [2] in the year 1995. They performed sound attenuation experiments on a sculpture made by Eusebio Sempere exhibited at Juan March foundation, Madrid. They measured sound attenuation in outdoor conditions for sound wave vectors perpendicular to cylinder's vertical axis. Sound attenuation peak at 1670 Hz was found, that was the formation of the first band gap.

The periodicity of the scatterers can be along one, two or three dimensions forming 1-D, 2-D and 3-D sonic crystals, respectively. The scatterers are responsible for multiple scattering of the reflected and incident sound waves. According to Bragg's law [3], the mechanism of sound attenuation is destructive interference of the reflected sound waves with the incident sound waves within the band gaps. If the interference is constructive then energy of the original wave gets transmitted through the sonic crystal, which forms the propagation bands. Bragg's law is written as

$$n\lambda = 2d \sin\theta \tag{1}$$

Here λ is the wavelength of the incident waves, d is the distance between the scatterers, θ is the angle the incident wave makes with the plane of scatterers and n is a positive integer. For first-order reflection and plane wave sources, the above equation gives the centre frequency, f_c as

$$f_c = \frac{c}{2a} \tag{2}$$

Here d is replaced by the lattice constant a and c is the speed of sound. In theory, band gaps should be at multiples of the above fundamental frequency. Similarly, another property known as the filling fraction affects the width of the band gap. Filling fraction is the fraction of the structure occupied by the cylinders [4]. For a square lattice arrangement having scatterers of radius r, the filling fraction is written as

$$F = \frac{\pi r^2}{a^2} \tag{3}$$

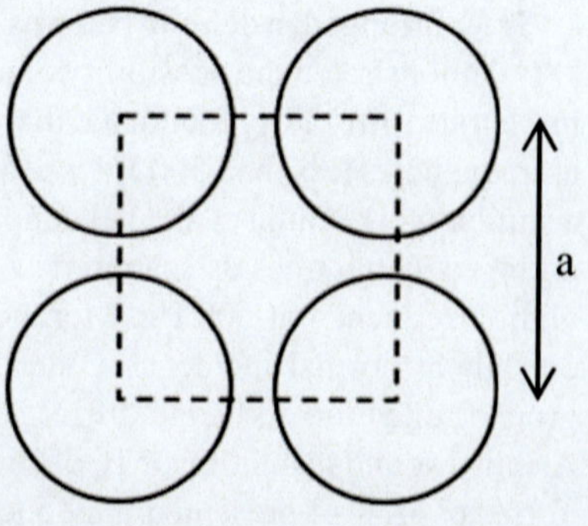

Fig. 1 Unit cell of a two-dimensional sonic crystal shown by the dotted line for square configuration

Gupta and Lim [5] numerically and experimentally studied the effect of the geometric parameters on a rectangular sonic crystal. The parameters were two lattice constants along two perpendicular directions and diameter of the scatterers. They found out the lattice constant along the direction of plane wave propagation to affect the position of the centre frequency and the other two parameters to affect the bandwidth. For periodic structures, an important theorem is given by Kittel [6], which gives the wave function of a particle inside a periodic arrangement written as

$$\psi(r) = u(r)e^{ikr} \tag{4}$$

Here, $\psi(r)$ is the Bloch wave function, $u(r)$ is a periodic function with same periodicity and symmetry as that of the lattice, k is the wave vector and r is the position. The exponential function represents a plane wave which gets modulated by the periodic function *u(r)*.

As sonic crystals are periodic, a group of scatterers can be stacked together to form the entire lattice. This smallest group of scatterers is known as a unit cell as shown in Fig. 1. Analysis of one-unit cell is sufficient for the analysis of the whole structure.

Researchers have been working on theoretical models for predicting the band gaps. Gupta et al. [7, 8] developed an one-dimensional model and a quasi two-dimensional model for band gap and sound attenuation calculations. They prepared a 1D Webster horn model by considering the pressure to be constant over the cross section of the waveguide. Then for the quasi 2D model they improved their previous 1D model by taking a parabolic pressure profile across the cross section. They compared the obtained results with finite element simulations. The finite element results came out to be on par with the theoretical models. Elford et al. [9] computed the acoustic band structure and transmission loss of sonic crystals with resonant array elements and their various combinations with the use of finite element method. Similarly, Gupta et al. [10] did a finite element simulation on 2D sonic crystals and found maximum sound transmission loss of 18 dB in the band gap region. Hoare et al. [4] performed analysis of 2D and 3D sonic crystals using Finite Difference Time Domain (FDTD) method.

Trees arranged in definite patterns can act as sonic crystal noise barriers. Rosa et al. [11] demonstrated the possibility of sound attenuation by periodically arranged trees in a lattice form. They found that there is an improvement in sound attenuation values at frequencies less than 500 Hz when compared with a typical green belt or forest. In a similar work, Gulia et al. [12] conducted a 3D computational study on periodically arranged Thuja trees as scatterers. A good amount of sound attenuation was found within frequency of 500 Hz. Morandi et al. [13] performed measurements on sonic crystals at normal incidence under EN 1793-6 standards, which allows to cancel ground reflection and edge diffraction by applying suitable windowing techniques. Average sound insulation of 10 dB was found in a frequency range of 550–1,000 Hz. Torres et al. [14] presented the concept of sonic crystal acoustic screens to control transportation noise. Their performance was found to be comparable to classical acoustic barriers.

In this paper, we have presented a finite element analysis of sonic crystals for highway noise reduction. We have used a total of 50 scatterers arranged in a 10×5 pattern. The sound sources in the highway are assumed to be planar in nature with frequencies ranging up to 8000 Hz. Frequency-domain analysis and eigenfrequency analysis were performed using COMSOL Multiphysics. Maximum sound attenuation around 27 dB was found in the first band gap.

2 Methodology

2.1 The Model

The entire model is constructed in COMSOL Multiphysics background. The radius of the scatterers and the lattice constant are 25 mm 90 mm respectively. So, Eq. (3) returns an approximate value of 0.25 as filling fraction. Air is taken as host material. The scatterers are assumed to be sound hard which is implemented as a sound hard boundary condition. A sound hard boundary takes the normal component of acceleration and velocity as zero at the boundaries. As the structure is periodic, periodic boundary conditions are added to the boundaries of the unit cell in both directions. Free triangular elements of maximum size 8.575×10^{-3} m is used for the mesh generation. This size of the triangular element comes by dividing the wavelength to 20, i.e. $\lambda/20$. The unit cell generated is presented below in Fig. 2.

The unit cell is extended to periodically arrange scatterers in a 10×5 grid as shown in Fig. 3. Plane wave background pressure field of 1 Pa is added to the left of the array.

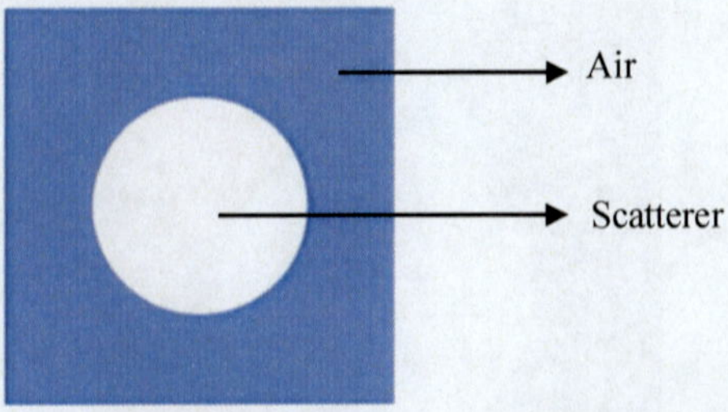

Fig. 2 Unit cell generated in COMSOL

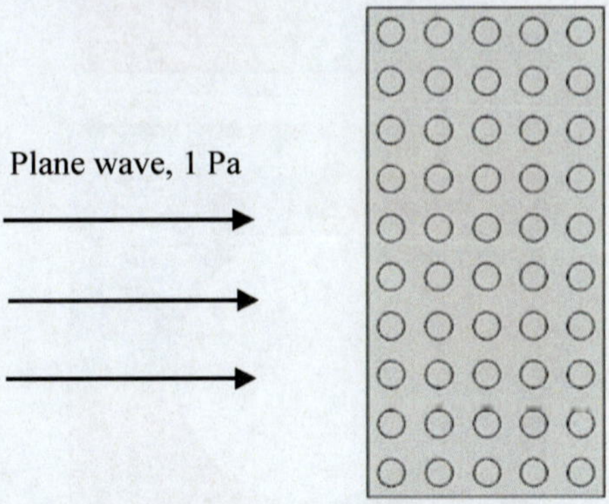

Fig. 3 10 × 5 array having background pressure field of 1 Pa

2.2 *Simulation*

An eigenfrequency study is performed on the unit cell of Fig. 2 for the first ten eigenfrequencies. A parametric sweep is done for the wave vector along the direction of the wave for a range of $0 \leq k_x \leq \pi/a$ that is the range of the irreducible part of the Brillouin zone [15]. The eigenfrequency study finds the band structure and eigenmodes of the unit cell. For the array, frequency-domain study is performed. The frequencies are varied between ranges of 100–8000 Hz with a step size of 10 Hz. The frequency-domain study returns the pressure distribution and the transmission loss across the array.

3 Results and Discussion

3.1 *Eigenmodes and Band Structure*

The eigenfrequency analysis produces the eigenmodes and band structure of the unit cell. As the structure is periodic in nature, eigenfrequency analysis of one-unit cell suffices the analysis for the whole structure. Figure 4 shows the first six eigenmodes at respective eigenfrequencies of 1400.8, 2224.3, 3904, 4110.8, 4390.1, and 4960.3 Hz

a) b) c) d) e) f)

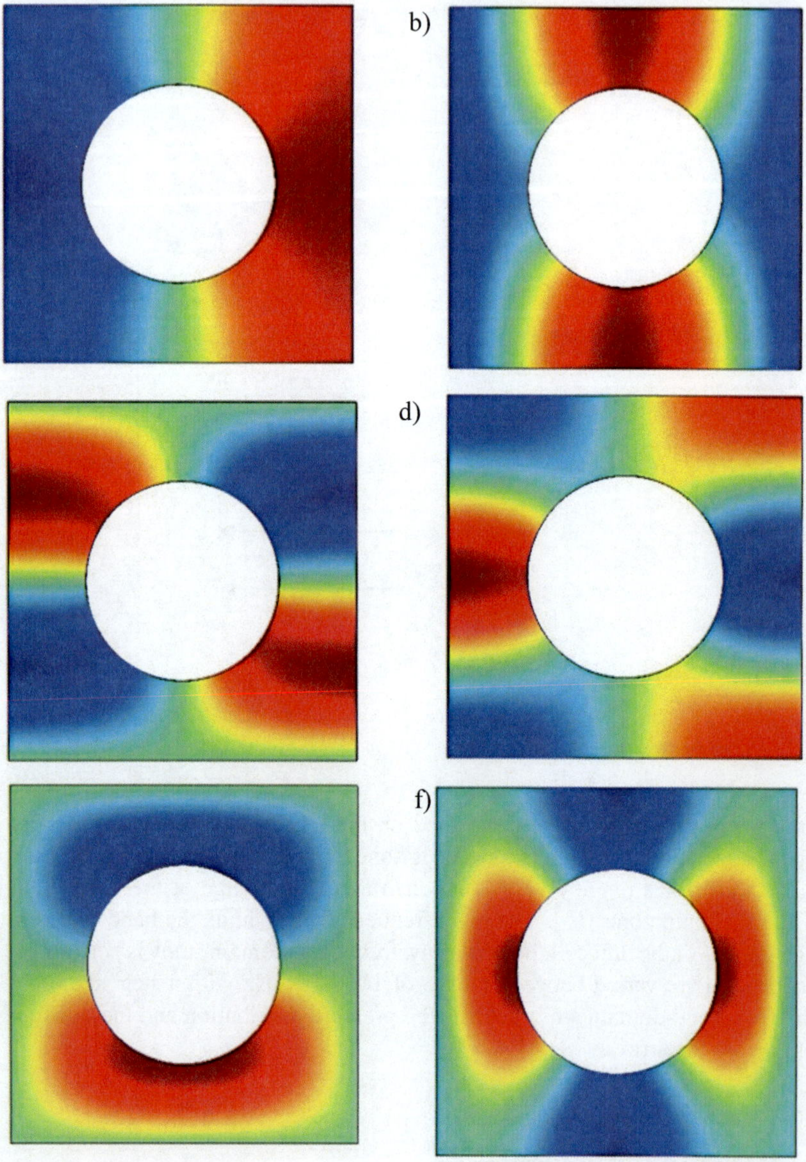

Fig. 4 First six eigenmodes showing the variation of acoustic pressure field inside a unit cell at k_x = 34.907 1/m

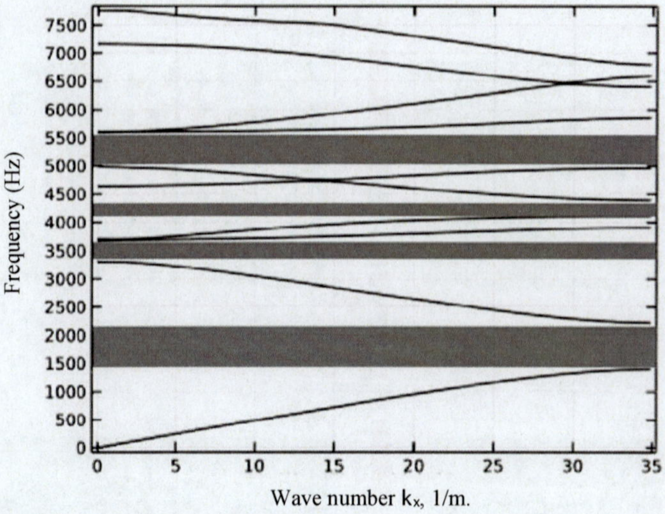

Fig. 5 Band diagram

when k_x is fixed at 34.907 1/m. The pressure in the following six eigenmodes varies between a maximum of 2.14 Pa to a minimum of −1.92 Pa.

Similarly, we the draw band diagram, which is a plot of eigenfrequencies at different k_x values. Figure 5 shows the band diagram. The grey areas are the band gaps within which sound attenuation takes place. The centre frequency of the first band gap is found to be around 1812 Hz, which is close to the centre frequency of 1906 Hz found from Eq. (2). Total of four band gaps is found within 8000 Hz. Corresponding sound attenuation values will be presented in the next section. Second and forth band gaps have centre frequencies that are multiples of the centre frequency of the first band gap. There is also a narrow third band gap around a centre frequency of 4200 Hz.

3.2 *Pressure Distribution and Transmission Loss*

The pressure variation across the sonic crystal at different frequencies of 1810 Hz, 3000 Hz, 5250 Hz and 7000 Hz are shown in Fig. 6a–d, respectively. Out of the four frequencies 1st and 3rd frequencies fall inside 1st and 4th band gaps. Sound attenuation can be clearly observed at those frequencies in Fig. 6a, c, respectively. For the other two frequencies, which lie outside the band gaps, sound wave passes through the crystals with not much attenuation.

Transmission loss occurs across the array of scatterers at the band gap regions. Transmission loss can be calculated from the exit and inlet pressure of the array.

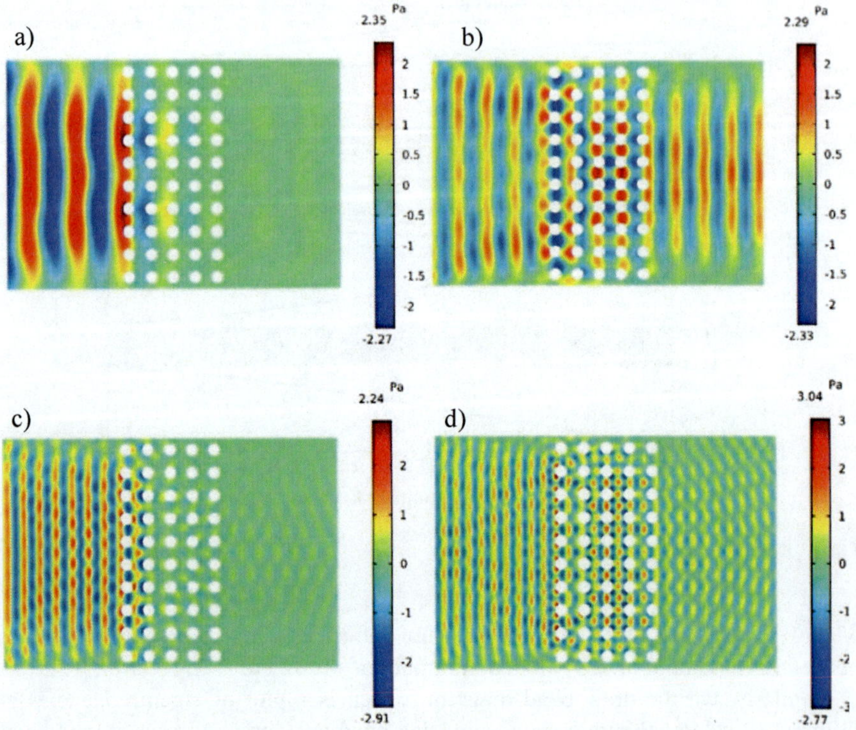

Fig. 6 Pressure variation at different frequencies of **a** 1810 Hz **b** 3000 Hz **c** 5250 Hz **d** 7000 Hz

$$TL = 20\log_{10}\frac{p_i}{p_o} \tag{5}$$

Here p_i and p_o are pressure at inlet and outlet of the array, respectively. Transmission loss in the first band gap is presented below in Fig. 7. Approximate transmission loss of 27 dB is found around 1730 Hz.

Transmission loss at other band gaps can be viewed in Fig. 8. Sound transmission losses can be accurately seen over the band gaps except for the third band gap, which shows transmission loss over a wider band than the actual width of the band gap.

4 Conclusions

In this paper, a finite element study over a sonic crystal grid of 10 × 5 has been done. Planar sound sources are considered for the highway noise over a frequency range of up to 8000 Hz. Four band gaps are found within 6000 Hz showing good sound transmission loss values. Maximum sound attenuation of 27 dB is found in the first band gap. Band gaps were calculated from the eigenfrequencies of the unit

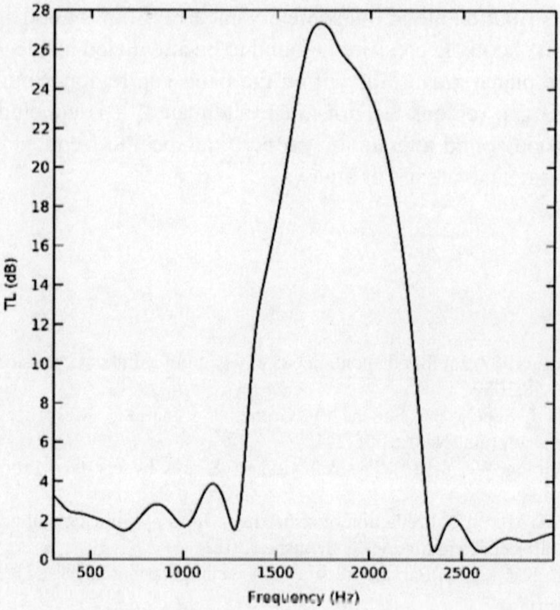

Fig. 7 Transmission loss in the first band gap

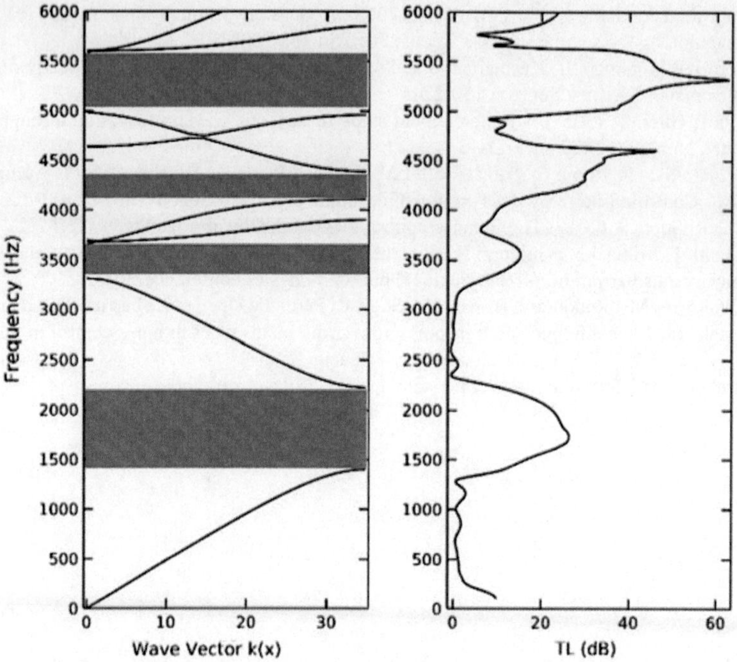

Fig. 8 Band diagram and corresponding transmission loss

cell. Pressure distribution along the sonic crystal array shows good agreement with the band diagram. Acoustic pressure is found to be attenuated inside the array, if the frequency of the planar wave falls within the band gap region. Sound frequencies outside the band gap regions are not much attenuated. It is concluded that, sonic crystals show good sound attenuation property at specific frequency bands which can be used to user's advantage in highways.

References

1. Yablonovitch E (1987) Inhibited spontaneous emission in solid-state physics and electronics. Phys Rev Lett 58:2059
2. Martinez Sala J, Sancho JV, Sanchez V, Gomez J, Llinarez J, Meseguer F (1995) Sound attenuation by sculpture. Nature 378:241
3. Bragg WH, Bragg WL (1913) The reflexion of X-rays by crystals. Proc R Soc Lond A 88(605):428–438
4. Hoare SH, D. T. Murphy, "Prediction of scattering effects by sonic crystal noise barriers in 2d and 3d finite difference simulations", Acoustics 2012
5. Gupta A, Lim KM (2012) Parametric study on rectangular sonic crystal. Appl Mech Mater 152–154:281–286
6. Kittel C (1996) Introduction to solid state physics. Wiley, New York
7. Gupta A, Lim KM, Chew CH (2011) Analysis of frequency band structure in one-dimensional sonic crystal using Webster horn equation. Appl Phys Lett 98:201906
8. Gupta A, Lim KM, Chew CH, ANN (2012) A quasi two-dimensional model for sound attenuation by the sonic crystals. J Acoust Soc Am 132:2909
9. Elford DP, Chalmers L, Kusmartsev FV, Swallowe GM (2011) Matryoshka locally resonant sonic crystal. J Acoust Soc Am 130:2746
10. Gulia P, Gupta A (2017) A finite element study of acoustic wave propagation through sonic crystal. Nonlinear Stud 24(1):3–13
11. Martınez-Sala R, Rubio C, Garcia-Raffi LM, Sanchez-Perez JV, Sanchez Perez EA, Llinares J (2006) Control of noise by trees arranged like sonic crystals. J Sound Vib 291:100–106
12. Gulia P, Gupta A (2016) Traffic noise control by periodically arranged trees. TRJ 2(2)
13. Morandi F, Miniaci M, Guidorzi P, Marzani A, Garai M (2015) Acoustic measurements on a sonic crystals barrier. In: 6th international building physics conference, IBPC
14. Peiró-Torres MP, Redondo J, Bravoc JM, Sánchez Pérez JV Open noise barriers based on sonic crystals. In: XII conference on transport engineering in advances in noise control in transport infrastructures, CIT 2016, 7–9 June 2016, Valencia, Spain
15. Kittel C (1996) Introduction to solid state physics. Wiley, New York

Acoustic Properties of Additive Manufactured Porous Material

Deepak C. Akiwate, Mahendra D. Date, B. Venkatesham, and S. Suryakumar

Abstract Acoustic porous materials are extensively used in many engineering applications like building, automobile, aviation, and marine. The health risk factor and environmental claims, associated with traditional materials such as glass wool, mineral fibers, and polymer foams demand for the alternative porous acoustic absorbing materials. Advances in additive manufacturing (AM) allow to manufacture complex structures and give an alternative method to produce porous materials. This study investigates the acoustic properties of porous sound-absorbing material produced by using additive manufacturing (AM) technique and explores the feasibility of AM to manufacture acoustic absorptive materials. For study, three samples with different aperture ratios were fabricated by AM technique, and their sound absorption coefficients were measured experimentally by using the impedance tube. The theoretical formulation for predicting normal sound absorption coefficient of sample with and without air gap was developed and compared with experimental results. The predicted absorption coefficient agrees well with measured results. The measured results indicate that the absorption coefficient of the structures fabricated through AM can be altered by varying aperture ratio and air gap behind the sample. This study reinforces the capability of AM for producing complex acoustic structures with better acoustic properties.

D. C. Akiwate · M. D. Date · B. Venkatesham (✉) · S. Suryakumar
Indian Institute of Technology Hyderabad, Telangana 502285, India
e-mail: venkatesham@iith.ac.in

D. C. Akiwate
e-mail: me14resch11001@iith.ac.in

M. D. Date
e-mail: mddate@iith.ac.in

S. Suryakumar
e-mail: ssurya@iith.ac.in

M. Singh and Y. Rafat (eds.), *Recent Developments in Acoustics*, Lecture Notes in Mechanical Engineering, https://doi.org/10.1007/978-981-15-5776-7_12

1 Introduction

Porous materials are widely used to absorb sound along the path in many engineering applications like building design, automobile, and aviation [1, 2]. These materials include glass wool, mineral fibers, polymer foams and fibers, and are well known as traditional sound absorbers due to their ease of use and flexibility to adopt complex shapes. However, these materials have environmental issues like non-degradability and non-recyclability [3, 4]. In addition, glass and mineral fibers have high health risk factors to human health. These environmental and health issues associated with traditional acoustic materials demand for alternative porous sound-absorbing acoustic materials. A good amount of literature is available on exploring natural materials for acoustic applications [5–7]. Aiming to the same, Zulkifli et al. [5] investigated acoustic properties of multilayer coconut coir fiber and found good absorption coefficient in the relatively higher frequency range. Ersoy and Kucuk [6] measured the acoustic absorption of industrial waste, developed during processing of tea leaves and observed good acoustic absorption between 500 and 3200 Hz. Similarly, Fatima and Mohanty [7] studied acoustic and fire-retardant properties of jute composite material which found application in many household and industrial appliances. Many studies have already reported about acoustic properties of various materials with different manufacturing methods. However, the recent advances in additive manufacturing (AM) enables to manufacture complex shapes which were difficult by using traditional methods [8]. The use of additive manufacturing method to produce acoustic porous materials and structures were discussed in literature [9–12]. Setaki et al. [10] prepared complex-shaped compact resonator configurations using AM technique, and the measured absorption coefficients of these configurations were validated with predicted results. Liu et al. [11] studied acoustic properties of microperforated panels manufactured with AM technique which were light and thin compared to traditional ones. In another study, Liu et al. [9] investigated the effect of perforation angle and backside air gap on the acoustic absorption coefficient of additive manufactured porous material. But, the combined effect of aperture ratio and back air gap on acoustic absorption was not studied yet. However, there are limited studies which explore additive manufacturing for acoustic applications.

The main objective of this study is to demonstrate the feasibility of additive manufacturing to produce sound-absorbing porous materials and to investigate the effect of aperture ratio as well as air gap behind the sample. The proposed method in this work can provide a basic framework to produce sound-absorbing materials using additive manufacturing and can be extended for more complex configurations specifically designed for desired frequencies.

2 Materials and Experimental Setup

2.1 Materials and Manufacturing of Sample

In this study, the sample dimensions chosen are 30.00 mm diameter and 20.00 mm length as shown in Fig. 1a. The angle of perforations and the diameter of the pores were kept constant as 15.00^0 and 0.80 mm, respectively. The aperture ratio of the samples was varied by changing the distance between the pores b, as shown in the Fig. 1b. Hole spacing (b) of 2.00, 4.00, and 6.00 mm were chosen for current study; their corresponding aperture ratios are 9.36%, 2.34%, and 1.06%, respectively. These samples will be henceforth referred as ADM1, ADM2, and ADM3. The samples were fabricated in a Polyjet-based AM system (Object30 Prime, Stratasys), with VeroBlue (RGD 840) resin as the raw material. A Polyjet-based system was chosen as it provides greater accuracy and finer feature resolution compared to the prevalent FDM-based AM systems. All samples were printed with 16.00-micron layer resolution, along the axial direction of the pores. Table 1 shows the detailed dimensions of manufactured samples used for the current study.

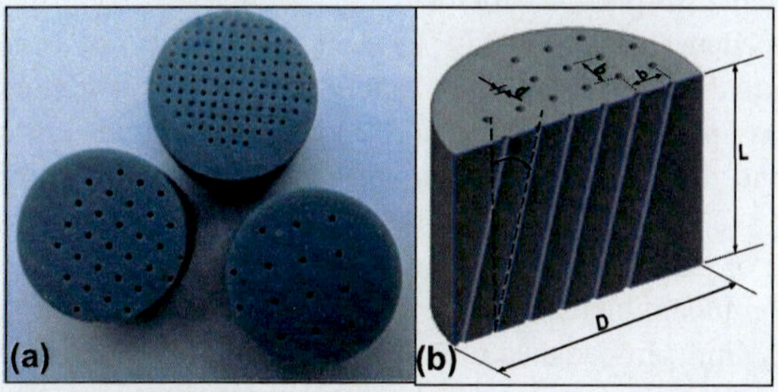

Fig. 1 Additive manufactured test specimens **a** images of samples **b** sectional view with geometric terminology

Table 1 Geometric parameters of additive manufactured porous samples

Sample	ADM1	ADM2	ADM3
Sample diameter, D (mm)	30	30	30
Hole diameter, d (mm)	0.8	0.8	0.8
Sample length, L (mm)	20	20	20
Hole spacing, b (mm)	2	4	6
Perforation angle, θ (degree)	15	15	15
Aperture ratio, P (%)	9.358	2.339	1.050

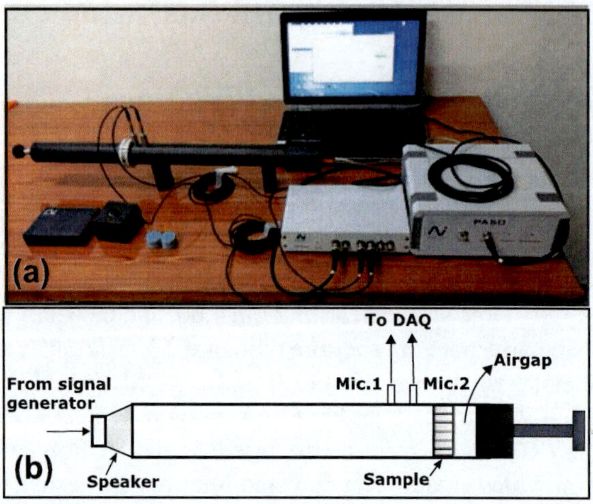

Fig. 2 **a** Experimental setup for measuring absorption coefficient **b** schematic of providing air gap in impedance tube

2.2 *Experimental Setup*

The normal sound absorption coefficient of the sample was measured by using two microphone impedance tube methods [13–15]. Impedance tube consists of a tube-like structure with the speaker at one end and sample holder at another end as shown in Fig. 2. The sample holder was capable of providing rigid termination as well as air gap behind the sample. The random signal was given as input to speaker and the plane wave propagation inside the tube was considered for measurements. The assumption of plane waves holds up to higher order cut-on frequency which is decided based on tube dimensions. The spacing between two microphones decides lower frequency limit. Impedance tube of 30 mm internal diameter was used with measurement frequency range from 800 to 6300 Hz. Pressure sensed by microphones were acquired by data acquisition system (DAQ) which was connected to computer interface. The signal generator was also connected to the DAQ to generate a random signal. The microphone switching method was used to minimize the measurement error due to phase mismatch between the two microphones, also this method enables the results independent of the gain of two microphone channels. ASTM 1050 standard was followed for measuring absorption coefficient [15].

3 Theoretical Analysis

The manufactured porous materials are consisting of pores similar to narrow tubes, inclined at an angle of 15° to the direction of wave propagation. The wave propagation in narrow tube depends on the geometrical parameters such as shape and size of the tube. The dissipation of sound in narrow tubes involves the viscous and thermal losses at the boundary of the wall, as the dimensions of the tube become comparable to the boundary layer thickness. These narrow tube viscous and thermal effects can be incorporated by estimating effective complex density and effective complex bulk modulus of the medium, respectively [16]. The surface impedance of porous material when backed by rigid termination, was estimated using viscothermal effects. Later, surface impedance was used for calculating the absorption coefficient. However, when the sample is backed by the backside air gap, it changes the complete physical system and it acts as a microperforated plate (MPP) backed by air cavity [17]. In this case, narrow tubes act as inertance while air gap acts as compliance. The formulation for both cases with and without air gap to predict absorption coefficient in normal incidence is as follows.

3.1 Porous Material Without Air Gap

The effective complex density ρ and complex bulk modulus K is given by [15];

$$\rho = \rho_0 \left(1 + \frac{\sigma\phi}{j\omega\rho_0} G_c(s)\right) \tag{1}$$

and

$$K = \frac{\gamma P_0}{\gamma - (\gamma - 1)F(B^2\omega)}, \tag{2}$$

where

$$F(B^2\omega) = \frac{1}{1 + \frac{\sigma\phi}{jB^2\omega\rho_0} G_c(Bs)} \tag{3}$$

and

$$G_c(s) = \frac{\frac{-s}{4}\sqrt{-j}\,\frac{J_1(s\sqrt{-j})}{J_0(s\sqrt{-j})}}{1 - \frac{2}{s\sqrt{-j}}\frac{J_1(s\sqrt{-j})}{J_0(s\sqrt{-j})}}. \tag{4}$$

Here P_0, γ, B are the ambient pressure, ratio of specific heats, and square root of Prandtl number, while s is the shear wave number given by $s = \bar{c}\sqrt{\frac{8\omega\rho_0}{\sigma\phi\cos^2(\theta)}}$. J_0 and J_1 are the Bessel functions of zero and first orders, while ρ_0 is the ambient density of air. The quantity $\sigma\phi$ which is the product of flow resistivity and porosity of the material, and given is by $8\eta/(\bar{r}^2)$. Here, η is the dynamic viscosity of air and $\bar{r}$ is the hydraulic radius of narrow tube which is the ratio of cross-sectional area to the perimeter of the tube. As the sample is supported by rigid termination, its surface impedance is given by

$$Z = \frac{-jZ_c\cot(kL')}{P}. \tag{5}$$

Here, Z_c and k are the characteristic impedance and wave number based on viscothermal effects discussed above. P is the aperture ratio of porous material while $L^{'}$ is the effective length of sample including an end correction of $0.85d$. The reflection coefficient from surface impedance can be calculated by $R = \frac{Z-Z_c}{Z-Z_c}$ and subsequently absorption coefficient by $\alpha = 1 - |R|^2$.

3.2 *Porous Material with Air Gap*

As discussed earlier, manufactured porous material backed by air gap acts as MPP with the cavity. The specific impedance of the MPP layer including end correction is given by [16]

$$Z = (R + jM)/\rho_0 c_0 = (r + jm), \tag{6}$$

where $r = \frac{32\eta L^{'}}{P\rho_0 c_0 d^2}\left[\sqrt{1+\frac{x^2}{32}} + \frac{\sqrt{2}}{8}\frac{xd}{L}\right]$, $m = \frac{L}{Pc_0}\left[1 + \frac{1}{\sqrt{9+\frac{x^2}{2}}} + \frac{0.85d}{L^{'}}\right]$, and $x = \frac{\sqrt{d}}{2}\sqrt{\frac{\omega\rho_0}{\eta}}$. Here, r and $m\omega$ are the normalized resistance and specific acoustic reactance of MPP. The term $L^{'}$ represents the effective length of tubes, given by $L/\cos(\theta)$. Therefore, the total normalized specific acoustic impedance of the MPP backed by air gap is given by

$$Z = r + j\left[m\omega - \cot\left(\frac{\omega H}{c_0}\right)\right]. \tag{7}$$

Here, H is the depth of air gap behind the sample. The absorption coefficient of a complete system with air gap can be estimated as

$$\alpha = \frac{4Re(Z_t)}{[1 + 4Re(Z_t)]^2 + [Im(Z_t)]^2}. \tag{8}$$

4 Results and Discussion

The sound absorption coefficients of additive manufactured samples ADM1, ADM2, and ADM3 were measured in impedance tube without air gap and results are shown in Fig. 3a. It is observed that all three samples have similar trend for absorption coefficient with respect to frequency. The sample ADM1 which is having highest aperture ratio and has the highest amplitude of absorption coefficient of 0.93 at a frequency of 3550 Hz, while sample ADM3 has lowest aperture ratio and amplitude of absorption coefficient of 0.16 at 3120 Hz. The frequencies of maximum absorption observed in all samples were well below of their quarter-wave tube effects. It is due to

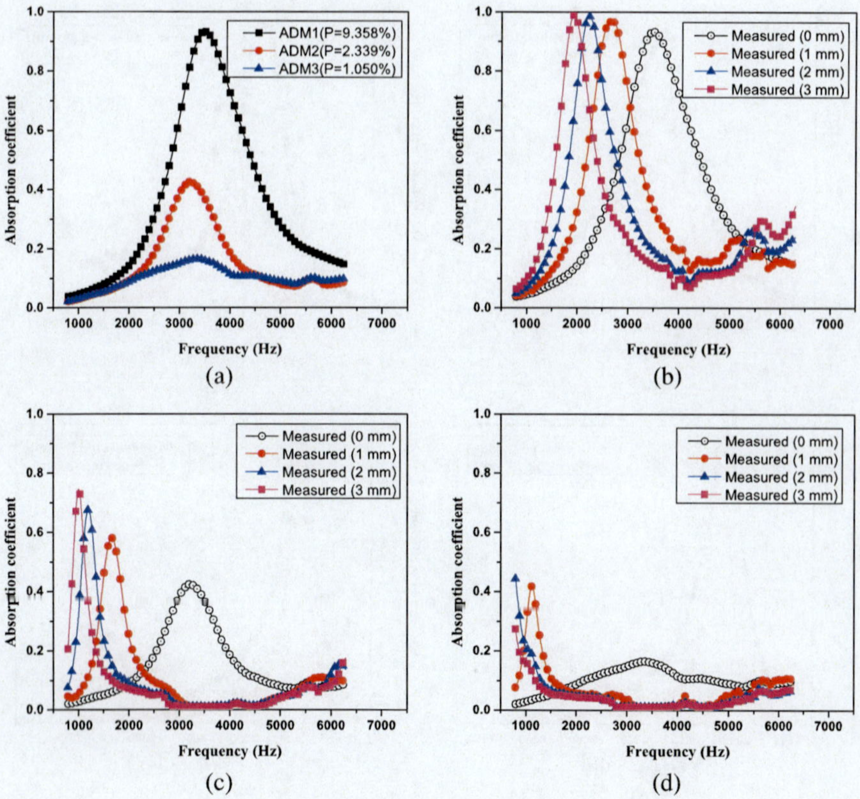

Fig. 3 Effect of design parameters on absorption coefficient of all additive manufactured samples for **a** aperture ratio **b** air gap for ADM1 **c** air gap for ADM2 **d** air gap for ADM3

viscothermal dissipation near the narrow tube walls which reduces the speed of sound in narrow tubes as compared to the speed in ambient air. So, it causes a reduction in the frequency value of peak absorption. It can be noted that as the aperture ratio increases the amplitude of absorption and the frequency of peak absorption increases. However, there will be the certain limit for aperture ratio because of diminishing of viscothermal effects. Therefore, an optimum aperture should be obtained to achieve maximum absorption at the desired frequency. All three manufactured samples were tested in impedance tube with air gap of 1, 2, and mm. The absorption coefficients of samples with these conditions are shown in Fig. 3b–d. The provision of air gap shows similar behavior in all samples regardless of their aperture ratio and the same was observed earlier by Liu et al. [9]. Providing air gap behind the sample modifies the complete physical system as compared to previous case. Because, the narrow tubes become open at both ends and behaves like acoustic inertance and air gap acts as acoustic compliance. Therefore, it creates the system analogues to Helmholtz resonator. Air gap significantly reduces the peak frequency of observation while at the same time the

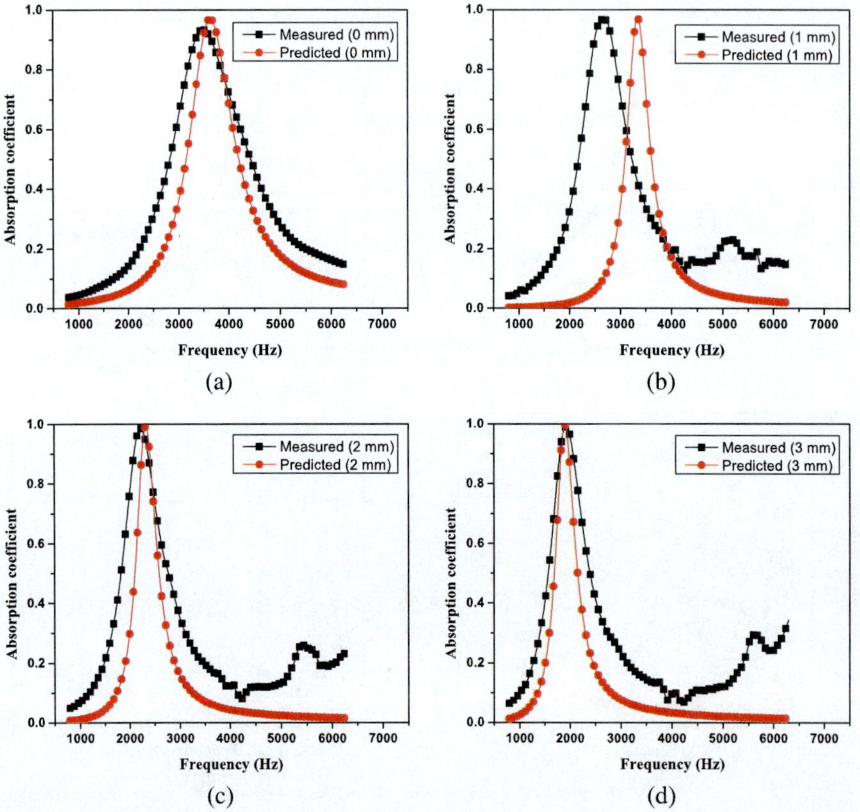

Fig. 4 Comparison of absorption coefficient of samples ADM1 with backside air gap of **a** 0 mm **b** 1 mm **c** 2 mm **d** 3 mm

amplitude of absorption coefficient increases with narrow band response compared to without air gap. This is one of the desirable phenomenon in terms of low-frequency noise control where most of traditional acoustic material may not be effective. The peak absorption coefficient with different air gaps can be observed in case of ADM1 and ADM2 while for ADM3 peak absorption can be observed only with 1 mm air gap and for another two cases it is below the lowest measurement frequency (<800 Hz) as shown in Fig. 3d. Thus, it is concluded that optimum aperture ratio and air gap can be employed to design the additive manufactured porous materials at desired frequency.

The measured absorption coefficient of the sample ADM1 with and without air gap was compared with theoretical formulation discussed in the previous section. The absorption coefficient of sample "ADM1" backed by rigid support was estimated by using viscothermal formulation as discussed in Sect. 3. Equations 1 and 2 were used to estimate effective medium properties while Eq. 5 was used for predicting equivalent impedance of the structure. Figure 4a shows the comparison of predicted and measured absorption coefficient of ADM1. There is good agreement between measured and predicted results. In case of the sample with air gap, MPP with air-gap formulation was used. Figure 4b–d shows the comparison of predicted and measured absorption coefficients of sample ADM1 with an air gap of 1, 2, and 3, respectively. There is a good agreement between amplitude as well as the frequency of maximum absorption, while for air gap of 1 mm there is a discrepancy in frequencies as shown in Fig. 4b. The predicted frequency of maximum absorption is higher than that of measured one because of the reactance due to air gap is much higher than sample reactance at lower frequencies, and will become equal at the resonance, i.e., at the frequency of peak absorption. This higher reactance of air gap shifts the predicted frequency of maximum absorption toward higher side. The reactance due to air gap reduces by increasing the air gap behind the sample. Hence, the predicted and measured results agree well for 2 and 3 mm air-gap cases. However, maintaining the exact 1 mm air gap behind the sample was challenging one and small deviations in air-gap measurement causes large deviations in predicted and measured results. Thus, the proposed theoretical methodology to predict absorption coefficient with and without air gap has been validated. The proposed formulation is applicable when the reactance due to air gap will be in the order of sample reactance. However, further studies required to understand the effect of narrow air gaps on frequency shift.

5 Conclusions

In this work, an alternative manufacturing method has been successfully demonstrated to fabricate artificial acoustic porous materials. These materials have good acoustic properties in medium frequency range and can be used for low-frequency range sound absorption with specific air gap behind the samples. The amplitude and frequency of maximum absorption can be tuned by changing aperture ratio of material and air gap behind it. The increase in aperture ratio for chosen range, increases the amplitude as well as frequency of peak absorption. However, increase in air gap behind sample increases amplitude of peak absorption and significantly reduces the frequency of maximum absorption. Appropriate combination of air-gap and aperture ratio can be used to achieve good absorption coefficient in mid to low-frequency range where most of traditional materials may not be effective. The proposed theoretical formulation for both configurations, i.e., with and without air gap has been validated with experimental results. The method presented in this work provides a basic framework to design and manufacture more complex structures with AM to obtain good acoustic performance in desired frequency range.

Acknowledgments Deepak Akiwate is a recipient of Prime Minister's Fellowship Scheme for Doctoral Research, a public-private partnership between Science and Engineering Research Board (SERB), Department of Science and Technology, Government of India and Confederation of Indian Industry (CII). The authors host institute for research is IIT Hyderabad and the partner company is Eaton Technologies Private Limited, Pune.

References

1. Bernek LL, L Vér I (2006) Noise and vibration control engineering: principles and applications. John Wiley and Sons, Inc., New York
2. Ingard U (2010) Noise reduction analysis (physics). Jones & Bartlett Publishers
3. Asdrubali F, Schiavoni S, Horoshenkov KV (2012) A review of sustainable materials for acoustic applications. Build Acoust 19(4):283–312
4. Joshi SV, Drzal LT, Mohanty AK, Arora S (2004) Are natural fiber composites environmentally superior to glass fiber reinforced composites? Compos Part A: Appl Sci Manuf 35(3):371–376
5. Zulkifli R, Mohd Nor MJ, Mat Tahir MF, Ismail AR, Nuawi MZ (2008) Acoustic properties of multi-layer coir fibres sound absorption panel. J Appl Sci 8(20):3709–3714
6. Ersoy Sezgin, Küçük Haluk (2009) Investigation of industrial tea-leaf-fibre waste material for its sound absorption properties. Appl Acoust 70(1):215–220
7. Fatima S, Mohanty AR (2011) Acoustical and fire-retardant properties of jute composite materials. Appl Acoust 72(2–3):108–114
8. Ian G, David R, Brent S (2014) Additive manufacturing technologies: 3D printing, rapid prototyping, and direct digital manufacturing. Springer, New York
9. Liu Z, Zhan J, Fard M, Davy JL (2016) Acoustic properties of a porous polycarbonate material produced by additive manufacturing. Mater Lett 181:296–299
10. Setaki F, Van Timmeren A, Turrin M (2016) New sound absorption materials: using additive manufacturing for compact size, broadband sound absorption at low frequencies. In: INTER-NOISE and NOISE-CON congress and conference proceedings, vol. 253, pp. 6190–6195. Institute of Noise Control Engineering
11. Liu Z, Zhan J, Fard M, Davy JL (2017) Acoustic measurement of a 3D printed micro-perforated panel combined with a porous material. Meas J Int Meas Confed 104:233–236
12. Akiwate DC, Date MD, Venkatesham B, Suryakumar S (2018) Acoustic properties of additive manufactured narrow tube periodic structures. Appl Acoust 136:123–131
13. Munjal ML (2014) Acoustics of ducts and mufflers, 2nd edn. Wiley, Chichester
14. Impedance tube test system ver. 1.03 Users manual, BSWA Tech. Co. Ltd, Beijing, P.R China (2010)
15. ASTM (2008) E1050-08 standard test method for impedance and absorption of acoustical materials using a tube, two microphones and a digital frequency analysis system. ASTM International, West Conshohocken, PA
16. Jean A, Noureddine A (2009) Propagation of sound in porous media: modelling sound absorbing materials 2e. Wiley, New York
17. Liu Z, Zhan J, Fard M, Davy JL (2017) Acoustic properties of multilayer sound absorbers with a 3D printed micro-perforated panel. Appl Acoust 121:25–32

Sound Transmission Characteristics Through Multi-panel Structures of Wooden Doors and Uncertainty Components in the Measurements

Kirti Soni, Mahavir Singh, and Yudhisther K. Yadav

Abstract This paper examines the sound transmission loss (STL) through composite panel door systems commonly used in the recording studios, operating theaters, libraries, interview rooms, offices, museums, etc. The sound insulation property of different combinations of door panel of materials such as wood/plywood, cement board, wool, tec sound, laminate is investigated. Reverberation chamber method is used to determine the sound transmission loss. This paper presents the results of a series of doors tests to estimate the sound transmission loss through different combinations of wooden door system. The main aim of this study is to design such a system using wooden door panels that can provide significant noise reduction. The other objective of this study is to evaluate typical doors characteristics including the effect of minor modifications that could increase the noise reduction provided by the door system.

Keywords Acoustics materials · Sound transmission loss · Sound transmission class

1 Introduction

People are very sensitive to noise. Noisy environment living produces stress that negatively impacts our life and also interfere with our sleep and makes us tired and short tempered. If a building is located close to a heavily trafficked road it is quickly exposed to noise pollution. Every truck, car or even tram generates a background noise that can significantly harm our quality of life. In hospitals, noise is often considered as one of the main areas of worry, because it is responsible for sleep loss in patients, high stress levels (both in patient and working staff) and as a result medical error increases [1–3]. Several studies reported that far greater amount of noise comes

K. Soni (✉) · M. Singh · Y. K. Yadav
A & V Metrology, PMM Division, CSIR-National Physical Laboratory, New Delhi 110012, India
e-mail: soniks@nplindia.org

M. Singh
e-mail: mahavir@nplindia.org

M. Singh and Y. Rafat (eds.), *Recent Developments in Acoustics*, Lecture Notes in Mechanical Engineering, https://doi.org/10.1007/978-981-15-5776-7_13

through Doors and windows, not the walls [4–11]. It is clear from the laboratory study that more than 90% of all the exterior noise comes in through doors and windows. Reduced noise pollution using increased sound insulation creates a healthy working home/office environment that increases the productivity. Few studies exist regarding the sound insulation of the doors. There are numerous environments such as recording studios, operating theaters, libraries, interview rooms, and museums where effective sound-proofing doors are required [12]. The main aim of this study is to design a perfect structure of door that can provide an economical and durable noise reduction solution for confidential speech privacy. The secondary goal of the study is to design thinnest door with high STC (Approx.50). Sound transmission class (STC) offers an estimate of the acoustic performance of a door in certain common airborne sound insulation applications. Wood-door configuration can achieve isolation noise levels equal to or higher than more massive construction such as concrete, but to take advantage of that potential, it is important to characterize the airborne sound transmission loss properties of wood-door configuration. About 45 sound transmission class (STC) value is required for confidential speech privacy in healthy office/home environment [12]. In total twelve door panels of different configurations tested in laboratory. Six door configurations tested, to show the dependence of the sound transmission on the space/acoustic panel position between Cement Fiber Board/Plywood, with/without laminate and with different thickness of tec sound. And also to design the minimum thickness of the door that could reliably achieve the target of STC 50. The effect of 500/1000GSM acoustic wool to fill the space between Cement Fiber Board/Plywood was also examined. Some other significant factors that affect sound transmission characteristics were also pointed out in the paper.

2 Description of the Measurement

The test results reported here were obtained in the Sound Transmission Suite Lab at the Acoustics & Vibration Metrology Section, CSIR-National Physical Laboratory, New Delhi, India. The sound transmission loss characteristics of the different wooden doors were measured by ISO 16283-1:2014 and ASTM Standard Classification E-413. Door samples were mounted in a removable test frame between two chambers, without rigid contact to either reverberation chambers. Door panel specimens were mounted in an opening (930 mm × 630 mm) between source room and receiving room. Both rooms are irregular in shape with no parallel surfaces and are equipped with stationary diffusers. The source and receiving rooms in the test suite have volumes of 257 and 271 cubic metres, respectively. Test signals are supplied to each room by duo-decahedral loudspeakers system with independent pseudorandom noise sources. Pink noise is used to measure decays in the receiving room. B & K Type 2270 Sound level metre in each room are used to obtain space averaged sound pressure levels and reverberation times. A sound transmission class rating (STC) was obtained for each door by ASTM E-413, Standard Classification for Determination of Sound Transmission Class.

The sound pressure level and reverberation time measurements are made for a 1/3-octave band with centre frequencies from 100 to 4000 Hz. The data are then used to calculate sound transmission loss for each frequency band as per specified standards. The sound transmission loss was calculated as Eqs. (1) and (2)

$$STL = L_s - L_r + 10\log_{10}\left(\frac{S}{A}\right), \tag{1}$$

and

$$A = \frac{0.161V}{T}, \tag{2}$$

where

L_s average sound pressure level in source room (dB)
L_r average sound pressure level in receiving room (dB)
S area of the test partition (m^2)
A total absorption area of receiving room (m^3/s)
V volume of receiving room (m^3)
T reverberation time (s).

The evaluated uncertainty in measurement is ±1.0 dB which is at a coverage factor $k = 2$ and which corresponds to a coverage probability of approximately 95% for normal distribution.

3 Results and Discussions

Different structures of doors (Fig. 1) based on applications were tested in CSIR-NPL Sound Transmission laboratory. The details of the measured transmission loss values for wooden doors are influenced by numerous parameters describing the doors such as: (a) plywood thickness, (b) Cement Fiber Board thickness (c) wool thickness/GSM), and (c) layer spacing i.e., air gap etc. Investigation of the data shows that these parameters have different effects on transmission loss values than on STC values. In some cases changes to a door structure, that would usually significantly change STC ratings, have only very small effects on transmission loss values and vice versa. It is important to specify doors samples in terms of their measured transmission loss values rather than to guess the effects of improvements to doors structures by various modifications to them. Sample 1 is four-layer structure and layer 1 is 12 mm Cement Fibre Board, layer 2 is 50 mm (Thick) × 50 mm (Width) Wooden Baton frame filled with 50 mm 500 GSM wool, layer 3 is 5 mm Thick Tec-sound, and layer four is 12 mm Thick Cement Fibre Board. STC value of Sample 1 is 46 while STC value of Sample 2 is 45 is somewhat similar to the Sample 1 in structure

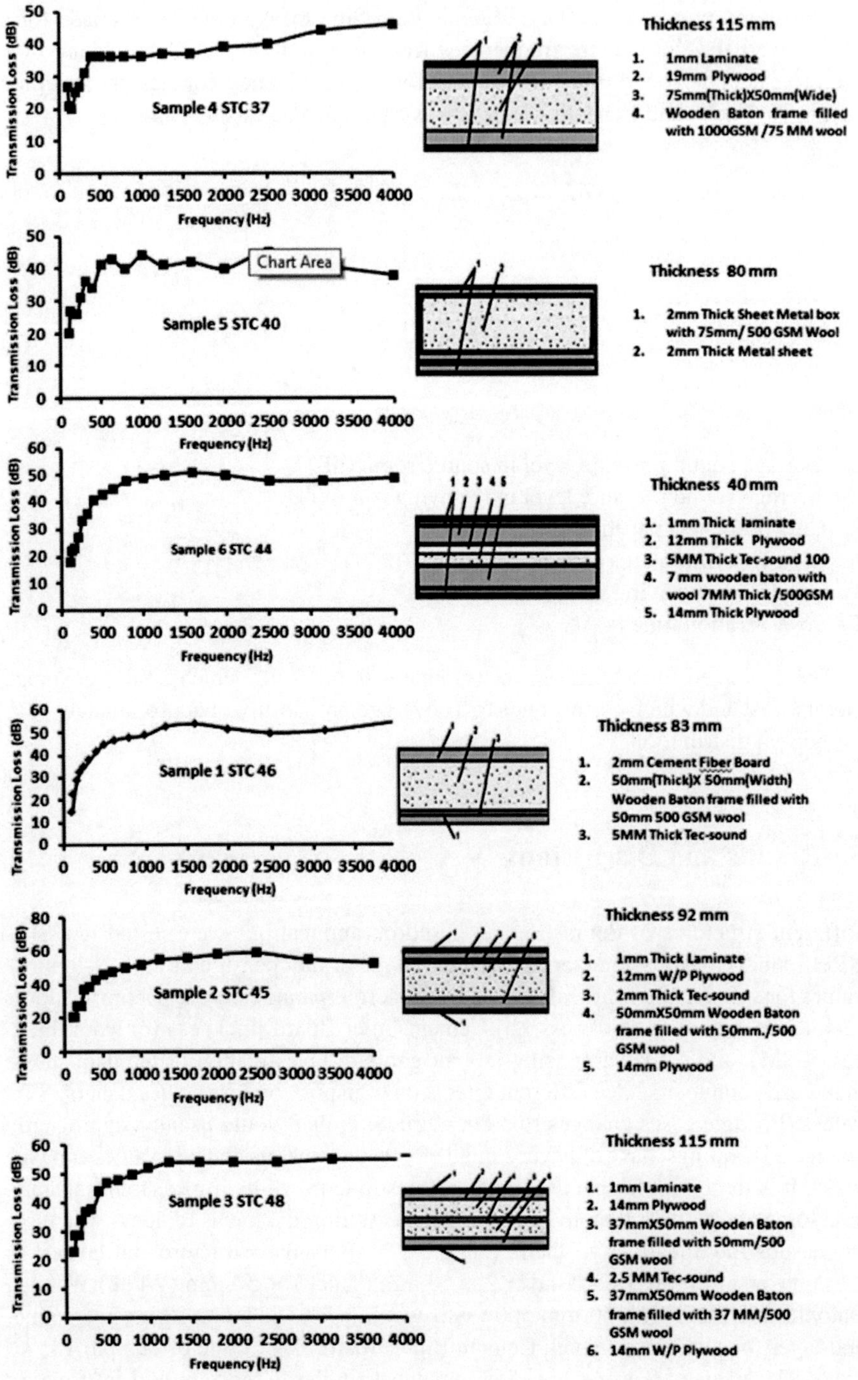

Fig. 1 Sound transmission loss of different samples (1-12)

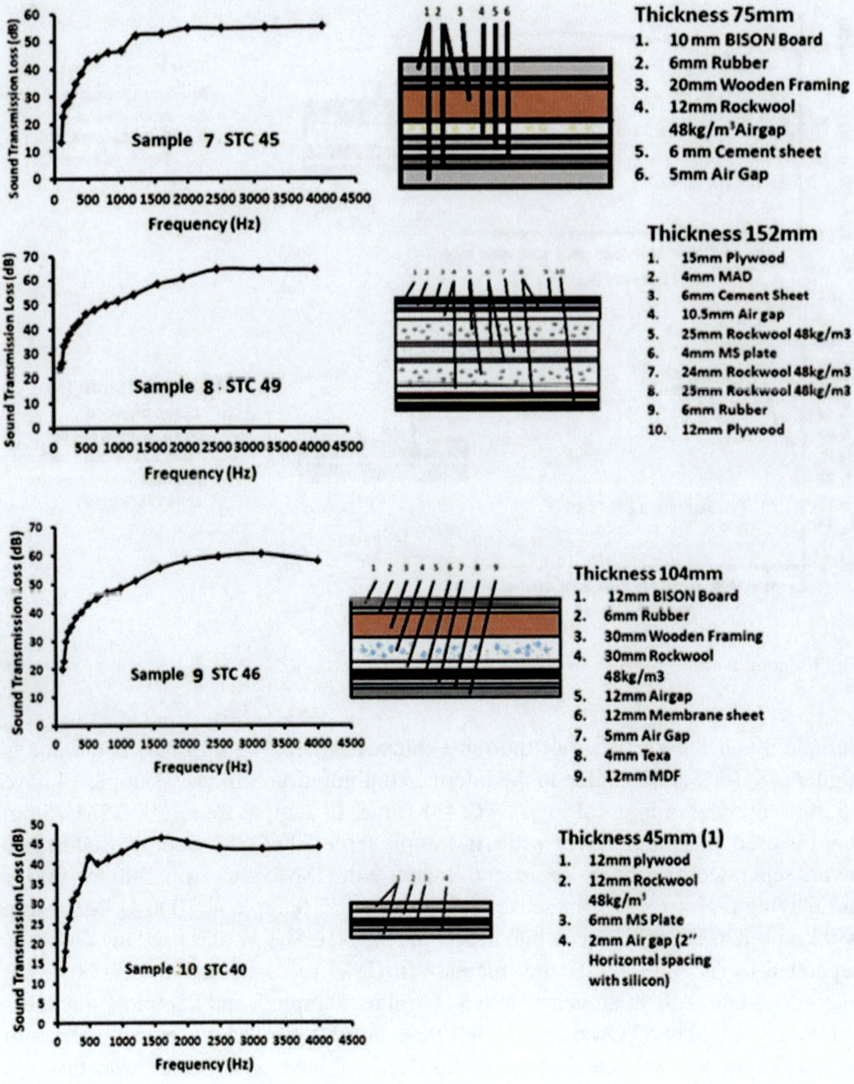

Fig. 1 (continued)

except 1 mm thick Laminate is used in upper and lower layer and 12 mm and 14 mm plywood are used in place of cement fiberboard. In Sample 1, 5 mm thick Tec-sound is used in the lower layer while in Sample 2 it is used in the upper part of the layer. It is clear that 1 mm thick laminate and upper or lower position of the thick Tec-sound is not highly affecting the STC value. Sample 6 thickness is just half the thickness of Samples 1 and 2, while STC value is almost similar 44. In Sample 6, 5 mm Tec-sound and only 7 mm wool are used in the centre part. It is obvious that 40 mm, 83 mm and 92 mm door samples are giving 44, 46, and 45 STC values. Similarly,

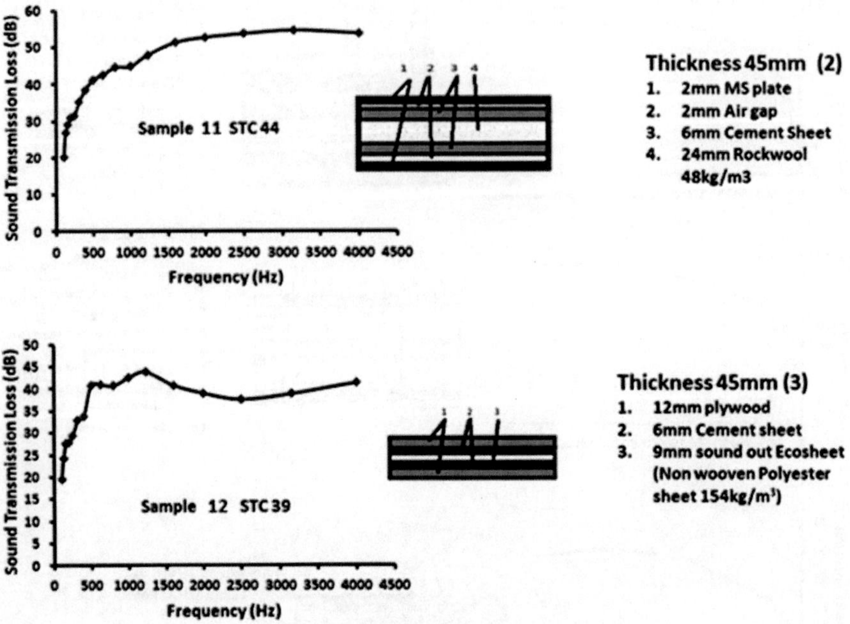

Fig. 1 (continued)

Sample 3 and Sample 4 are of the same thickness, i.e., 115 mm but STC value is higher (48) for Sample 3 due to the internal configuration.Similarly sample 14 have 152mm thickness represent high STC(49) value. In sample four 1000GSM/75mm wool is used as a single layer while in sample three 500 GSM wool is used in two layers separated by 2.5mm Tec-sound. Increase in GSM value from 500 to 1000 is not playing major role for increase in STC value. In Sample 4, 1000 GSM/75 mm wool is used as a single layer while in Sample 3, 500 GSM wool is used in two layers separated by 2.5 mm Tec-sound. Increase in GSM value from 500 to 1000 is not playing a major role in an increase in STC value. Sample 5 and Sample 1 thickness is almost same, but STC value is higher for Sample 1, i.e., 46 as cement fiberboard is used in place of the metal sheet. Sample 7 to 12 have STC value lower than 40. The best configuration which is giving highest STC in all samples is Sample 3, and Sample 6, is also better than other as its thickness is only 40 mm. However, STC value is 44. In Sample 4 if plywood thickness change from 14mm to 10mm and an air gap of 2mm is introduced in the center of the sample than higher STC value (around 50) can be achieved.

3.1 Comparisons of TL(avg) and STC Values for the Different Types of Doors

The first analysis of the data was to examine the relationship between the mean TL(AVG) values of each of the door types and the corresponding mean STC values. The TL(AVG) values are plotted versus the corresponding STC values in Fig. 2. Each data point represents the average results of one of the 18 types of wooden doors corresponding to the values. There is a statistically significant relationship ($R^2 = 0.987$) between the *TL(AVG)* and *STC* for the 18 values. However, the scatter about the mean trend is quite small. That is, for a given STC value there is a substantial range of possible *TL(AVG)* values.

3.2 Uncertainty Components Occurring in Measurements of Sound Transmission Loss

If the measured or predicted noise level depends on many input values, then the final result is a function of many influences (Eq. 3) [14]

$$L = f(Y + Y + Y + Y + \cdots + Y) \tag{3}$$

where L = the measured, which is determined by directly measured quantities Y , Y , Y … Y (input variables). One and all of them carry some standard uncertainty u and combined standard uncertainty uc, under assumption that the individual influences in Eq. (3) are independent. The uncertainty of measurement comprises many sources and many components. The ISO guide divides these components into two classes: type A or type B estimations, depending on the method used to estimate their numerical values. Type A evaluation of standard uncertainty is obtained from statistical analysis of the results of series of experimental measurements, like standard deviations. Type B quantities are those for which there are no experimental data from a set of measurements to statistically evaluate their standard uncertainties, but probability distributions based on experience or other information, like calibration certificates, manufacturer's data, or the result of a previous uncertainty evaluation [14]. The sound transmission loss of a sound insulating material is measured (using Eq. 1) in the laboratory through the measurement of the one-third octave band levels $\boldsymbol{L}$ and $\boldsymbol{L}$ at the source and receiver chambers, respectively, and reverberation time T of the latter. The mean value was calculated from $\boldsymbol{n}$ measurements is taken as the estimated measurement result. In this case, the standard uncertainty of parameter is equal to the experimental standard deviation s of the mean of a series of independent observations.

$$u = \frac{s}{\sqrt{n}} \tag{4}$$

Where, $u=$ uncertainty of parameter
$s=$ standard deviation of parameter of n observation
$n=$ number of observations

The derivatives with respect to L_s and L_r are respectively, 1 and −1. The derivatives with respect to *T, A, V* are, respectively,

$$\frac{\partial L}{\partial T}=\frac{10}{\ln 10}\frac{1}{T};\ \frac{\text{dB}}{s}$$

$$\frac{\partial L}{\partial A}=\frac{10}{\ln 10}\frac{1}{A};\ \frac{\text{dB}}{\text{m}^2}$$

$$\frac{\partial L}{\partial V}=-\frac{10}{\ln 10}\frac{1}{V};\ \frac{\text{dB}}{\text{m}^3}$$

Hence, the combined standard uncertainty in STL will be

$$u_c=\sqrt{\left(\left(\frac{\partial L}{\partial L_s}\right)^2 u_{Ls}^2+\left(\frac{\partial L}{\partial L_r}\right)^2 u_{Lr}^2+\left(\frac{\partial L}{\partial T}\right)^2 u_T^2+\left(\frac{\partial L}{\partial A}\right)^2 u_A^2+\left(\frac{\partial L}{\partial V}\right)^2 u_V^2\right)};\ \text{dB} \quad (5)$$

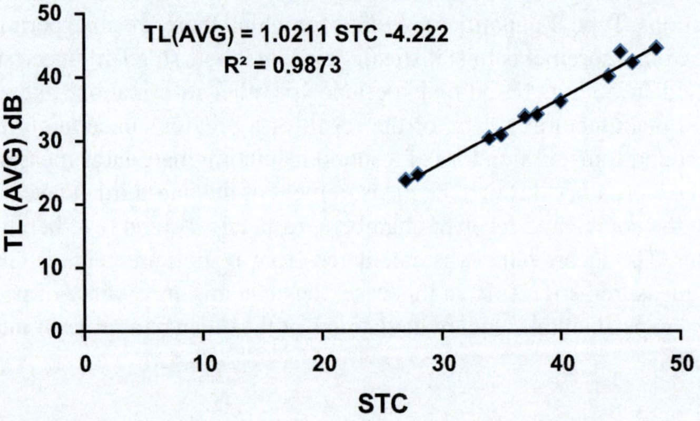

Fig. 2 Scatter plot between sound Transmission class (STC) and Transmission loss (TL)

Table 1 Uncertainty components in the acoustic measurements and analysis

Frequency (Hz)	$(u(L1)/sqrt(n))^2$	$(u(L2)/sqrt(n))^2$	$(u(T)/sqrt(n))^2$	$u(S)^2$	$u(V)^2$	$(\delta L/\delta L1)^2$	$(\delta L/\delta L2)^2$	$(\delta L/\delta T)^2$	$(\delta L/\delta S)^2$	$(\delta L/\delta V)^2$	CSU
100	0.0766	0.6538	0.0187	0.0000	0.0082	1.0000	1.0000	1.5170	55.0677	0.0003	0.8711
125	0.0670	0.0458	0.0206	0.0000	0.0082	1.0000	1.0000	1.2887	55.0677	0.0003	0.3733
160	0.0094	0.2653	0.0183	0.0000	0.0082	1.0000	1.0000	1.2365	55.0677	0.0003	0.5453
200	0.0406	0.0752	0.0027	0.0000	0.0082	1.0000	1.0000	1.4916	55.0677	0.0003	0.3462
250	0.0166	0.0512	0.0040	0.0000	0.0082	1.0000	1.0000	2.1287	55.0677	0.0003	0.2761
315	0.0166	0.0200	0.0011	0.0000	0.0082	1.0000	1.0000	1.9419	55.0677	0.0003	0.1969
400	0.0470	0.0778	0.0023	0.0000	0.0082	1.0000	1.0000	1.6162	55.0677	0.0003	0.3586
500	0.0380	0.0393	0.0016	0.0000	0.0082	1.0000	1.0000	1.3883	55.0677	0.0003	0.2822
630	0.0206	0.0467	0.0020	0.0000	0.0082	1.0000	1.0000	1.6353	55.0677	0.0003	0.2657
800	0.0174	0.0627	0.0006	0.0000	0.0082	1.0000	1.0000	1.6257	55.0677	0.0003	0.2847
1000	0.0094	0.1340	0.0001	0.0000	0.0082	1.0000	1.0000	1.5257	55.0677	0.0003	0.3789
1225	0.0094	0.2967	0.0008	0.0000	0.0082	1.0000	1.0000	1.6844	55.0677	0.0003	0.5544
1600	0.0094	0.2033	0.0008	0.0000	0.0082	1.0000	1.0000	1.8576	55.0677	0.0003	0.4629
2000	0.0146	0.8193	0.0008	0.0000	0.0082	1.0000	1.0000	2.3940	55.0677	0.0003	0.9142
2500	0.0130	1.6085	0.0002	0.0000	0.0082	1.0000	1.0000	3.0489	55.0677	0.0003	1.2736
3150	0.0214	1.2027	0.0002	0.0000	0.0082	1.0000	1.0000	3.7675	55.0677	0.0003	1.1068
4000	0.0160	0.3018	0.0003	0.0000	0.0082	1.0000	1.0000	4.4117	55.0677	0.0003	0.5648

4 Concluding Remarks

In this study, the evaluation of STC for the composite panels in the aspect of thickness, density, size and with different materials has been achieved using reverberation chamber method. It is observed from the study that for higher STC, higher thickness is not required as Samples 3 and 4 have same thickness but have different STC values. However, using minimum thickness panel (Sample 6), higher STC value can be achieved. Also, the inner configuration of the panel also plays an important role in achieving high STC value. In Samples 1, 2, and 3, high STC (48) value is observed for Sample 3 because of the sound partition of the wool in the center.

Acknowledgments Part of this work [Project No. CNP 170432] was supported by Soundworks, Bengaluru. The Authors are grateful for this support.

References

1. Aaron JN, Carlisle CC, Carskadon MA, Meyer TJ, Hill NS, Millman RP (1996) Environmental noise as a cause of sleep disruption in an intermediate respiratory care unit. Sleep 19:707–710
2. Topf M (1985) Noise-induced stress in hospital patients: coping and non auditory health outcomes. J Human Stress 11:125–134
3. Topf M, Dillon E (1988) Noise-induced stress as a predictor of burnout in critical care nurses. Heart and Lung 17:567–574
4. Andro B et al (2008) Prediction of sound transmission through automotive door seal systems. J Acoust Soc Am 123(5):3534
5. Hongisto et al (2000) Sound insulation of doors- Part 2: comparison between measurement results and predictions. Jo Sound and Vibr 230:149–170
6. Hongisto (2000) Sound insulation of doors-Part 1: prediction models for structural and leak transmission. J Sound Vibr 230:133–148
7. Bradley JS, Birta JA (2001a) A simple model of the sound insulation of gypsum board on resilient supports. Noise Control Engg J 49:216–223
8. Bradley JS, Birta JA (2001b) On the sound insulation of wood stud exterior walls. J Acoust Soc Am 110:3086–3096
9. Braganca L, Patr´ıcio J (2004) Case study: comparison between the acoustic performance of a mixed building technology building and a conventional building. Build Acoust 11(1):79–90.
10. Rasmussen B, Rindel JH (2010) Sound insulation between dwellings—descriptors applied in building regulations in Europe. Appl Acoust 71(3):171–180
11. Warnock AC (1993) Sound transmission through slotted concrete blocks with attached gypsum board. J Acoust Soc Am 94:2713–2720
12. Bradley JS, Gover BN (2011) Selecting walls for speech privacy IRC report RR-314
13. ASTM E413 (1983) Standard classification for determination of sound transmission class. Am Soc Test Mater
14. ISO 10140-2, Acoustics—laboratory measurement of sound insulation of building elements—part 2: measurement of airborne sound insulation

Prediction of Acoustical Performance with and Without Airflow Resistivity of Fibrous Natural Materials

Manish Raj, Shahab Fatima, and N. Tandon

Abstract Traditional sound-absorbing materials for noise control are rockwool and glasswool which serve the purpose pretty well but are accompanied by some serious health hazards during their manufacturing or processing or application. Hence researchers and practitioners are coming up with new materials, particularly from natural sources, which apart from being cheap and readily available, have good noise absorption properties along with minimum effect on environment and health hazards. When sound is incident on these materials, some portion is reflected back (due to impedance mismatch), some portion is dissipated within the material (due to thermal and viscous loss in the pores) and the remaining is transmitted to the medium behind the material. Sound absorption is defined as dissipation plus transmission; hence, any parameter which increases any of these two effects will increase the sound absorption. Just like the synthetic sound-absorbing materials, these natural materials are also porous in nature which creates sufficient interest to explore them for potential acoustical applications. Several researchers have conducted acoustical investigations on materials which are used in some other form such as bamboo, jute, kenaf, hemp, ramie, sisal, coconut coir, tea leaf fibre, betel nut fibre and also on some waste materials like recycled rubber, cigarette butt, bagasse, oil palm empty fruit bunch fibre. In this chapter, a mathematical model has been developed to predict the normal sound absorption coefficient of jute and waste cotton. It can also be predicted by estimating the airflow resistivity of a material and feeding that into established models but the amount of error encountered in this approach is often large. So, with an objective to minimise the error, characteristics impedance and complex wave number of the materials have been experimentally estimated. With the minimum sum of squares of

M. Raj (✉) · S. Fatima · N. Tandon
Center of Automotive Research & Tribology (Formerly ITMMEC), Indian Institute of Technology Delhi, New Delhi 110016, India
e-mail: manish.raj@itmmec.iitd.ac.in; manishraj.iitdelhi@gmail.com

S. Fatima
e-mail: fatima@itmmec.iitd.ac.in

N. Tandon
e-mail: ntandon@itmmec.iitd.ac.in

M. Singh and Y. Rafat (eds.), *Recent Developments in Acoustics*, Lecture Notes in Mechanical Engineering, https://doi.org/10.1007/978-981-15-5776-7_14

error curve fitting approach, a MATLAB code has been developed to obtain the modified coefficients in Delany–Bazley (D–B) model. These equations have been used to predict the sound absorption coefficients of Jute fibres and waste cotton, which is compared with the actual ones obtained by conducting experiments on impedance tube. The modified model was found to be of the greater degree of accuracy than the original D–B model. Further, the airflow resistivity term was eliminated in the Matlab code and the model was used to predict the normal sound absorption coefficient at different frequencies. The results of this model were compared with the result of the original D–B model and experimentally estimated normal absorption. It was also found to be closer to the actual behaviour than the one predicted by D–B model.

Keywords Natural fiber · Acoustical mathematical modelling · Sound absorption coefficients · Jute · Waste cotton

1 Introduction

Traditional acoustical materials, like rockwool, glasswool, etc., need large amount of external energy for its production, which increases its overall carbon footprint. Apart from this, they are also found to be hazardous to the personnel involved with it. Owing to these limitations, mainly since 2000s research has focused towards natural and recycled materials to be explored as potential acoustic materials which are environmental friendly and less hazardous [1]. Over the years, these materials have been tested in fibrous form, non-woven or as a composite.

Bamboo was one of the earlier natural materials to be explored for acoustical application. It was inferred that its thickness has a direct effect on SAC over all frequency range and shifting the air space depth to a higher value shifted the peak value of SAC towards lower frequency. Increase in the density of the fibre increased the number of available pores for viscoelastic damping, which resulted in higher SAC. This trend, however, prevailed only in medium and higher frequency range. Increased fibre diameter had a negative effect on SAC. On comparison with glass wool with same parameters, it fetched almost similar properties and these results boost the claim as it to be a valid replacement for glass wool [2]. Rubber is used as vibration isolators due to its good damping properties and as noise is absorbed by viscoelastic damping, it motivated researchers to explore this material as a potential acoustical material. Using recycled rubber particle will serve the dual purpose of waste disposal with noise absorption. It was observed that smaller particle fetches more attenuation in the low-frequency range and increase in thickness increased the maximum value of SAC [3]. The natural material obtained from dried and chopped harvested tea leaves which yield biodegradable fibre was tested between 500 and 6300 Hz, the maximum value of SAC reached was around 0.7 at 5000 Hz. Another natural material, woven cotton cloth for which the SAC value was increasing linearly with frequency, was used as backing. The thickness of the backing has a direct effect on the SAC value and absorption at low-frequency increased [4]. Coir is obtained from the outer shell of coconut which is used as a material for making rope in rural areas. It was observed that for noise absorption, it performs well at medium and high frequencies, but its behaviour at low frequencies needs to be improved. While using the material fibres in association with perforated plates, the SAC at low-frequency improved but the performance at medium frequency deteriorated. Same thing happens when the fibre is backed by an air gap. It was observed that perforated plates with coir fibres in the association of air gap increase the SAC at low frequency. The additional conclusion was adding the perforated plates reduced the thickness of the air gap needed to be maintained behind the fibre for the same amount of SAC [5]. The low-frequency response is a very important property, and traditionally, the performance of natural materials has not been much appreciable at this, but cotton was found to be one of the best with SAC up to 0.9 at 1000 Hz. Sisal fibre was found to be a poor absorber within all range of frequencies with their peak SAC around 0.5 at 5000 Hz [6]. Wool fibre acted as poor sound insulator at low and medium frequencies but performed well in the higher frequency range with their SAC of 0.8 at frequencies greater than 4000 Hz [7]. Very promising behaviour was shown while investigation on the acoustic properties of Kenaf and sheep wool by varying the thickness of the sample. These materials performed as a better sound-absorbing material with their NRC value reaching 0.7 and 0.7, respectively [8, 9]. Noise performance test was conducted with waste cigarette butts in used and unused condition and it has shown the SAC as high as 0.8 at frequencies greater than 1000 Hz [10]. All these results have been reported by experimental results. It is quite well known that the instrumentation involved in the field of noise engineering is extremely costly and time consuming. Hence, an alternate approach which predicts a result with acceptable degree of accuracy by application of empirical relationship is discussed in the next section.

Nomenclature	
SAC	Sound absorption coefficient
NRC	Noise reduction coefficient
AFR	Airflow resistivity
Z_b	Characteristics impedance
j	$\sqrt{-1}$
K_b	Propagation constant
σ	Airflow resistivity
Z_s	Normalised acoustic surface impedance
α	Absorption coefficient
l_s	length of sample
K	Bulk modulus of air = 1.42e5
C_0	Speed of sound in air = 342 m/s
f	Frequency

(continued)

(continued)

Nomenclature	
ω	Angular frequency
ρ_0	Density of air = 1.213 kg/m^3
a, b, c, d, e, g, i, k	Power law constants
D–B	Delany–Bazley

2 Basis of Mathematical Model Formulation

The development of empirical relationship in the field of noise control revolves around the concept of how much energy sound is able to dissipate as it travels through the pores of a material. The pores of an absorbing material can be of various complex shapes but for the ease of understanding and mathematical formulations; it is considered to be cylindrical. In this concept of lossy cylinder, the mass of the air in pores acted like an inductor (L), the property of compressibility of the air resembles a capacitor (C), the viscous loss at the wall of the pores resemble a resistor (R) and the thermal loss associated in the process resembles a small shunt resistor (G) [11].

Later, Delany and Bazley [12] introduced power law equation in terms of characteristics impedance and propagation constant equations by curve fitting approach which gives a rough idea of the absorption to be expected by a porous material. In this, the porosity of the material was assumed to be very close to unity and thus the only material property governing the absorption was AFR, σ. The equation was given in terms of $\left(\frac{f}{\sigma}\right)$. This equation was later modified up to a greater degree of accuracy by Miki [13] by using the same experimental data. Further modification was done by Komatsu [14] in which he changed the variables of power law from $\left(\frac{f}{\sigma}\right)$ to $\left(2 - \log\frac{f}{\sigma}\right)$. These models consider the flow resistivity as the only variable to be considered as an input to predict the SAC value with porosity assumed to be close to unity which is not always a very realistic assumption with respect to natural materials. With passage of time, other authors like Allard and Champoux [15] have incorporated other material properties like porosity, tortuosity and characteristics length to be a variable while estimating the SAC. This was further modified by Kino [16] by mathematical modifications which predicted the SAC to a greater degree of accuracy. This increase in accuracy was not fairly compensated by the amount of costs involved in instrumentation [17]. So, to save the extra cost, some researchers have shown that if we can just measure the characteristic impedance and propagation constant of a material, it is very much possible by mathematical modifications to predict their SAC value even without any knowledge of their AFR [18]. The general equations of power law are given as [19],

$$Z_b(f) = \rho_0 C_0 \left\{ 1 + a\left(\frac{f}{\sigma}\right)^b - jc\left(\frac{f}{\sigma}\right)^d \right\} \quad (1)$$

$$K_b(f) = \frac{2\Pi f}{C_0} \left\{ 1 + e\left(\frac{f}{\sigma}\right)^g - ji\left(\frac{f}{\sigma}\right)^k \right\} \quad (2)$$

$$Z_s = -jZ_b \cot(K_b l_s) \quad (3)$$

$$\alpha = 1 - \left[\frac{Z_s - \rho_0 C_0}{Z_s + \rho_0 C_0}\right]^2 \quad (4)$$

3 Materials and Methods

Jute is a natural material mainly cultivated in the Gangetic plains of southeast Asia. Owing to its good aspect ratio, it is used for making ropes and garments [20]. It has shown a good acoustical behaviour [21] along with good mechanical properties [22]. Cotton is used worldwide for making pillows and mattresses for human comfort. These items start to degrade with their usage and turn into waste when unable to provide sufficient amount of comfort. As cotton has been proved to be good sound-absorbing material [23], the properties of waste cotton need to be explored to provide a cheap and environment friendly alternate acoustical material. So, samples of these two materials of diameter and thickness 100 mm are fabricated and their AFR is measured. The results obtained are tabulated in Table 1. This AFR is fed into the D–B model to predict the sound absorbtion coefficient of these materials, a plot of which is shown in Figs. 1 and 2.

The actual trend is measured by using a four microphone impedance tube, and the predicted trend and actual trend is compared as shown in Figs. 1 and 2. It is inferred that there is a significant difference between the actual and predicted trend. Apart from this, the phenomenon of coincidence dip [24] is not visible in the D–B model. Hence we need to minimise the error between predicted and actual trend. This can be achieved by modifying the coefficients of the power law.

Table 1 Details of fabricated materials

SI. No.	Material	Bulk density (kg/m^3)	AFR
1	Jute fibre	300	20,087
2	Waste cotton	140	49,050

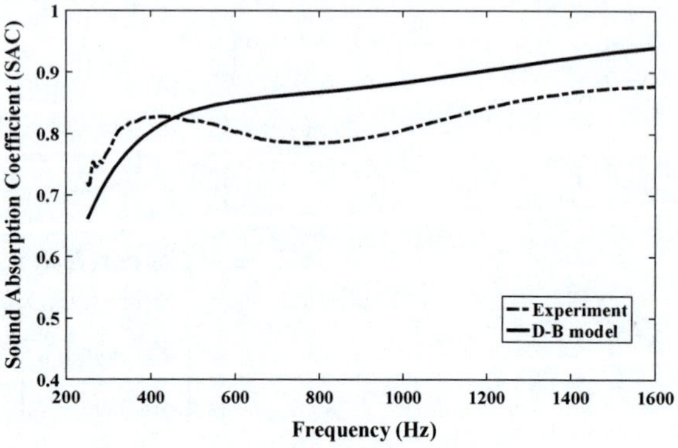

Fig. 1 Sound absorption coefficient of Jute by experiment and D–B model

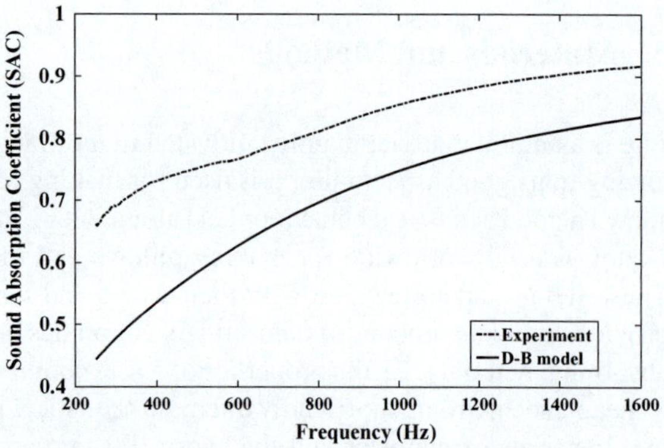

Fig. 2 Sound absorption coefficient of waste cotton by experiment and D–B model

4 Modified Model with AFR

The concept of power law was introduced in the 70s when computing power was severely constrained; hence, a single generic model of D–B, which was originally for synthetic material, was also in application for SAC prediction for other materials with near unit porosity. The trend of actual acoustical behaviour may be well above the predicted one or vice versa. Hence, even the conservative design approach is not much fruitful. With advancement in computing power, the authors firmly believe, it is time when the models should be materials specific. It has been tried by some other researchers too for coconut coir [25] and cotton shoddy [17]. To minimise the error between the actual and predicted SAC values, the characteristics impedance and propagation constant were measured in the impedance tube set-up (Figs. 3 and 4) and using the curve fitting approach with minimum sum of squared error, the constants in the power law equations were estimated by coding in MATLAB.

Characteristic impedance is the sum of resistive impedance (real part) and reactive impedance (imaginary part). Hence, the term Z_b is equivalent to $R - jX$, which on equating the real part of characteristic impedance becomes

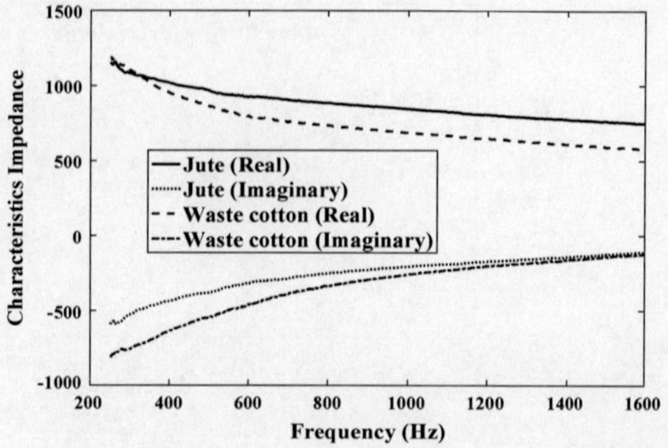

Fig. 3 Characteristic impedance

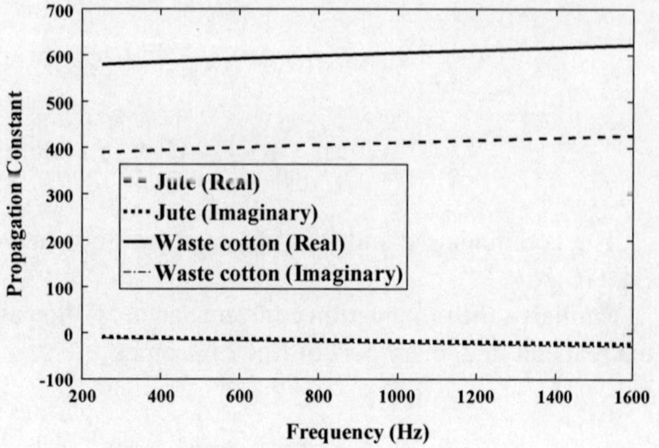

Fig. 4 Propagation constant

$$R = \rho_0 C_0 \left\{ 1 + a \left(\frac{f}{\sigma} \right)^b \right\} \tag{5}$$

$$\frac{R}{\rho_0 C_0} - 1 = a \left(\frac{f}{\sigma} \right)^b \tag{6}$$

$$\log\left(\frac{R}{\rho_0 C_0} - 1 \right) = b \log(f) + \log\left(\frac{a}{\sigma^b} \right) \tag{7}$$

This is a linear equation and the constants can be estimated with slope and intercept of the line. The plot is shown in Fig. 5 is for jute. Same step is extended to estimate the other coefficients for jute and waste cotton.

$$X = \rho_0 C_0 \left(c \left(\frac{f}{\sigma} \right)^d \right) \tag{8}$$

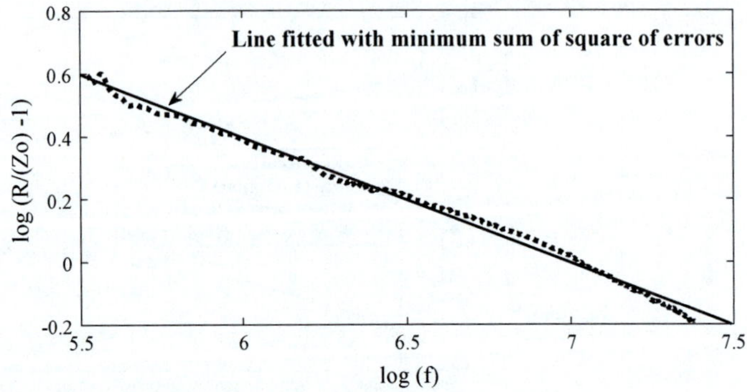

Fig. 5 Fitted line with minimum sum of square of errors

$$\frac{X}{\rho_0 C_0} = f^d \frac{c}{\sigma^d} \tag{9}$$

$$\log\left(\frac{X}{\rho_0 C_0}\right) = d\log(f) + \log\left(\frac{c}{\sigma^d}\right) \tag{10}$$

The coefficients c and d is obtained by slope and intercept of fitted line in the Eq. (10).

Similarly, the propagation constant can be written as $\alpha - j\beta$, which on comparing the real and imaginary part of Eq. 2 becomes

$$\alpha = \frac{2\pi f}{c_o}\left(1 + e\left(\frac{f}{\sigma}\right)^g\right) \tag{11}$$

$$\left(\frac{c_o\alpha}{2\pi f} - 1\right) = f^g \frac{e}{\sigma^g} \tag{12}$$

$$\log\left(\frac{c_o\alpha}{2\pi f} - 1\right) = g\log(f) + \frac{e}{\sigma^g} \tag{13}$$

The coefficient e and g is obtained by slope and intercepts of fitted line of Eq. (13). Similarly, on comparing the imaginary part,

$$\beta = \frac{2\pi f}{c_o}\left(i\left(\frac{f}{\sigma}\right)^k\right) \tag{14}$$

$$\left(\frac{c_o\beta}{2\pi f}\right) = f^k \frac{i}{\sigma^k} \tag{15}$$

$$\log\left(\frac{c_o\beta}{2\pi f}\right) = k\log(f) + \frac{i}{\sigma^k} \tag{16}$$

Table 2 Coefficients of power equation

Coefficients	D–B model [26]	Jute (fitted)	Waste cotton (fitted)
a	0.586	0.3113	0.0304
b	−0.75	−0.4003	−0.7871
c	0.768	0.0320	0.0091
d	−0.73	−0.8983	−1.0722
e	0.0857	1.1284	0.6962
g	−0.70	−0.9814	−0.9830
i	0.1749	0.2374	0.2038
k	−0.59	−0.4678	−0.4570

and by fitting a line to the Eq. (16) the coefficients *i* and *k* is obtained by its slope and intercept.

The coefficients for jute and waste cotton as obtained from the steps mentioned above are obtained and tabulated in Table 2.

5 Modified Model Without AFR

As shown in Eq. 7, $\log\left(\frac{R}{\rho_0 C_0} - 1\right) = b\log(f) + \log\frac{a}{\sigma^b}$; the coefficient b can be found by the slope of fitted line. To estimate the coefficient *a*, AFR of the material needs to be measured. In absence of the required instrumentation, we can still predict the absorption behaviour by mathematical modifications, $b = \text{slope}$, $\text{intercept} = \log\left(\frac{a}{\sigma^b}\right)$, so, $a = \left(\sigma^b\right)e^{\text{intercept}}$ [18].

Putting this value of a in the original equation, $R = \rho_0 C_0\left\{1 + a\left(\frac{f}{\sigma}\right)^b\right\}$, it is observed that there is a common item, (σ^b) in both numerator and denominator which gets cancelled out and the equation becomes $R = \rho_0 C_0\left\{1 + e^{\text{intercept}}\left(f^b\right)\right\}$ which is independent of the coefficient a [18]. Readers are encouraged to refer reference [18] for a detailed explanation of these steps. Hence, in this approach one coefficient is obtained directly from the slope of fitted line and the other is modified in terms of the intercept in such a way that the term containing AFR is cancelled out. Following the same steps, a MATLAB code has been developed to predict the absorption trend for jute and waste cotton. As observed from Figs. 6 and 7, the developed model agrees fairly well with the one observed experimentally on considering the overall performance in the range of frequencies. Apart from reduced error, it also gives a fair idea about the dip in absorption. Hence, it is better to use material specific models rather than the generic ones. The model without AFR performs very close to the one with AFR. The performance is so close that the difference practically vanishes when we consider behaviour at frequencies greater than 1000 Hz. This can be very much

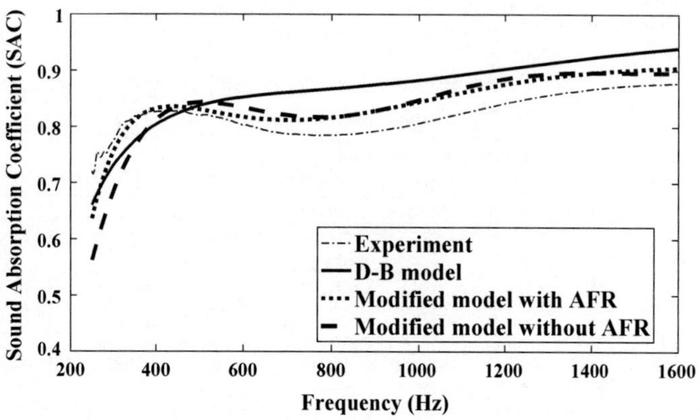

Fig. 6 Sound absorption coefficient of jute

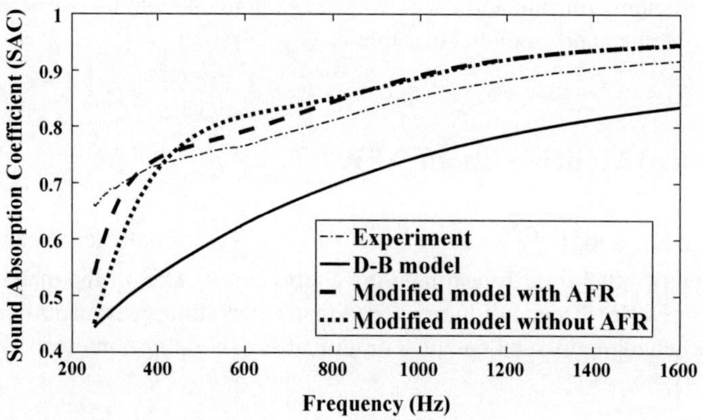

Fig. 7 Sound absorption coefficient of waste cotton

helpful while characterising an acoustical absorbing material, intended to reduce medium and high frequency noise.

6 Conclusions

Material specific models predict the absorption behaviour with greater degree of accuracy when compared with experimental results. The models developed with and without AFR of a material predicts the absorption behaviour fairly well when compared with the experimental results. The difference between these two models is converging in nature and it vanishes and practically the models overlap each other for a frequency greater than 500 Hz for jute and 700 Hz for waste cotton. In a broad sense, at a frequency greater than 1000 Hz, there is no difference between the results predicted by the model with and without AFR. The experimental results have shown that jute and waste cotton can turn out to be great acoustical absorber and can be considered as a valid replacement for synthetic acoustical absorbers currently in application.

Acknowledgments The authors are thankful to Department of Science and Technology-Government of India, for funding this project. They are also thankful to Prof. A. R. Mohanty and IIT-Kharagpur for sharing their experimental facility with the authors.

References

1. Asdrubali F, Schiavoni S, Horoshenkov K (2012) A review of sustainable materials for acoustic applications. Build Acoust 19:283–312
2. Koizumi T, Tsujiuchi N, Adachi A (2002) The development of sound absorbing materials using natural bamboo fibers. High Perform Struct Compos 59:157–166
3. Hong Z, Bo L, Guangsu H, Jia H (2007) A novel composite sound absorber with recycled rubber particles. J Sound Vib 304:400–406
4. Ersoy S, Küçük H (2009) Investigation of industrial tea-leaf-fibre waste material for its sound absorption properties. Appl Acoust 70:215–220
5. Hosseini Fouladi M, Nor MJM, Ayub M, Leman ZA (2010) Utilization of coir fiber in multilayer acoustic absorption panel. Appl Acoust 71:241–249
6. Oldham DJ, Egan CA, Cookson RD (2011) Sustainable acoustic absorbers from the biomass. Appl Acoust 72:350–363
7. Ballagh K (1996) Acoustical Properties of Wool. Appl Acoust 48:101–120
8. Hao A, Zhao H, Chen JY (2013) Composites: Part B Kenaf/polypropylene nonwoven composites: the influence of manufacturing conditions on mechanical, thermal, and acoustical performance. Compos B Eng 54:44–51
9. Berardi U, Iannace G (2015) Acoustic characterization of natural fi bers for sound absorption. Appl Acoust 94:840–852
10. Gómez Escobar V, Maderuelo-Sanz R (2017) Acoustical performance of samples prepared with cigarette butts. Appl Acoust 125:166–172
11. Flanagan JL (2013) Speech analysis and perception, 2nd edn. Springer, Berlin, pp 23–86
12. Delany ME, Bazley EN (1969) Acoustical properties o f fibrous absorbent materials. Appl Acoust 3:105–116
13. Miki Y (1990) Acoustical properties of porous materials-Modifications of Delany–Bazley models. J Acoust Soc Jpn (E) 11:19–24
14. Komatsu T (2008) Improvement of the Delany–Bazley and Miki models for fibrous sound-absorbing materials. Acoust Sci Technol 29:121–129
15. Allard JF, Champoux Y (1992) New empirical equations for sound propagation in rigid frame fibrous materials. J Acoust Soc Am 91:3346–3353
16. Kino N (2015) Further investigations of empirical improvements to the Johnson-Champoux-Allard model. Appl Acoust 96:153–170
17. Manning J, Panneton R (2013) Acoustical model for Shoddy-based fiber sound absorbers. Text Res J 83:1356–1370
18. Jones PW, Kessissoglou NJ (2015) Simplification of the Delany–Bazley approach for modelling the acoustic properties of a poroelastic foam. Appl Acoust 88:146–152
19. Fatima S, Mohanty AR (2016) Acoustical characterization of bulk natural fibrous material using flow resistivity. In: Paper presented at 45th international congress and exposition on noise control engineering INTER-NOISE 2016, Hamburg, Germany, 21–24 August 2016

20. Mohanty A, Fatima S (2016) Noise control using green materials. SV Sound Vib Instrum Issue 13:13–15
21. Fatima S, Mohanty AR (2011) Acoustical and fire-retardant properties of jute composite materials. Appl Acoust 72:108–114
22. Bhowmick M, Mukhopadhyay S, Alagirusamy R (2012) Mechanical properties of natural fibre-reinforced composites. Text Prog 44:85–140
23. Arenas JP, Crocker MJ (2010) Recent trends in porous sound-absorbing materials. Sound Vib Mater Ref Issue, 12–17
24. Bhattacharya MC, Guy RW, Crocker MJ (1971) Coincidence effect with sound waves in a finite plate. J Sound Vib 18:157–169
25. Ramis J, Del Rey R, Alba J et al (2014) A model for acoustic absorbent materials derived from coconut fiber. Mater Constr 64:9–14
26. Rwawiire S, Tomkova B, Militky J, Hes L, Kale BM (2017) Acoustic and thermal properties of a cellulose nonwoven natural fabric (barkcloth). Appl Acoust 116:177–183

Dependence of Macro-Scale Response of Fibrous Materials on Polygonal Arrangement of Fibers

Sagar Deshmukh and Sripriya Ramamoorthy

Abstract Absorption characteristics of fibrous material inherently depend on the microscopic shapes and the dimensions of the fibers. Periodic Unit Cell (PUC) modeling approach is used for the optimization of arrangements of fibers. Periodic hexagonal and square arrangement of fibers are considered here for study. Five Johnson and Champoux-Allard (JCA) parameters and the transmission loss are computed to evaluate the effect of these two structure configurations. Steady Stokes and electric boundary value problem has been solved for estimation of the airflow resistivity, tortuosity and viscous characteristic length, while porosity and thermal characteristic length are estimated directly from mesh geometry. This study underlines the effects due to the change in fiber arrangements on to absorption characteristics of 50 mm thick sample size of the rigid porous medium. It is observed that for equal centric diameter of fibers hexagonal configuration yields better transmission loss compared to square configuration of fibers over frequency range of 0–8 kHz.

Keywords PUC · Transmission loss · Porosity · Airflow resistivity · Tortuosity · Thermal characteristic length · Viscous characteristic length

Nomenclature

ϕ	Porosity
σ	Airflow resistivity
α_∞	Tortuosity
Λ	Viscous characteristic length
Λ'	Thermal characteristic length

S. Deshmukh · S. Ramamoorthy (✉)
Department of Mechanical Engineering, Indian Institute of Technology Bombay, Mumbai 400076, India
e-mail: ramamoor@iitb.ac.in

S. Deshmukh
e-mail: sagar.deshmukh@iitb.ac.in

M. Singh and Y. Rafat (eds.), *Recent Developments in Acoustics*, Lecture Notes in Mechanical Engineering, https://doi.org/10.1007/978-981-15-5776-7_15

V_f Volume of fluid domain
V Total volume of porous media
$\boldsymbol{v}$ Velocity field
p Pressure
$\boldsymbol{E}$ Electric field
μ Dynamic viscosity

1 Introduction

Fibrous porous materials such as glass wool as well as reticulated porous materials such as polyurethane foams are used for passive noise control in many applications in automotive, aerospace and building industries. These materials attenuate the airborne sound waves by increasing air resistance and thus reducing the amplitude of sound wave [1].

Several attempts have been made to compute absorption characteristics of porous materials such as empirical model of Delany and Bazley [2] based on airflow resistivity of medium for range of fibrous materials. Similar empirical relationships for estimation of absorption coefficient through measurement of complex surface impedance and complex propagation constant for multi-layer reticulated polyurethane foam has been proposed by Dunn and Davern [3]. However, these models impose restrictive requirement on the ratio of frequency to static flow resistivity ($0.01 < f/\sigma < 1$) and assumption of near unit porosity. Thus several geometrical parameters have been evolved to relate absorption characteristics of porous materials with its microscopic shape and size. The Johnson and Champoux-Allard (JCA) model [4–6] is based on Biot theory [7, 8] of wave propagation in fluid-saturated porous media. According to Biot theory of wave propagation, two longitudinal waves (each one in solid and fluid phase) and one transverse wave (in solid phase) propagates through porous medium. Under the acoustic excitation, solid frame of porous material can be assumed to be rigid for some foams and hence single longitudinal wave propagates through the fluid medium. Thus porous medium can be seen as equivalent fluid with effective density (ρ) and bulk modulus (K). These effective properties account for the losses caused by viscous and thermal effects and depends on five parameters, viz., flow resistivity (σ), porosity (ϕ), tortuosity (α_∞), viscous characteristic length (Λ) and thermal characteristic length (Λ'). Several experimental and numerical studies have been carried out for estimation of JCA parameters [8–10]. Recently, Periodic Unit Cell (PUC) modeling of microstructural configuration for estimation of JCA parameter was proposed by Perrot et al. [10]. In this method, steady Stokes boundary value problem is solved for estimation of pressure and velocity field across the fluid domain to compute the flow resistivity of the porous medium, while electric boundary value problem is solved for electric field to compute tortuosity and viscous characteristics length of the porous medium.

Various attempts have been made to relate absorption characteristic of the porous materials with microscopic parameters of porous medium [11, 12]. Perrot et al. [10] used bottom-up approach for optimization of shape, size and throat length for the hexagonal arrangement of fibers. The present work is an attempt to compare the polygonal arrangement of fibers for the enhancement of absorption characteristics. Outline of the paper is as follows: In Sect. 2, square and hexagonal arrangements of fibers have been chosen to estimate JCA parameters. In Sect. 3, the predicted JCA parameters are used to evaluate the transmission loss using [16]. In Sect. 4, results of velocity and electric fields are presented and the transmission loss for the two configurations is also compared. Section 5 concludes the paper.

2 Numerical Calculations

2.1 Model Geometry

In addition to the 2D hexagonal arrangement of fibers [10] square arrangement of fibers is considered for study as shown in Fig. 1. Both of these arrangements are commonly seen in most of the fibrous materials. The optimum circular radius of 32 μm and centric diameter of 268 μm of fibers are considered from Perrot et al. [10] as shown in Fig. 1.

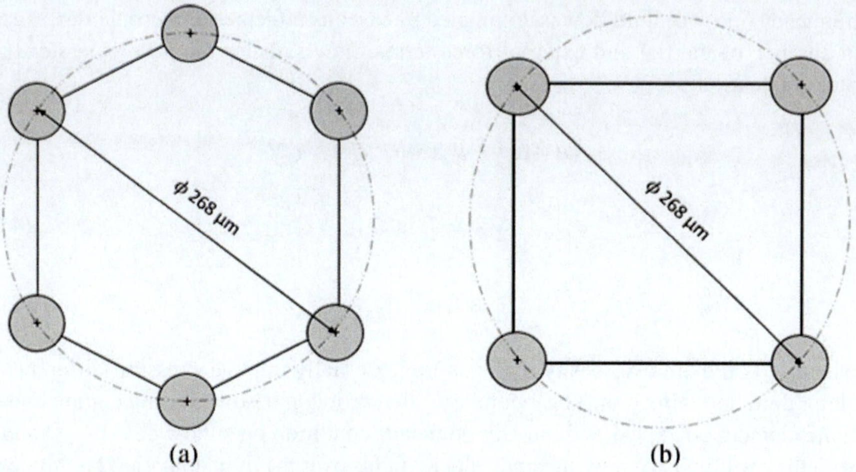

Fig. 1 Structural arrangement of fibers **a** hexagonal **b** Square

2.2 *Estimation of JCA Parameters*

2.2.1 Porosity (ϕ)

Porosity can be estimated from geometric model by ratio of volume of fluid domain to total volume of fluid as well as solid phase of the porous medium.

$$\phi = \frac{V_f}{V} \tag{1}$$

2.2.2 Air Flow Resistivity (σ)

Navier–Stokes equations for fluid flow governs the motion of fluid in domain and can be seen as Newton's second law of motion for fluids. Generalized form of Navier–Stokes equation for compressible Newtonian fluid is as follows [13]:

$$\rho\left(\frac{\partial \boldsymbol{v}}{\partial t} + \boldsymbol{v} \cdot \nabla \boldsymbol{v}\right) = -\nabla p + \mu \Delta \boldsymbol{v} + F \tag{2}$$

where $\rho\left(\frac{\partial \boldsymbol{v}}{\partial t} + \boldsymbol{v} \cdot \nabla \boldsymbol{v}\right)$ accounts inertial forces due to fluid motion, ∇p represents pressure difference across fluid domain, $\mu \Delta \boldsymbol{v}$ corresponds to viscous forces whereas F constitutes the effect of externally applied forces. In static conditions or at very low frequency ($\omega \rightarrow 0$) fluid flow is dominated by viscous effects over inertial ones, thus in absence of inertial and external force terms, flow satisfies the following steady Stokes equation:

$$\mu \Delta \boldsymbol{v} - \nabla p + \hat{e} = 0 \text{ in } \Omega_f \tag{3}$$

$$\nabla . \boldsymbol{v} = 0 \text{ in } \Omega_f \tag{4}$$

$$\boldsymbol{v} = 0 \text{ on } \partial \Omega_{sf} \tag{5}$$

where μ is dynamic viscosity and $\hat{e}$ is the externally applied pressure difference along fluid flow direction. This boundary value problem is solved using commercial finite element code [15] with no-slip boundary condition on solid walls (Eq. 5) and with the periodicity of pressure and velocity fields over the fluid domain (Ω_f). Model symmetry (shown in Fig. 2) is taken into account to reduce computation time of the simulation.

From average velocity over fluid domain, flow resistivity of medium can be estimated using following equation:

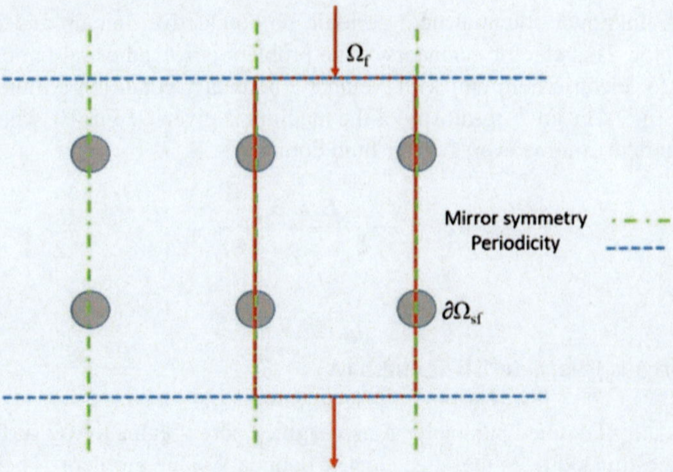

Fig. 2 Use of symmetry for square arrangement of fibers

$$\sigma = \frac{\hat{e}}{\phi < v >_f} \tag{6}$$

The symbol $< >_f$ designated as volume average over fluid domain.

2.2.3 Tortuosity (α_∞)

At high frequencies ($\omega \rightarrow \infty$), inertial effect dominates over the viscous one. In this case viscosity of the fluid can be neglected and fluid can be assumed to be a perfect fluid. This incompressible perfect fluid flow problem coincides with electric conduction problem by assuming that the porous medium is filled with conducting fluid and having insulated solid phase (skeleton) [14]. This analogy can be conceived by assuming the change in electric potential corresponds with pressure difference, fluid inertial term corresponds with current density and fluid density corresponds to the electric resistivity of the conducting fluid. This assumption remains valid until wavelength of sound is large enough compared to microscopic dimensions of pores. This phenomenon is governed by the following set of equations:

$$\boldsymbol{E} = -\nabla \boldsymbol{q} + \boldsymbol{e} \text{ in } \Omega_f \tag{7}$$

$$\nabla.\boldsymbol{E} = 0 \text{ in } \Omega_f \tag{8}$$

$$\boldsymbol{E}.\mathbf{n} = 0 \text{ on } \partial\Omega_{sf} \tag{9}$$

where $\boldsymbol{q}$ is unknown potential field periodic on conductive domain and $\mathbf{n}$ is unit normal vector. This electric boundary value problem is solved using finite element code [15] for electric field ($\boldsymbol{E}$) with Neumann boundary condition at fluid surface interface (Eq. 9). From $\boldsymbol{E}$ tortuosity of the medium is given by Eq. 10, where $< >_f$ symbol refers to volume average over fluid domain.

$$\alpha_\infty = \frac{< \boldsymbol{E}.\boldsymbol{E} >_f}{< \boldsymbol{E} >_f < \boldsymbol{E} >_f} \tag{10}$$

2.2.4 Viscous Characteristic Length (Λ)

Johnson et al. [4] defined parameter Λ as weighted pore volume to wet surface ratio and can be estimated from the same electric boundary value problem as below:

$$\Lambda = 2\frac{\iiint E.EdV}{\iint E.EdA} \tag{11}$$

where area integral is over fluid–solid interface area.

2.2.5 Thermal Characteristic Length (Λ')

This can be calculated directly from geometry as doubled the ratio of volume of the fluid domain over area of fluid–solid interface [5].

$$\Lambda' = 2\frac{\iiint dV}{\iint dA} \tag{12}$$

where area integral is over fluid–solid interface area.

2.3 *Equivalent Density and Bulk Modulus*

From the Johnson and Champoux and Allard model [4–6] is effective density and effective bulk modulus of the equivalent fluid representing the porous medium can be expressed as function of angular frequency (ω):

$$\rho(\omega) = \rho_0\alpha_\infty\left(1 + \frac{\phi\sigma}{j\omega\rho_0\alpha_\infty}\left(1 + j\frac{4\omega\rho_0\mu\alpha_\infty^2}{\sigma^2\phi^2\Lambda^2}\right)^{1/2}\right) \tag{13}$$

and

$$K(\omega) = \frac{\gamma P_0}{\gamma - (\gamma - 1)\left(1 + \frac{8\mu}{j\omega\rho_0 B^2 \Lambda'^2}\left(1 + j\frac{\omega B^2 \rho_0 \Lambda'^2}{16\mu}\right)^{\frac{1}{2}}\right)^{-1}} \quad (14)$$

where P_0 is barometric pressure, γ, B^2, ρ_0 and μ are specific heat ratio, Prandtl number, density and dynamic viscosity of saturating air, respectively. From effective density $\rho(\omega)$ and bulk modulus $K(\omega)$ one can easily calculate characteristic impedance and complex wave number [5],

$$Z_c = \sqrt{\rho K} \quad (15)$$

and,

$$\gamma = \omega\sqrt{\frac{\rho}{K}} \quad (16)$$

2.4 Estimation of Transmission Loss

Transmission loss (TL) usually defined as the decrease in acoustic intensity when an acoustic wave propagates through the absorptive medium. Thus TL can be considered to be a benchmark parameter for comparing absorption characteristics of two different materials or configurations.

Therefore transmission loss of porous material for both configurations has been estimated over frequency range of 0–8 kHz using dedicated acoustic finite element based software [16]. The porous material of 50 mm thickness has been modeled and placed in middle of tube of having 1000 mm length as shown in Fig. 3. A long length of the tube ensures the propagation of only plane waves for sample under test, while the outlet of the tube is modeled to be infinitely long to ensure no reflection occurring from rear side of the tube. Thus transmission loss can be calculated by using following relation:

$$\mathrm{TL} = 10\log_{10}\left(\frac{W_o}{W_i}\right) \quad (17)$$

where W_0 and W_i are incident powers on the inlet and the outlet surface of the tube.

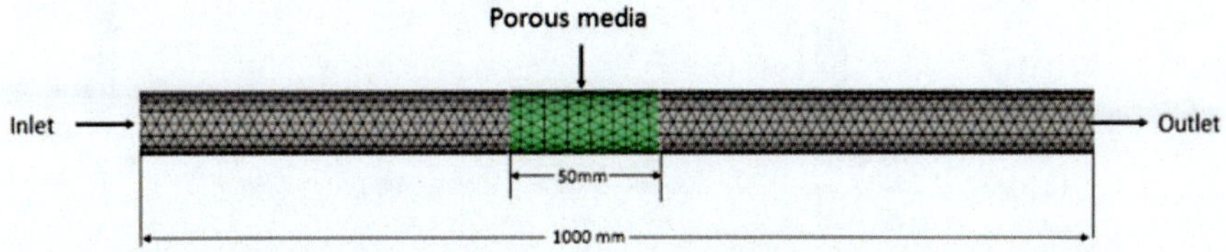

Fig. 3 Meshed model with porous material at center of tube

3 Results and Discussions

Steady Stokes and electric boundary value problem has been solved for velocity and electric field across the fluid domain. Velocity and electric field contours are as shown in Figs. 4 and 5. Though the pressure drop across the both PUC in downstream direction is 0.7 Pa, change in velocity profile does occur due to the different positions of fibers over domain as shown in Fig. 4 whereas in electric analysis due to the nonslip condition, current (or fluid) flow path is more concentrated along circumference of the fiber due to minimal fluid viscosity in high-frequency domain as shown in Fig. 5.

The values of all the five JCA parameters for hexagon and square arrangements of fibers are estimated using the Eqs. (1), (6), (10), (11) and (12) and shown in Table 1. The predicted JCA parameters as well as absorption coefficients for hexagonal fiber arrangement matched well with the results from Perrot et al. [10].

These five parameters for both the configurations are further used to define the rigid porous medium in [16] and transmission loss is estimated for desired frequency range of 0–8 kHz. Figure 6 shows the plots of transmission loss versus frequency for

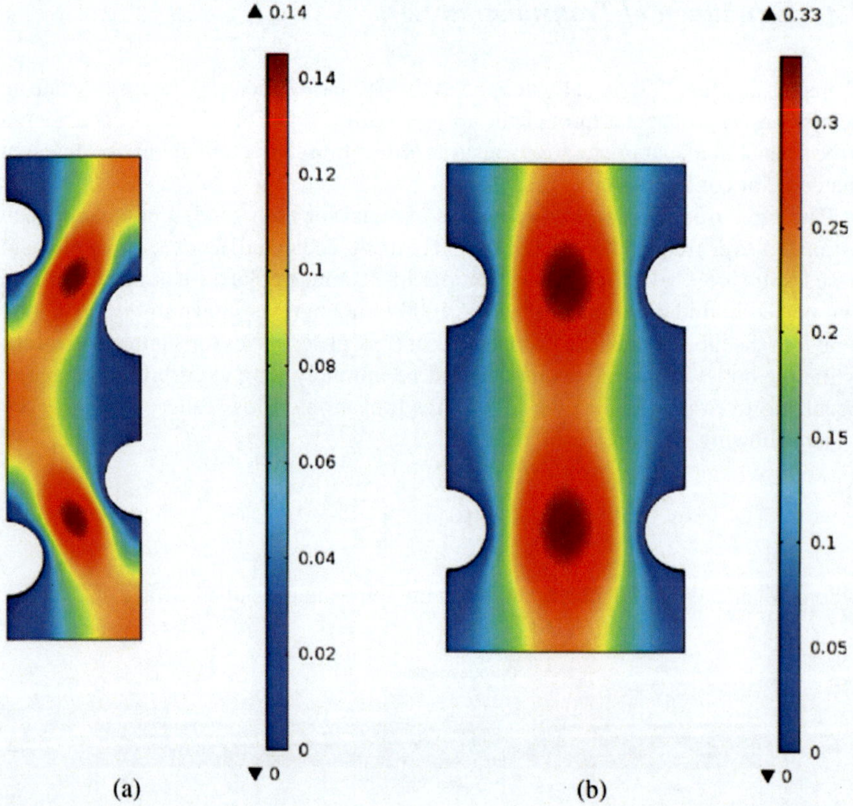

Fig. 4 Velocity field across the fluid domain for **a** Hexagonal **b** Square arrangement of fibers

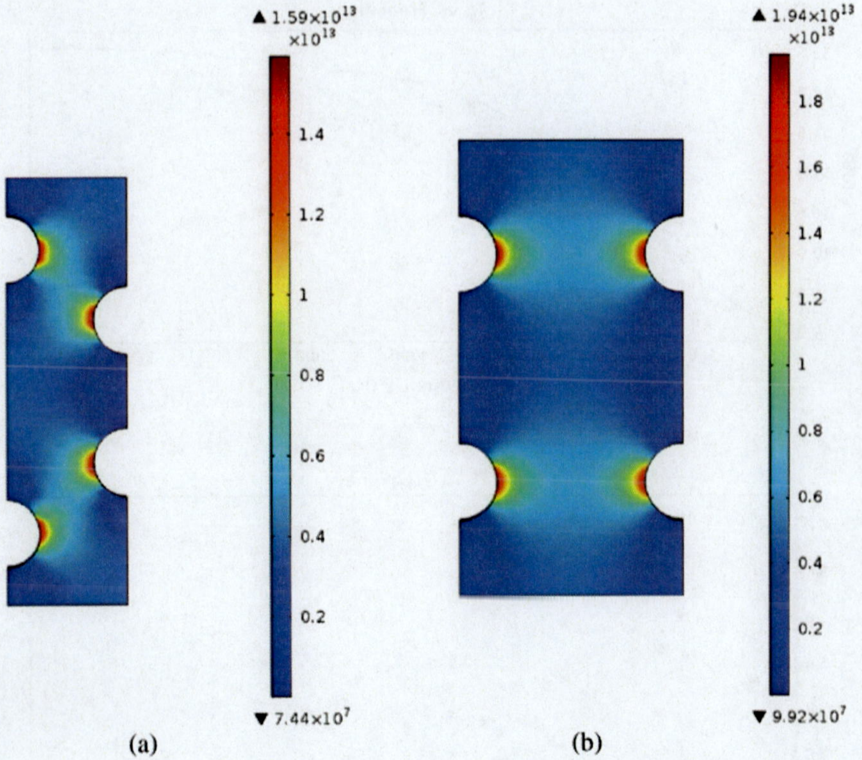

Fig. 5 Electric field across the fluid domain for **a** Hexagonal **b** Square arrangement of fibers

Table 1 Macroscopic JCA parameters for circular cross section having radius 32 μm and throat length 70 μm

Parameter	Units	Hexagon	Square
Porosity (ϕ)	%	86.36	91.03
Airflow resistivity (σ)	Pa-s/m^2	26524	11381
Tortuosity (α_∞)	–	1.069	1.05
Viscous Characteristic Length (Λ)	μm	112.60	176.96
Thermal Characteristic Length (Λ')	μm	200.00	324.73

both hexagonal and square arrangements of fibers. From plots, it can be seen that TL in hexagonal arrangement is substantially more compared to square arrangement, which is expected also due to less number of fibers per centric diameter of 268 μm.

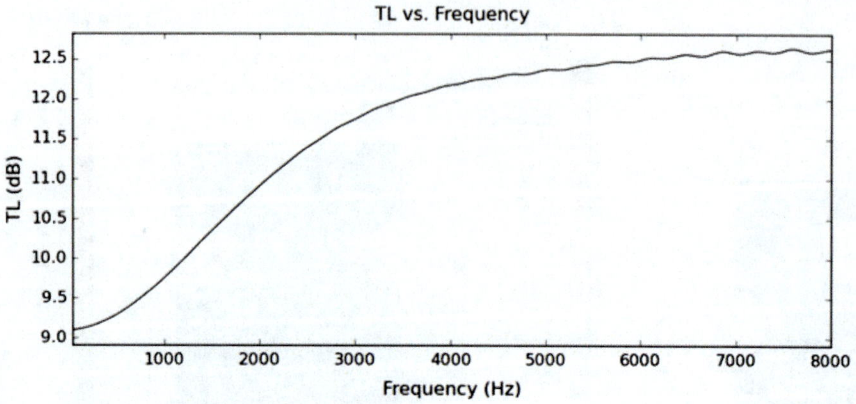

(a)

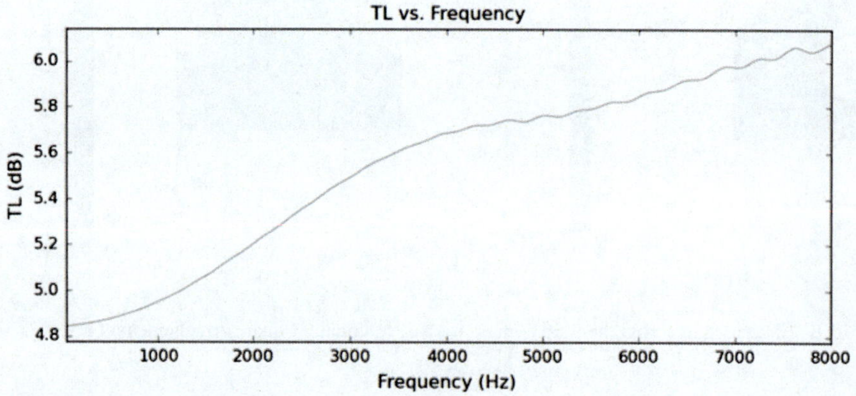

(b)

Fig. 6 Transmission loss versus Frequency for **a** Hexagonal **b** Square arrangements of fibers

4 Conclusion

JCA parameters for both configurations have been estimated, and result confirms that arrangement of fiber affects the absorption characteristics of the porous medium. Transmission loss for hexagonal arrangement increases proportionally with frequency up to 5 kHz and remains almost steady over remaining range of frequency. However, it gradually increases by a small value for square configuration as shown in Fig. 6. Plots show that the hexagonal arrangement of fibers performs well over square arrangement of fibers, transmission loss of hexagonal fiber is almost twice that of square one. Hence for given optimized radius and throat thickness of fibers, polygonal arrangements can be optimized for better absorption characterization of porous material. From Table 1 and Fig. 6, it can be concluded that better absorption characteristics correspond to minimum thermal and viscous characteristic length.

Acknowledgments This work was supported by IMPRINT INDIA grant no. 6367 from MHRD and DST, India.

References

1. Ashby MF, Evans AG, Fleck NA, Gibson LJ, Hutchinson JW, Wadley HG (2000) Metal foams: a design guide. Butterworth-Heinemann
2. Delany ME, Bazley EN (1969) Acoustical properties of fibrous absorbent materials. Appl Acoust 3:105–116
3. Dunn IP, Davern WA (1986) Calculation of acoustic impedance of multi-layer absorbers. Appl Acoust 19(3):321–334
4. Johnson DL, Koplik J, Dashen R (1987) Theory of dynamic permeability and tortuosity in fluid-saturated porous media. J Fluid Mech 176:379–402
5. Allard JF, Atalla N (1993) Modelling sound absorbing materials: propagation of Sound in Porous Media. Elsevier Applied Science, New York
6. Champoux Y, Allard JF (1991) Dynamic tortuosity and bulk modulus in air-saturated porous media. J Appl Phys 70:1975–1979
7. Biot MA (1956) Theory of propagation of elastic waves in a fluid-saturated porous solid. J Acoust Soc Am 28:168–191
8. Atalla Y, Panneton R (2005) Inverse acoustical characterization of open cell porous media using impedance tube measurements. Can Acoust 33(1):3–10
9. Leclaire P et al (1996) Determination of the viscous and thermal characteristic lengths of plastic foams by ultrasonic measurements in helium and air. J Appl Phys 2009–2012
10. Perrot C, Chevillotte F, Panneton R (2008) Bottom-up approach for microstructure optimization of sound absorbing materials. J Acoust Soc Am 124(2):940–948
11. Cortis A et al (2003) Influence of pore roughness on high-frequency permeability. Phys Fluids:1766–1775
12. Gasser S et al (2005) Absorptive properties of rigid porous media: application to face centered cubic sphere packing. J Acoust Soc Am:2090–2099
13. Versteeg HK, Malalasekra W (2007) An introduction to computational fluid dynamics: The finite method. Pearson Education Ltd., New York
14. Brown RJ (1980) Connection between formation factor for electrical resistivity and fluid-solid coupling factor in Biot's equation for acoustic waves in fluid-filled porous media. Geophysics 1269–1275
15. COMSOL Multiphysics, 5.3 Documentation
16. MSC ACTRAN, 17.0, user guide vol. 1 and vol. 2

Acoustic Streaming in a Porous Media

Neetu Srivastava and Yasser Rafat

Abstract Srivastava shown in Srivastava (Powder Technol 320:108–113, 2017, Lect Notes Mech Eng 19:149–155) that the velocity of the steady flow outside the boundary layer was independent of viscosity but was dependent on permeability. Through this paper, we tried presenting practical implication of the theory developed by Srivastava (Powder Technol 320:108–113, 2017, Lect Notes Mech Eng 19:149–155). It was found that the flow may be described using the non-dimensional parameters. Results were plotted for the volumetric flow rate analysis of the problem.

Keywords Fluidized bed · Standing wave · Boundary layer

1 Introduction

Some phenomena in the field of microsystem technologies lead to streaming flows which are vital for both its positive effects (e.g., microfluidic mixing, fluid pumping) and its negative effects (e.g., particle/cell focusing). These effects generated due to the complex streaming pattern can be solved by perturbation method. In this work, acoustic streaming fields in 2D parallel plate have been structured and the effects of permeability on a boundary-driven acoustic streaming field are investigated. As a result large influences on streaming velocities and streaming patterns are observed. Srivastava [1, 2] investigation reveals that the position of nodes in standing wave is a function of permeability as well as effective viscosity. Hence, using perturbation theory a solution is derived near the boundary layer in the vicinity of rigid wall. The significant effect of permeability in the boundary layer zone can be predicted from the problem. These derived results have been supported by one practical example. Non-dimensional flow parameters are used to describe the flow.

N. Srivastava (✉)
Department of Mathematics, Amrita School of Engineering, Amrita Vishwa Vidyapeetham, Bengaluru, India
e-mail: s_neetu@blr.amrita.edu

Y. Rafat
Department of Mechanical Engineering, Aligarh Muslim University, Aligarh, Uttar Pradesh, India

M. Singh and Y. Rafat (eds.), *Recent Developments in Acoustics*, Lecture Notes in Mechanical Engineering, https://doi.org/10.1007/978-981-15-5776-7_16

Lord Rayleigh [3] has developed the theoretical work for the study of acoustic streaming. He demonstrated his work through Kundt's tube. Westervelt [4] continued this analysis by analyzing the streaming velocity due to acoustical disturbances. Further, the related theories were reviewed by Nyborg [5]. Qi [6] did the theoretical study to analyze the acoustic streaming in a circular tube. Vainshtein [7] analyzed Rayleigh streaming and Couette flow to investigate the effect of streaming on shear flow. The treatment has ignored the analysis of acoustofluidics devices involving porous media. This phenomenon is of great practical interest for a wide range of applications such as sonoprocessed fluidized bed reactors. Aiming towards maximizing the efficiency of each particular process, this manuscript is mainly focusing on the presentation of theoretical development of acoustic streaming through the fluidized bed. Fluidized bed in the flow structures follows the Brinkman equation [8]. The coefficient of effective viscosity μ_e is different from μ, the coefficient of viscosity of clear fluid in the flow governing equation. The value of μ_e is different from μ in all the cases and was predicted by Lundgren [9]. Experimental values for μ_e for the steady flow through a wall bounded by a porous medium are measured by Givler and Altobelli [10] and found that the ratio $(\mu_e/\mu) = \gamma^2$ varies from 5.1 to 10.9 (Fig. 1).

In this paper, we have followed the lines of Landau and Lifschitz [11] to discuss the Rayleigh streaming through fluidized. According to Landau and Lifschitz the acoustic streaming (Rayleigh Streaming) is particularly pronounced under the condition $\lambda \gg h \gg \delta$, where λ is the wavelength, h is the characteristics length scale, and δ is the viscous penetration depth. In case of porous media (for example, sonoprocessed fluidized bed between the parallel plates), theoretical study [12] reports that the properties of Rayleigh streaming are seen when the characteristic pore size R is either less than or equal to boundary layer thickness, i.e., $R \leq \delta$, and in addition to this the oscillation amplitude is large as compared to the characteristic pore size. Valverde report has modified Landau and Lifschitz condition for the occurrence of

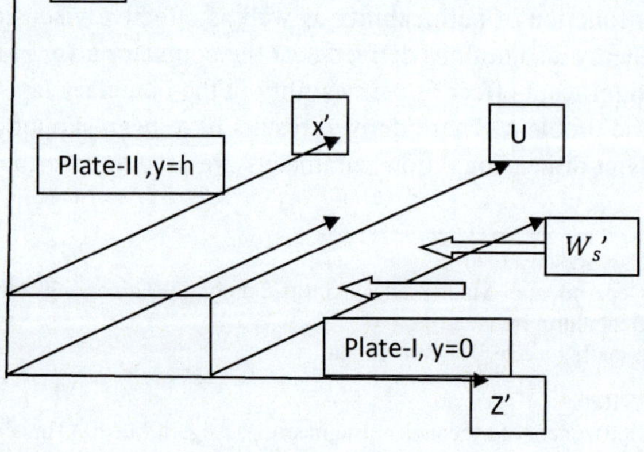

Fig. 1 Flow domain with co-ordinates

Rayleigh streaming in presence of porous media as $\lambda \gg h \gg \delta \geq R$. Brinkman equation represents the flow in porous region. Under the Landau and Lifschitz condition, we have used the method of successive approximation employed by Schlichting [13]. The results have a general application to the process efficiency of acoustofluidics devices involving porous media immersed in viscous fluids where the typical size of the constituent particle is small as compared to the boundary layer thickness.

2 Perturbed Flow Region Between in the Channel

The standing wave generated in the medium enclosed between two parallel walls describes simple Couette flow as one of the plates in the flow structures is at rest and the other is moving with velocity U and satisfies

$$\frac{U}{c} \ll 1. \tag{1}$$

where c is the wave velocity. This condition is sufficient to ignore the compressibility of the main flow. Defining a plane standing wave:

$$W = \mathrm{Re}\left(w_0 \cos Az' \cos \omega t'\right), \; -\infty < x < \infty \tag{2}$$

where $A = \frac{\omega}{c}\sqrt{\left[1 + i\left(\frac{\sigma^2}{Ro}\right)\right]}$, external flow velocity is defined by $W\left(z', t'\right)$, t' is the time and y' is the normal co-ordinate. On the x' variable, none of the flow parameters depends. Taking into an account an assumption that the appearance of acoustic streaming in the flow region is due to the small distance between the plate, the flow equation can be written as

$$\gamma^2 \mathrm{CurlCurl\ Curl}(\vec{v}) = -\frac{1}{\kappa}\mathrm{Curl}(\vec{v}) \tag{3}$$

where $\vec{v} = \left[0, v_y, v_z\right]$,$\sigma$ is the conductivity and μ is the dynamic viscosity. The relevant boundary conditions are

$$\left.\begin{array}{l} u' = 0, v' = 0, w' = w_s' \text{ at } y' = 0 \\ u' = U, v' = 0, w' = w_s' \text{ at } y' = h \end{array}\right\} \tag{4}$$

where w_s' is defined by Srivastava [3]. Let us define stream function, ψ', where

$$v_z = \frac{\partial \psi'}{\partial y'}, \; v_y = -\frac{\partial \psi'}{\partial z'} \tag{5}$$

Introducing the following non-dimensional variables: $y = \frac{2y'}{h} - 1$, $z = 2nz'$,

$$\psi = \frac{32c_a}{3w_0^2 h}\psi'. \tag{6}$$

Substituting Eqs. (5) and (6) in Eq. (3), we can re-write the Eq. (3) with boundary layer approximation as

$$\frac{\partial^4 \psi}{\partial y^4} - \alpha^2 \frac{\partial^2 \psi}{\partial y^2} = 0. \tag{7}$$

where $\alpha = \frac{\sigma}{\gamma}$. To solve the Eq. (7) we can expand the stream function as

$$\psi = \psi_0 + \alpha^2 \psi_1 + \ldots\ldots\ldots\ldots\ldots\ldots\ldots \tag{8}$$

To eliminate the constants from the solution we use the boundary conditions (4) and can write the solution as

$$\psi_0 = \left(y^3 - y\right) \sin Az \tag{9}$$

Solution to the first approximation for the Eq. (9) can be written as

$$\psi_1 = \left(\frac{y^5}{20} - \frac{y^3}{10} + \frac{y}{20}\right) \sin Az \tag{10}$$

where $A = \frac{\omega}{c}\sqrt{\left[1 + i\left(\frac{\sigma^2}{Ro}\right)\right]}$.

3 Results and Discussion

In this section, results for the current study are plotted using **MATHEMATICA 7**. In this section, we have presented some of the contour plots related to the current model.

Figures 2 and 3 represent the y-component of the velocity. This figure indicates that the y-component of velocity will be modified with the effect of $= \frac{\sigma}{\gamma}$. The y-component of velocity differs significantly for the values computed at $\alpha = 1.33$, $\alpha = 64$ and Ro $= 20$. As we reduce the $\gamma = \frac{\mu_e}{\mu}$, y-component of velocity will increase as well as the number of vortices will also increase which will result in mass transfer enhancement. Figures 5 and 6 are plotted for the streamline patterns of current study for the different values of Darcy to the gamma. A huge difference is found in the magnitude of acoustic streaming patterns for two different values of α. These figures indicate that decrease in γ also increases the number of vortices in streaming patterns, which means that decrease in γ can be used to enhance the mass

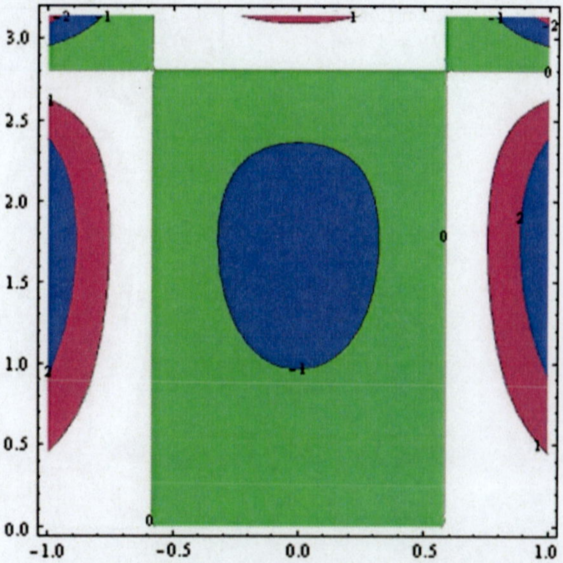

Fig. 2 y-Velocity component for $\alpha = 1.33$ ($\sigma = 8/\gamma = 6$), Ro $= 20$

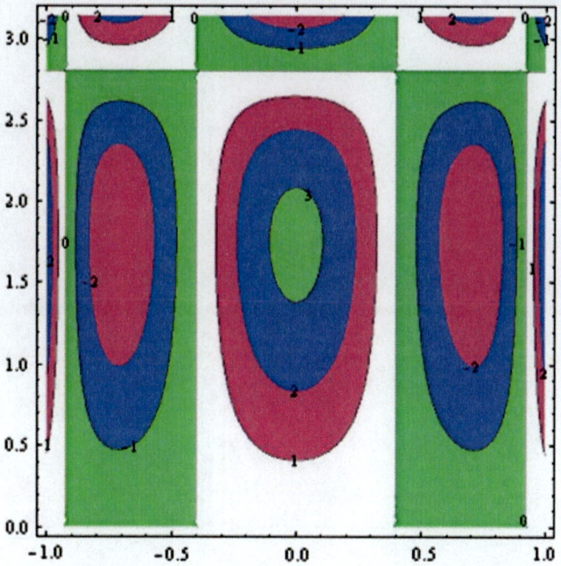

Fig. 3 y-Velocity component for $\alpha = (64)$ ($\sigma = 8/\gamma = 1$) and Ro $= 20$

transfer. The number of vortices could significantly enhance the mass transfer in acoustofluidic devices which has a potential application in diffusion-based systems.

Figures 4 and 5 are plotted for the Roshko number Ro $= 20$, $\gamma = 1$ and for the two different values of Darcy number. These figures indicate that the periodicity patterns are slightly changed for the different values of Darcy number (Fig. 6).

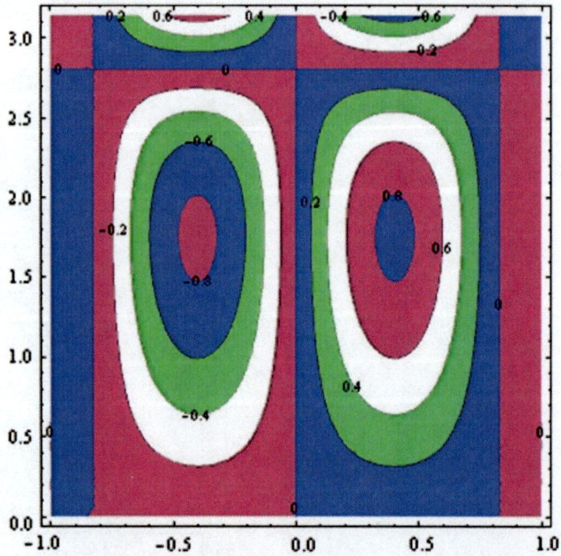

Fig. 4 Stream function for $\alpha = (64)$ $(\sigma = 8/\gamma = 1)$ and Ro = 20

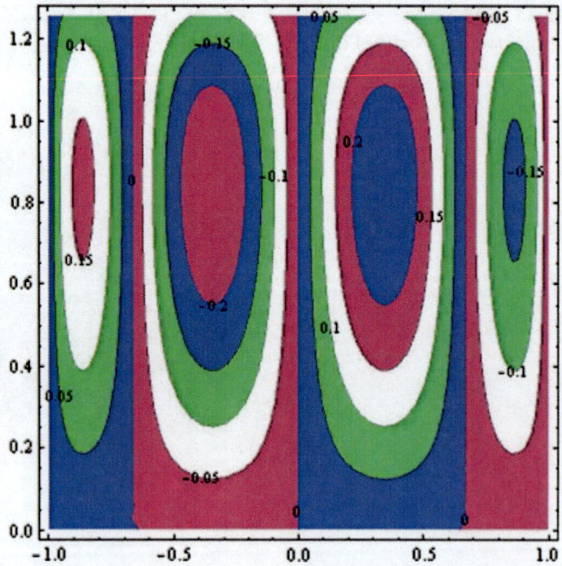

Fig. 5 Stream function for $\alpha = 6$ $(\sigma = 6/\gamma = 1)$ Ro = 20

Figures 7 and 8 represent the z-component velocity for gamma = 1 means the effective viscosity = viscosity of the fluid and for two different values of Darcy number. In many situations $\mu_e = \mu$, i.e., $\gamma^2 = 1$. These patterns show that the velocity will get minutely affected with a change in Darcy number.

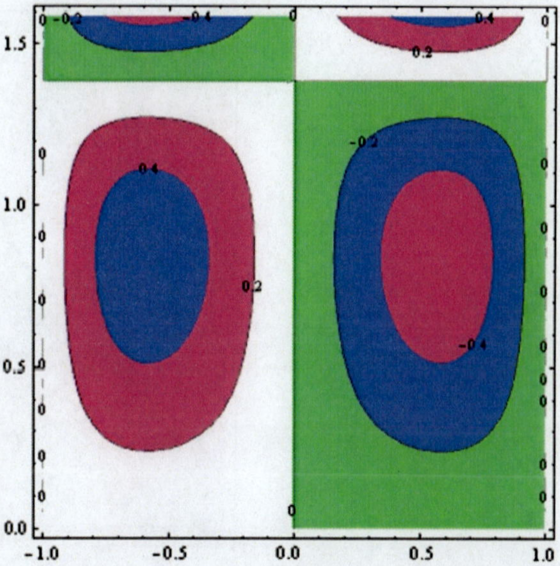

Fig. 6 Stream function for $\alpha = 1(\sigma = 6/\gamma = 6)$ Ro = 20

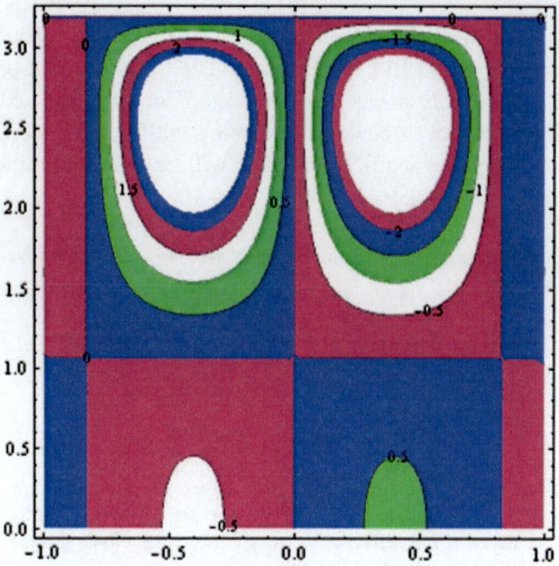

Fig. 7 Velocity component for $\alpha = 8$ ($\sigma = 8/\gamma = 1$), Ro = 20

4 Conclusion

The boundary layer theory is developed in the fluidized bed comprises fluid-saturated porous media. Development of acoustic streaming within the interstices will affect the streaming patterns, periodicity, and number of vortices as well as stagnation

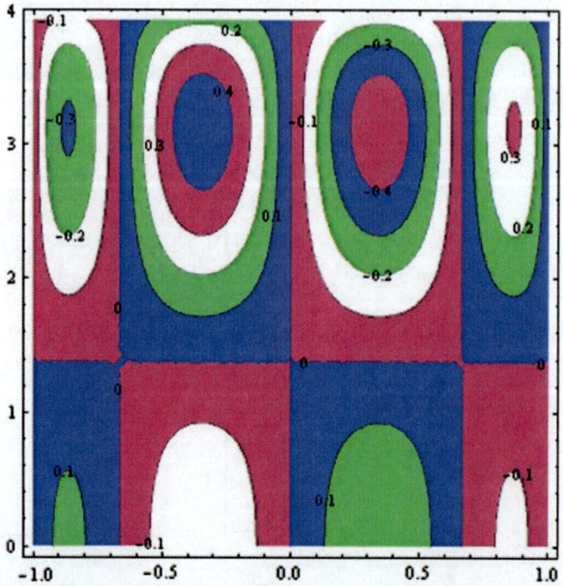

Fig. 8 Velocity component for $\alpha = 5$ ($\sigma = 5\ /\gamma = 1$) Ro $= 20$

point which in result will affect the mass transfer. Examples with the relevant engineering applications are gas fluidized bed reactors of small particles subjected to high-intensity acoustic wave (sonoprocessed fluidized beds). In both the cases, the development of acoustic streaming within interstices may have a significant effect on the process efficiency. This will be a very significant delivery in case of drag reduction models.

Acknowledgements Authors would like to thank the University authorities for providing the necessary help.

References

1. Srivastava Neetu (2017) Rayleigh type streaming effect on magneto-hydrodynamic characteristics of fluidized bed particles. Powder Technol 320:108–113
2. Srivastava N Effect of permeability on the Rayleigh-type acoustic streaming. Lecture Notes Mech Eng Ch.19:149–155
3. Landau LD, Lifschitz EM (1993) Fluid mechanics, Course of theoretical physics, Vol.6, 2nd edn. Oxford: Pergamon Press
4. Westervelt PJ (1953) The theory of steady rotational flow generated by a sound field. J Acoust Soc Am 25:60–67
5. Nyborg WL (1958) Acoustic streaming due to attenuated Plane waves. J Acoust Soc Am 25:68–75
6. Qi Q (1993) The effect of compressibility on Acoustic streaming near a boundary for a plane traveling wave. J Acoust So Am 94:1090–1098

7. Rayleigh L (1884) On the circulation of air observed in Kundt's tube, and on some acoustical problems. Philos Trans Series I 175:1–21
8. Lundgren TS (1972) Slow flow through stationary random beds and suspension of sphere. J. Fluid Mech 51:273–299
9. Brinkman HC (1947) Calculation of viscous forces exerted by a flow in fluid on a dense swarm of particles. Appl Sci Res Al:27–36
10. Givler RC, Altobelli SA, A determination of the effective viscosity for the Brinkman Forchheimer flow model. J Fluid Mech 258 (1994):355–370
11. Schlichting H (1955) Boundary layer theory. Pergamon
12. Valverde JM (2013) Acoustic streaming in the gas fluidized beds of small particles. Soft Matter 9:8792–8814
13. Vainshtein P (1995) Rayleigh streaming at large Reynolds number and its effect on shear flow. J Fluid Mech 285:249–269

Acoustic Noise and Vibration

Experimental Studies on Twelve Bladed Fan Vibration Considering Flexibility of Shaft

Shashank Shekhar Singh and Ahmad Ali Khan

Abstract A fan rotor system can generate a large amplitude blade-passing frequency (BPF) component if the gap between rotating fans and stationary diffusers is not equal. A series of experiments was conducted in this study using Spectra Quest's Machinery Fault Simulator (MFS) to observe the behavioral changes of the BPF and its harmonics before and after installing an obstruction nearby an axial fan. In order to incorporate the flexibility of the shaft a modification in the existing setup (MFS-LITE) has been designed and fabricated. The whole study is divided into two parts: in the first part readings were taken without considering the effect of flexibility of shaft and for performing this, the shaft of smaller length is taken; and in the second part readings were taken with a longer shaft in order to incorporate the effect of flexibility during experimentation. The result shows that the obstruction had an obvious and consistent effect on the 1BPF and 2BPF components for the axial fan rotor system.

Keywords Blade-passing frequency (BPF) · Machinery Fault Simulator (MFS) · Flexibility · Fan

1 Introduction

During the operation of fluid turbo-machinery, such as pumps, fans, or turbines, fluid-dynamic perturbations are produced and can lead to vibration and noise emission. Typical fluid-dynamic excitations in such machines are generally associated with pure-tone (rotation frequency, blade-passing frequency) [1], and broadband frequency components. The rotation frequency is due to small misalignments, unbalance, or manufacturing imperfections of the impeller. The broadband phenomena are usually caused by flow turbulence and cavitation. In some cases, excitation at the

S. S. Singh (✉)
Department of Mechanical Engineering, JECRC, Jaipur 302005, Rajasthan, India
e-mail: shashank.me@jecrc.ac.in

A. A. Khan
Department of Mechanical Engineering, ZHCET, AMU, Aligarh 202002, Uttar Pradesh, India

M. Singh and Y. Rafat (eds.), *Recent Developments in Acoustics*, Lecture Notes in Mechanical Engineering, https://doi.org/10.1007/978-981-15-5776-7_17

blade-passing frequency is dominant. This study will focus on the issues about the fan rotors and the blade-passing frequencies.

Each time a blade passes a point in space or an obstruction, an impulse force fluctuation is experienced by the fluid or solid-body at the point. If a fan with N blades is running at rpm, then the number of impulses experienced per second is rpm.

The frequency is called the blade-passing frequency (BPF). This frequency is inherent in pumps, fans and compressors, and normally does not present a problem. However, large amplitudes of the BPF component and its harmonics can be generated in a pump or fan rotor system if the gap between rotating fans and stationary diffusers is not equal all the way around. These different gaps will cause the air flow rate of the pump or fan to vary, which makes the static and dynamic pressures of the blades change as well [2]. As a response to the fluctuations of these pressure loads, the larger amplitude of the BPF component will be generated. The BPF component or its harmonics sometimes can coincide with a system natural frequency, and that causes high vibration [3]. High BPF vibration can be generated if impeller wear ring seizes on shaft, or if welds fastening diffuser vanes fail. In addition, high BPF vibration can be caused by abrupt bends in pipe or duct, obstructions which disturb flow, damper settings, or if pump or fan rotor is positioned eccentrically within housing.

1.1 Detecting Fan Problems Using Vibration Analysis

Fans are subjected to operational forces generated by their operating speed, static pressures, and system arrangement. These operational forces cause forced vibration and may originate from the rotating parts themselves. Unbalanced fan wheels and drive sheaves are examples of forces that cause force and couple unbalance. Axial load on fan bearings is generated primarily by the negative static pressure at the inlet of the fan wheel. The relationship between a fan operating conditions (flow quantity, pressure, speed, and arrangement) generates forces that cause vibration and problems that reduce the expected life of the bearings and fan components.

1.2 Summary of Common Fan Vibration Problems (Detected Using Vibration Analysis)

1. Unbalance of Overhung Fan Rotating Parts (Fan Wheel or Drive Sheave): Overhung rotor unbalance will cause high 1X RPM in both the axial and radial directions. Axial readings tend to be in-phase and steady, whereas radial phase readings might be unsteady. Overhung rotors have both force and couples unbalance, each of which will likely require correction. Thus, correction weights must always be placed in 2 planes to counteract both force and couple unbalance.

2. Unbalance of Center-Hung Fan Wheel: Force unbalance will be in-phase and steady. Amplitude due to unbalance will increase by the square of speed (a 3X speed increase = 9X higher vibration). 1X RPM is always present and normally dominates the spectrum. This can be corrected by the placement of only one balance correction weight in one plane at center of gravity (CG) of the rotor. Approximately 0° phase difference should exist between outboard and inboard horizontal, as well as between the outboard and inboard vertical direction.
3. Belt Drive Problems and Drive Misalignment: Sheave misalignment is a very common problem with belt-driven fans. When it is present, the vibration at fan speed is typically the highest vibration peak in spectra taken on the motor. It can be confirmed by measuring the phase in the axial direction at fan speed. Misalignment will cause phase differences at or approaching 180° between the motor and fan inboard bearings (closest to the pulleys) with phase comparisons made at fan speed.
4. Fan Stall Condition: This unstable condition can cause the ductwork to vibrate excessively and produce a "howling" sound. In many cases sub-synchronous frequencies will occur at approximately 66–75% of the fan RPM that will oscillate in amplitude and will produce sub-synchronous sidebands around 1X blade pass frequency (where blade pass frequency = blades × RPM).
5. Fan Surge Condition: Surge is a condition resulting from improper fan selection or operation and is recognized by the "whoosh" sound as air surges back through the fan. The frequency is sub-synchronous typically at 33–50% of fan running speed.
6. Fan Bent Shaft Condition: Bent shaft problems cause high axial vibration with axial phase differences tending toward 180° on the same rotor. Dominant vibration normally occurs at 1X rpm if bent near shaft center, but at 2X rpm if bent near the coupling. Dial indicators can be used to confirm a bent shaft.
7. Misalignment Condition: Fan shaft misalignment is a condition typically found in direct drive overhung fans. Misalignment is a condition where the centerlines of two connected shafts typically at the coupling do not coincide. Parallel misalignment is a condition where the shaft centerlines are parallel, but not aligned. The radial vibration spectra will show 2X rpm higher than 1X and a 3X higher than normal. Axial 1X and 2X will be low with 180° out of phase across the coupling in the radial and axial direction. Angular misalignment will show high axial 1X and some 2X and 3X RPM 180° out of phase axial across the coupling [2].
8. Belt Wear: Belt wear can be detected by the presence of frequencies not only at belt rpm, but also at multiples of belt rpm (harmonics). Typically, when belt wear is present, the peak at 2X belt rpm will exceed the peak at 1X belt rpm. If belt wear is significant, it will usually produce many harmonics of belt speed. Note that belt speed will always be lower than either the motor or fan speed (rpm).
9. Fan Bearing Problems: Peaks (with harmonics) at non-synchronous frequencies are a symptom of rolling element bearing wear. Short bearing life for fans is typically the result of poor bearing selection for the application, such as

excessive loads, poor lubrication, or high temperatures. If the model number and manufacturer of the bearings are known, then the specific fault frequencies for the outer race, inner race, rolling elements, and cage can be determined. These fault frequencies for such bearings are found within tables of most predictive maintenance software today.

10. Air Flow and Air Turbulence Problems: Fans will normally have a peak in the spectrum at "blade pass" or "vanes pass" frequency. The increased amplitude of the peak indicates the possibility of clearance problems at the discharge of the scroll or flow restrictions causing high discharge pressure. Severe flow turbulence will generate sub-synchronous peaks and a raised noise floor.

2 Experimental Studies

2.1 Spectral Analysis

Spectral analysis is the core of the diagnosis of rotating machinery. Spectral data are usually taken as velocity data, which should be measured on all bearings on the fan, in all three directions, horizontal, vertical, and axial. The purpose of the spectral analysis is to identify the frequencies causing the machine to vibrate. If all vibration amplitude levels are within acceptable limits, then the machine would be accepted as normal. However, if any of the spectral components has high amplitude, then spectral analysis is used to correlate the frequency of the high amplitude vibration to a machine frequency.

The measurements include time waveform measurement, phase measurement, and measurement of the operational deflection shape (ODS) [4].

2.2 Machinery Fault Simulator

Spectra Quest's Machinery Fault Simulator (MFS) is an innovative tool to study the signatures of common machinery faults without compromising production schedule or profits [5]. The bench-top system has a spacious modular design featuring versatility, operational simplicity, and robustness. Each component is machined to high tolerances so it can be operated without conflicting vibration. Then, various faults can be introduced either individually or jointly in a totally controlled environment, making the MFS a good tool for learning machinery diagnosis.

2.3 Vibra-Quest

An innovative data acquisition and analysis solution for noise and vibration—Vibra-Quest (VQ) is an integrated solutions package for rotating/reciprocating machinery fault diagnosis, structural dynamics analysis and design, and acoustical analysis. It provides tools to solve noise and vibration problems, from experimental design to solution strategy development.

2.4 Experimental Setup

The experimental apparatus with a fan kit for understanding the characteristics of the fan vibration phenomenon. The apparatus consists of a shaft with length of 1000 mm and diameter of 19.05 mm. The shaft is coupled with a flexible coupling to minimize the effect of the high-frequency vibration generated by the 0.5 HP motor. A three-phase AC induction motor is connected to a variable speed control unit for achieving variable speeds. The motor can be run in the speed range of 0–3600 rpm. The shaft is supported by two identical ball bearings fitted into the support housings.

3 Results and Discussion

All experimental data were collected and analyzed by Spectra Quest's Vibra-Quest system. The Vibra-Quest system contains portable multi-channel data acquisition front-end and powerful analysis functions. For each kind of fan test, the results before and after installing the obstruction have been compared using the comparison panel of the Vibra-Quest system. In the present study, a 12 bladed axial fan has been considered and the experiment has been carried out for two different speeds for both solid and flexible shaft conditions. The experimental results obtained are explained as under.

3.1 12-Blade Axial Fan

(a) 1000 rpm: 1X = 16.67 Hz, 1BPF = 200 Hz, 2BPF = 400 Hz

The spectra of 12-balde fan running at 1000 rpm without and with the obstruction, when we are not considering the flexibility of shaft, has been shown. The figure clearly indicates that for the test with obstruction both the 1BPF and 2BPF have larger amplitudes compared to the test without obstruction. Due to the obstruction, the flow rate of the air is restricted. As it is obvious the pressure increases consequently, as

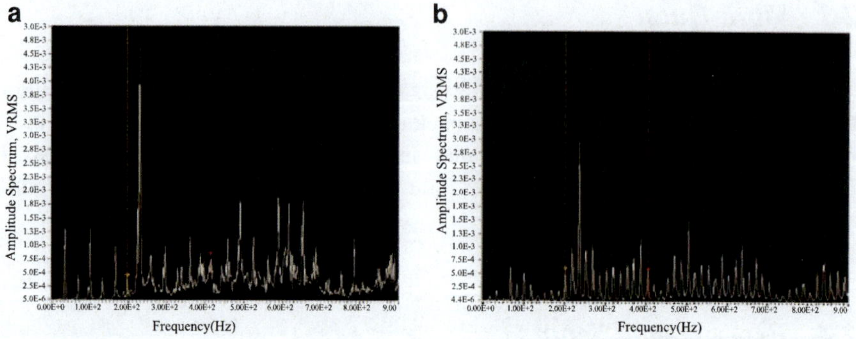

Fig. 1 **a** Spectra for 12-blade axial fan running at 1000 rpm before installing the obstruction, **b** Spectra for 12-blade axial fan running at 1000 rpm after installing the obstruction

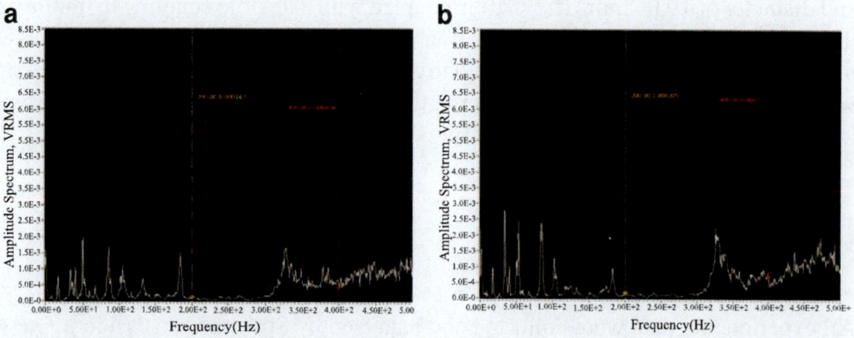

Fig. 2 **a** Spectra for 12-blade axial fan running at 1000 rpm before installing the obstruction, considering flexibility of shaft, **b** Spectra for 12-blade axial fan running at 1000 rpm after installing the obstruction, considering flexibility of shaft

the flow rate decreases. This explains why the amplitudes of the BPF component and its harmonics increase (Figs. 1 and 2).

(b) 2000 rpm: 1X = 33.33 Hz, 1BPF = 400.00 Hz, 2BPF = 800.00 Hz

Similar results at 2000 rpm test are observed, as shown in figures. Here also blade-passing frequencies increases with obstruction for non-flexibile shaft as well as for flexible shaft (Figs. 3 and 4).

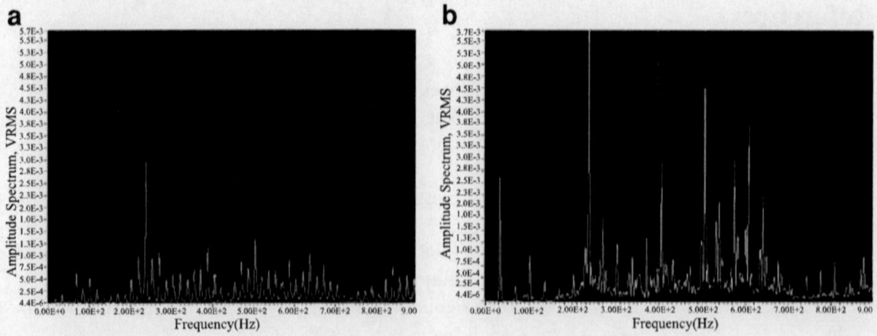

Fig. 3 **a** Spectra for 12-blade axial fan running at 2000 rpm before installing the obstruction, **b** Spectra for 12-blade axial fan running at 2000 rpm after installing the obstruction

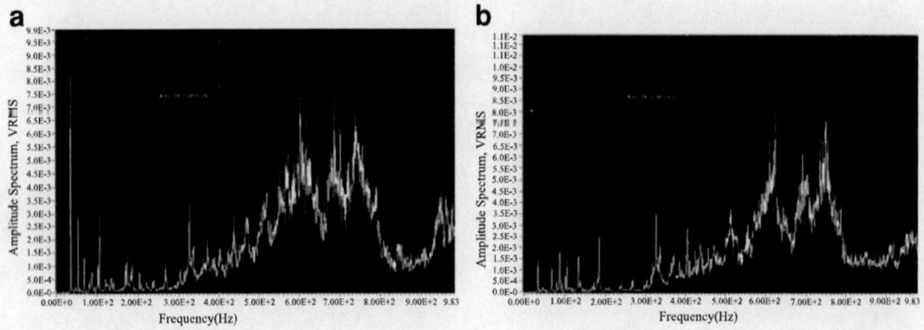

Fig. 4 **a** Spectra for 12-blade axial fan running at 2000 rpm before installing the obstruction, considering flexibility of shaft, **b** Spectra for 12-blade axial fan running at 2000 rpm after installing the obstruction, considering flexibility of shaft

4 Conclusion

Large amplitudes of the BPF and harmonics can be generated in pump and fan rotor systems if the gap between rotating fans and stationary diffusers is not equal all the way around. In this study, the BPF behavioral changes were simulated and analyzed using Spectra Quest's Machinery Fault Simulator and the Vibra-Quest software package. A series of experiments were conducted to observe the behavioral changes of 1BPF and 2BPF components before and after installing the obstruction near an axial fan. The results show that the obstruction has an obvious and consistent effect on the 1BPF, 2BPF for the 12-blade axial fan, as obstruction is not all side so they affect the flow of air for both rigid and flexible shaft.

References

1. Dileep N, Anusha K, Satyaprathik C, Kartheek B, Ravikumar K (2013) Condition monitoring of FD-FAN using vibration analysis. Int J Emerg Technol Adv Eng 3(1)
2. Asad AZ, Ramesh Kumar GR (2013) Condition monitoring of centrifugal blower using vibration analysis. Int J Multidiscip Sci Eng 4(5)
3. Leso N, Puttonen J, Porkka E (2011) The effect of foundation on fan vibration response. J Struct Mech 44(1):1–20
4. Zhou Shaoping, Zhang Jie, Yongsheng Su (2008) Vibration analysis and fault diagnosis of the fan unit and support structure. J Pressure Equipment Syst 6:45–48
5. Xie F, Li Z, Ganeriwala S (2007) Vibration signal analysis of fan rotors technology. Spectra-Quest Inc

Dynamic Analysis for Healthy and Defective Gears: A Review

Swati Gautam, Sidra Khanam, and N. Tandon

Abstract There are ample literature addressing the dynamic analysis for gears. In this paper, some important literature on dynamics of both healthy gears and gears with defects have been reviewed. Dynamic analysis of gears is significant from the condition monitoring point of view. Single degree of freedom dynamic model for spur gears has been detailed first. Dynamic study for worm gear with force analysis has been reported. Also, the effect of defects like crack and spall on dynamic behavior of gears studied by researchers has been included in this review. The study of defect model can be utilized in improving the efficiency of the system and preventing the failures in industries as well.

Keywords Vibration · Healthy and defective gears · Dynamic models

Nomenclature

c_1, c_2	Damping coefficients of first and second tooth pair in mesh, respectively
c_m	Mesh damping coefficient
$D(t)$	Instantaneous defect width along the contact line
$e(t)$	Static transmission error
e_1, e_2	Displacement excitations due to gear error
F_0	External torque load
F_x, F_y, F_z	Three orthogonal force components

S. Gautam (✉) · N. Tandon
ITMMEC, IIT Delhi, New Delhi 110016, India
e-mail: swatiindia4@gmail.com

N. Tandon
e-mail: ntandon@itmmec.iitd.ac.in

S. Khanam
ZHCET, Aligarh Muslim University, Aligarh 202002, Uttar Pradesh, India
e-mail: sidra.khanam10@gmail.com

M. Singh and Y. Rafat (eds.), *Recent Developments in Acoustics*, Lecture Notes in Mechanical Engineering, https://doi.org/10.1007/978-981-15-5776-7_18

$F(t)$	Defect morphology
$G(t)$	Time-periodic defect excitation generated by tooth shape deviations and errors
I_1, I_2	Mass moments of inertia of gear no. 1 and 2, respectively
J_w, J_g, J_4	Moment of inertias of the worm, worm gear, and the flywheel, respectively
K_w, K_g	Elastic constants of the worm shaft and the worm gear shaft, respectively
k	Total mesh stiffness of the gear pair
k_0	The mesh stiffness per unit contact length
k_1, k_2	Stiffness of first and second tooth pair in mesh, respectively
k_a	Axial compressive stiffness
k_m	Mean mesh stiffness of one pair contacted with healthy teeth
k_t	Stiffness of a single tooth
k_f, k_b, k_s	Fillet-foundation, bending, shear stiffness, respectively
k^i	Mesh stiffness of the ith tooth pair
k_{tooth}	Total mesh stiffness due to teeth deformation of N tooth pairs
k_{s_crack}, k_{b_crack}	Shear and bending stiffness of cracked tooth, respectively
m_e	Equivalent mass
T_1, T_2	Input and output torques, respectively
$P(\eta)$	Defect depth along the contact line
R_1, R_2	Base radii of gear
r_w, r_g	Pitch radii of the worm's helical tooth and the worm gear, respectively
W_0	Static load
W_f, W_n	Friction force, normal force
ϕ_1, ϕ_2	Angular displacements of gear number 1 and 2, respectively
ϕ', ϕ''	First- and second-order derivatives of ϕ, respectively
$\theta_1, \theta_2, \theta_3, \theta_4$	The rotation angles of worm shaft, worm's body, worm- gear body, and flywheel, respectively
ϕ_n	The pressure angle
λ	The worm's lead angle
λ_1, λ_2	Correction coefficients of the fillet-foundation stiffness for driving and driven gears, respectively

1 Introduction

Gears are toothed wheels, which transmit motion and power from one shaft to another due to successive engagement of teeth. Gears can be broadly classified as spur, helical, worm, and bevel. For obtaining a constant angular velocity ratio for all positions of the gear, the common normal at the point of contact between a pair of teeth must always pass through the pitch point. This condition is known as the law of gearing. Gears have wide utilities ranging from daily-life to industrial applications.

Mathematical models are utilized to study the dynamics of gear. Governing equations are formulated to represent the physical phenomenon of meshing between gear teeth. Effect of defects like tooth breakage, spall, crack, etc. can be suitably incorporated by modifying the governing equations. By solving these governing equations using numerical methods, simulated results can be obtained to understand the mechanism. Moreover, prediction of response for different defect sizes is also possible. Hence, dynamic modeling is important from condition monitoring perspective, and therefore, the aim of this paper is to review various gear dynamic models.

1.1 *Modeling for Healthy Gears*

Yang and Sun [1] presented a rotary model for the dynamics of spur gear pairs. The authors assumed a simple dynamic model for a gear pair. A mathematical model along with consideration of important factors such as backlash, damping during impact, and elasticity may yield good result.

Ozguven and Houser [2] proposed one such model for a spur gear pair which included effects like variable mesh stiffness and mesh damping, profile modifications, backlash, and gear errors. This study suggested that in high precision spur gears the effect of pitch errors on the dynamic forces is of less importance with respect to the excitation caused by mesh stiffness variation. To analyze the dynamics for spur gears, Ozguven [3] proposed a 6 DOF nonlinear model considering mesh stiffness with time variation.

A three-degree freedom dynamic model was developed by Kahraman and Singh [4]. Their model took into account the nonlinearities caused due to radial clearances in the radial rolling element bearings and backlash between spur gears. Chung and Shaw [5] prepared a mathematical model, simulated it numerically and performed an experiment, and compared it with the numerically obtained results. They found good agreement in experiments and numerical results. They also studied the moment of inertia of flywheel, rigidity of shaft, friction, and rigidity of gear tooth.

Dresig et al. [6] presented a vibratory model for the worm gear that covered the essential geometry, mass, stiffness, and damping parameters as well as friction. They observed under which conditions self-excited chatter vibrations arise in worm gears with a flexible drive shaft and an elastic gear contact—even for a constant friction coefficient. Sun et al. [7] developed a new method to calculate the bending deformation of helical tooth. First, they divided the gear section into several copies along tooth width and they calculated each section's moment of inertia. Kar and Mohanty [8] studied various time-varying parameters which were responsible for generating vibration and noise in helical gear. The variation in contact line was considered as the most important factor in their work.

1.2 Modeling of Gears with Defects

Several researchers have worked in the domain of gears with defects in them. Some of those works are discussed in this paper. Badaoui et al. [9] presented an extended gear dynamic model and some advanced signal processing techniques. They developed a numerical procedure to simulate the dynamics of gears with local tooth damages such as pitting or spalling. Li et al. [10] studied tooth root crack estimation using embedded modeling. An embedded model integrating a physical-based model of the gear box was established in this work. The authors found a parametric form of truncated Fourier series for meshing stiffness.

Choy et al. [11] studied the effects of surface pitting, wear, and partial tooth fracture on the vibrations of a gear transmission system. They developed an analytical model and found out the effects of wear and surface pitting by simulation of magnitude and phase changes in the gear mesh stiffness. Choy et al. [12] employed vibration signatures for damage detection in a gear transmission system. They simulated the dynamics of a gear transmission system numerically considering the effect of single and multiple tooth defects.

In a planetary gear, what should be the effect of manufacturing errors on its dynamic behavior has been studied by Chaari et al. [13]. A spectral iterative methodology has been used to find out the dynamic response. The excitation was induced by gear mesh stiffness. Parey and Tandon [14] developed a 6 DOF dynamic model of a gear system with the incorporation of a local defect in tooth. Their model consisted of two spur gears, two shafts, and two inertias for representation of load and prime mover and bearings. They considered the influences of time-varying mesh stiffness and damping, excitation because of gear errors, backlash, and profile modifications in their model. Parey and Tandon [15] prepared an impact velocity model. Their model could establish a relation between measurable vibration signal and the defect size on the gear tooth flank. They verified their analytically developed model with experiments. By use of empirical mode decomposition process, authors decomposed experimental vibration signals. In a single stage gearbox having spur gears inside, the influence of tooth crack on the vibration signal was studied by Wu et al. [16]. A lumped parameter model was used for simulation of the vibration signals for meshing gear pair.

Chaari et al. [17] developed an analytical method for quantification of the reduction in gear mesh stiffness because of two defects in gear tooth, viz. spall and breakage. Chaari et al. [18] formulated analytically the mesh stiffness of gear considering the effect of bending, contact deflections, and fillet-foundation. They modeled gear mesh stiffness considering the influence of crack in tooth. They concluded the influence of this fault as a decline in the gear mesh stiffness when the defective tooth is under meshing condition. Ma et al. [19] presented an improved analytical model (IAM) for the time-variable meshing stiffness (TVMS) evaluation for spur gear with crack defect. Details of their work have been presented in the subsequent section of this paper. What should be the load sharing between the worm threads and gear teeth at a particular instant of meshing, for this a methodology has been developed by

Simon [20]. By doing this, author evaluated the distribution of load along the teeth and transmission errors in various types of worm gears. Ma et al. [21] focused on the nonlinear dynamics and vibration phenomenon of a pair of spur gears with localized fault as spall. A dynamic model for a pair of gears with incorporation of spall was established. Also, authors studied time-varying mesh stiffness for investigation of the influence of spall on dynamics of gear.

Calculation methods for friction excitations and contact stiffness in helical gears were proposed by Jiang et al. [22]. Their calculation took into consideration the effect of variation of length of contact line with time. Taking into consideration the change in the position of meshing and the loss of length of contact line induced by tooth spalling defect, they obtained the friction force which is variable with respect to time, by subtracting the length of spalling defect at the mesh position from the length of the normal contact line, and they obtained the contact stiffness by Hertz contact algorithm. Litvin et al. [23] proposed an improved version for a worm gear system. The proposed gear drive has been shown less transmission errors, reduced sensitivity to alignment errors and improved bearing contact.

Vecer et al. [24] described important condition indicators and examined the capability of some important condition indicators for description of the extent of wear by use of vibration signals. Combination and data analysis of three measuring techniques of rotating machine condition monitoring were studied by Loutas et al. [25]. Recently, work on neural network based fault diagnosis of worm gears has also been reported [26].

Section 2 discusses analytical formulation for spur and worm gear sets in healthy condition, i.e., without defects.

2 Models for Healthy Gears

2.1 *Spur Gear*

A dynamic model proposed by Ozguven and Houser [2] comprises two disks, one spring and one damper. The disks replicate the inertias of both the gears. The spring and the damper represent the meshing stiffness and damping, respectively. The nonlinearity of mesh stiffness, damping, and gear error excitations have been considered.

The equations of motion of the system shown in Fig. 1, as given by Ozguven and Houser [2] can be written as

$$\begin{aligned} &I_1\phi_1' + R_1c_m(R_1\phi_1' - R_2\phi_2') - R_1c_1e_1' - R_1c_2e_2' \\ &+R_1k_m(R_1\phi_1 - R_2\phi_2) - R_1k_1e_1 - R_1k_2e_1 = T_1 \end{aligned} \quad (1)$$

$$I_2\phi_2' + R_2c_m(R_2\phi_2' - R_1\phi_1') - R_2c_1e_1' - R_2c_2e_2'$$

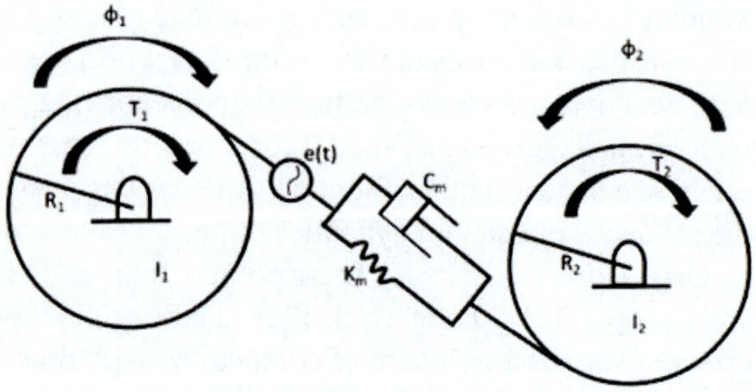

Fig. 1 Two DoF model for spur gear pair [2]

$$+R_2k_m(-R_1\phi_1 + R_2\phi_2) - R_2k_1e_1 - R_2k_2e_2 = -T_2 \tag{2}$$

where I_1 and I_2: mass moments of inertia of gear number 1 and 2, ϕ_1 and ϕ_2: angular displacements of gear number 1 and 2, $\phi'$and ϕ'': 1st and 2nd order derivatives of ϕ, respectively, c_m: mesh damping coefficient, c_1 and c_2: damping coefficients of first and second tooth pair in mesh, k_m: gear meshing stiffness, k_1 and k_2: stiffness of first and second tooth pair in mesh, T_1 and T_2: input and output torque, respectively, e_1 and e_2: displacement excitations due to gear errors.

The variable motion x may be given as

$$x = R_1\phi_1 - R_2\phi_2 \tag{3}$$

which reduces Eqs. (1) and (2) to

$$m_e x'' + c_m x' - c_1 e_1' - c_2 e_2' + k_m x - k_1 e_1 - k_2 e_2 = W_0 \tag{4}$$

where the equivalent mass m_e and the static load W_0 are given as

$$m_e = \frac{I_1 I_2}{(I_1 R_2^2 + I_2 R_1^2)} \quad W_0 = \frac{T_1}{R_1} = \frac{T_2}{R_2} \tag{5}$$

2.2 Worm Gear

Chung and Shaw [5] did the dynamic study for worm and worm wheel. Figure 2 shows the schematic for worm-gear set geometry comprising of worm, worm wheel, and flywheel used in the dynamic analysis.

Subsequent sections detail about force analysis in all the three directions and equations of motion of the gear system shown in Fig. 2.

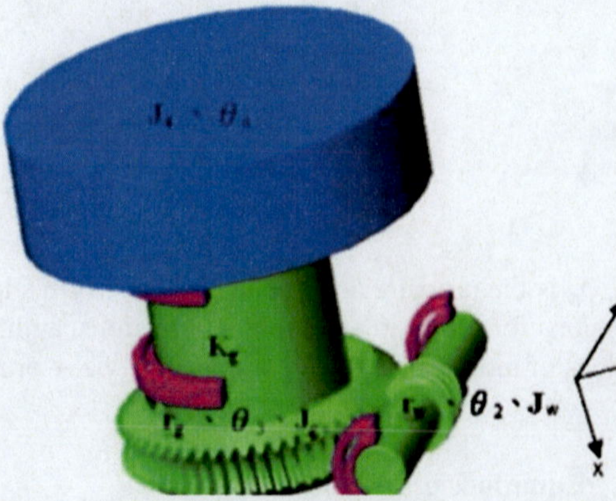

Fig. 2 Worm and worm gear geometry [5]

2.2.1 Force Analysis

First of all, a force analysis was done for the worm and worm wheel as below [5]:

Figure 3 represents the free body diagram for worm and worm gear, respectively, also the normal force W_n and frictional force W_f between the tooth of the worm ad that of the wheel have been decomposed and projected into three orthogonal force components F_x, F_y, and F_z and may be mathematically expressed as below:

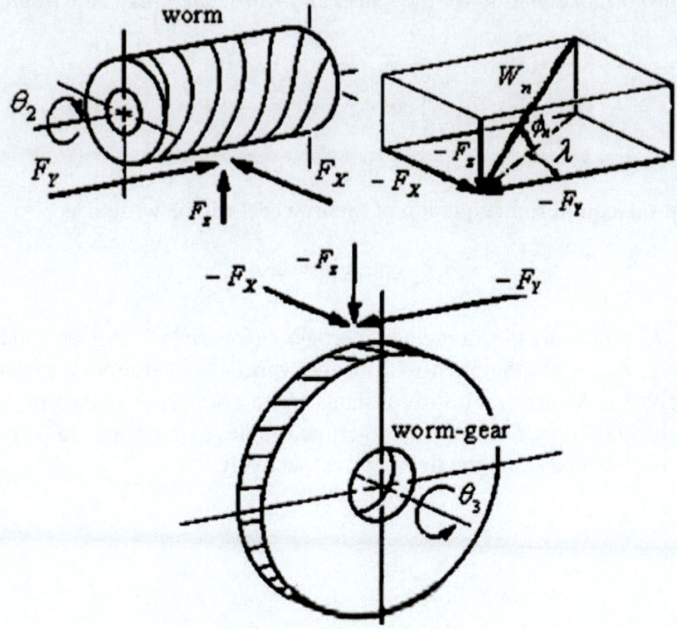

Fig. 3 Decomposition of contact forces in worm gear set [5]

$$F_x = W_n \cos\phi_n \sin\lambda + W_f \cos\lambda \tag{6}$$

$$F_y = W_n \cos\phi_n \cos\lambda - W_f \sin\lambda \tag{7}$$

$$F_z = W_n \sin\phi_n \tag{8}$$

where ϕ_n is the pressure angle, and λ is the worm's lead angle. After obtaining the expressions for forces in all the three directions, authors [5] proceeded to formulate equations of motions for three components, viz. worm, worm wheel, and flywheel.

2.2.2 Equations of Motion

The equilibrium equations of motion are formulated for three free bodies. These three free bodies are (1) the worm and worm shaft, (2) worm gear and shaft, and (3) flywheel. The equilibrium equation for the worm and worm shaft is written as

$$\begin{aligned} J_w\ddot{\theta}_2 &= -F_X(\theta_2, \theta_3, \dot{\theta}_2, \dot{\theta}_3)r_w - K_w(\theta_2 - \theta_1) \\ &= -(W_n(\theta_2, \theta_3, \dot{\theta}_2, \dot{\theta}_3)\cos\phi_n \sin\lambda)r_w - (W_f(\theta_2, \theta_3, \dot{\theta}_2, \dot{\theta}_3)\cos\lambda)r_w \\ &\quad - K_w(\theta_2 - \theta_1) \end{aligned} \tag{9}$$

The equilibrium equation for the wheel and wheel shaft may be written as

$$\begin{aligned} J_g\ddot{\theta}_3 &= -F_Y(\theta_2, \theta_3, \dot{\theta}_2, \dot{\theta}_3)r_g - K_g(\theta_3 - \theta_4) \\ &= -(W_n(\theta_2, \theta_3, \dot{\theta}_2, \dot{\theta}_3)\cos\phi_n \cos\lambda)r_g - (W_f(\theta_2, \theta_3, \dot{\theta}_2, \dot{\theta}_3)\sin\lambda)r_g \\ &\quad - K_g(\theta_3 - \theta_4) \end{aligned} \tag{10}$$

Finally, the equilibrium equation of the flywheel can be written as

$$J_4\ddot{\theta}_4 = -K_g(\theta_4 - \theta_3). \tag{11}$$

where J_w, J_g, and J_4 are the moment of inertia of the worm, worm gear, and flywheel, respectively; $r_{w,}$ $r_{g,}$ are the pitch radii of the worm's helical tooth and worm gear, respectively; K_w, K_g are the elastic constants of the worm shaft and worm-gear shaft, respectively; and $\theta_1, \theta_2, \theta_3, \theta_4$ are the rotational angles of the worm shaft, the worm's body, worm-gear body, and the flywheel, respectively.

3 Models for Gears with Defects

The following sections describe the analytical formulation for gears with two defects, respectively, crack and spall.

3.1 *Crack Model*

Ma et al. [19] proposed an improved analytical method (IAM) for the evaluation of the time-varying mesh stiffness (TVMS) for spur gear cracks. Authors considered fillet-foundation effect in the evaluation of TVMS. Finally, modeling of cracked-gear rotor system was done using finite element methods and equations of motion were developed as described in subsequent sections.

3.1.1 TVMS Calculation of Cracked-Gear Pairs Considering the Effect of Fillet-Foundation Deformation

Figure 4 illustrates mesh stiffness calculations using two methods i) traditional analytical method and improved analytical method.

For traditional analytical model (TAM), the overall mesh stiffness of the gear pair with double-tooth engagement is modeled as springs connected in parallel. The stiffness is computed by summing up the tooth pair stiffnesses as

$$k^i = 1 \Big/ \left(\frac{1}{k_{hi}} + \frac{1}{k_{t1}} + \frac{1}{k_{f1}} + \frac{1}{k_{t2}} + \frac{1}{k_{f2}} \right), \quad k = \sum_{i=1}^{N} k^i \tag{12}$$

where k^i is the mesh stiffness of the ith tooth pair; k_{hi} is the local contact stiffness; k_t is the stiffness of single tooth; k is the total mesh stiffness; N is the number of meshing tooth pairs.

Single tooth stiffness k_t consists of the bending stiffness k_b, shear stiffness $k_{s,}$ and axial compressive stiffness k_a which may be expressed as

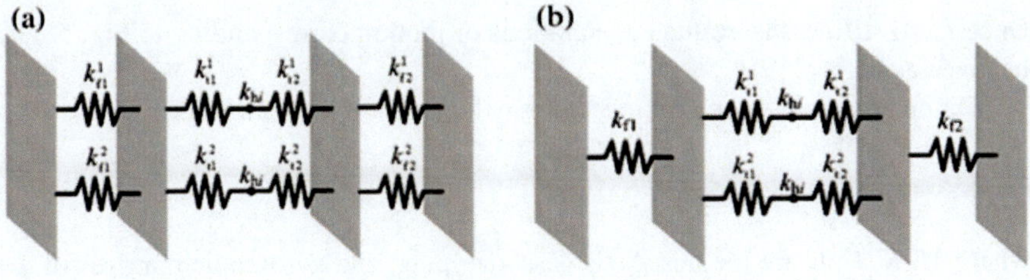

Fig. 4 **a** Traditional analytical model and **b** Improved analytical model [19]

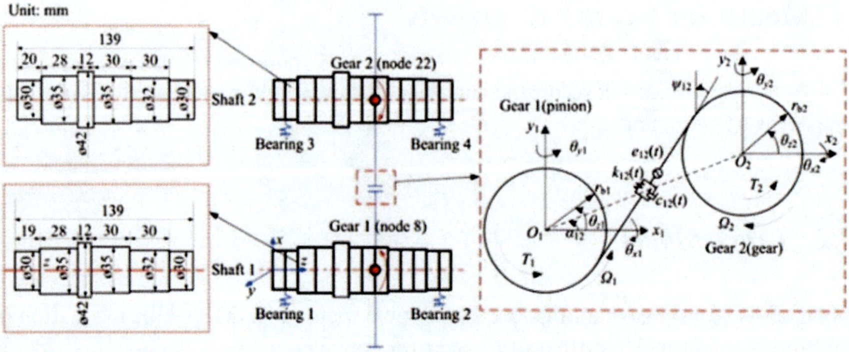

Fig. 5 Cracked-gear rotor system model with dimensions [19]

$$k_t = 1 \bigg/ \left(\frac{1}{k_b} + \frac{1}{k_s} + \frac{1}{k_a} \right) \tag{13}$$

For a tooth with crack, only bending and shear stiffnesses change, then

$$k_t = 1 \bigg/ \left(\frac{1}{k_{b\_crack}} + \frac{1}{k_{s\_crack}} + \frac{1}{k_a} \right) \tag{14}$$

where k_{b_crack} and k_{s_crack} are the bending and shear stiffness of cracked tooth.

For the IAM, the overall mesh stiffness k may be given as

$$k = 1 \bigg/ \left(\frac{1}{\lambda_1 k_{f1}} + \frac{1}{k_{tooth}} + \frac{1}{\lambda_2 k_{f2}} \right) \tag{15}$$

where λ_1, λ_2 are the correction coefficients of the fillet-foundation stiffness for driving gear and driven gear, respectively; and k_{tooth} is the mesh stiffness due to teeth deformation of N tooth pairs in contact. After evaluating TVMS, the equations of motion for the entire system shown in Fig. 5 can be written.

3.1.2 Equations of Motion

Once mesh stiffness is evaluated, equations of motion corresponding to Fig. 5 may be given as below [19].

Motion equation of the gear system may be given in the matrix form as

$$\boldsymbol{M}\ddot{\boldsymbol{u}} + (\boldsymbol{C} + \boldsymbol{G})\dot{\boldsymbol{u}} + \boldsymbol{K}\boldsymbol{u} = \boldsymbol{F}_{\boldsymbol{u}} \tag{16}$$

where $\boldsymbol{M}$, $\boldsymbol{K}$, $\boldsymbol{C}$, $\boldsymbol{G}$ are the mass, stiffness, damping, and gyroscopic matrices of the global system; $\boldsymbol{u}$ and $\boldsymbol{F}_u$ are the displacement and external force vectors of the global system.

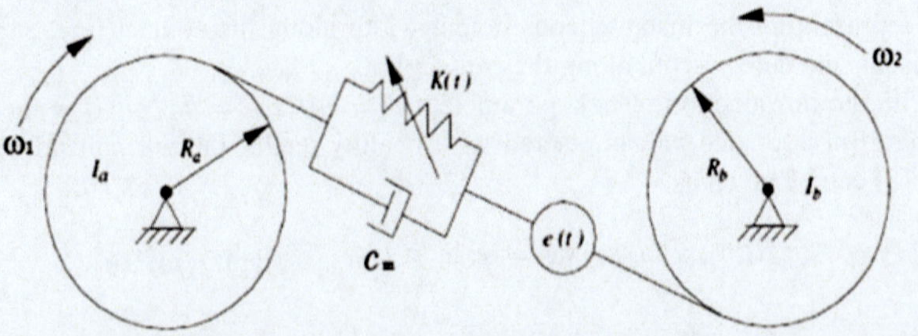

Fig. 6 Model of spur gear pair [21]

3.2 *Spall Model*

Ma et al. [21] studied spur gear pair dynamics having spalling defect. They evaluated stiffness which is variable with time and also analyzed the effect of spalling defect on mesh stiffness.

3.2.1 Model for the Gear System

Figure 6 is the schematic diagram for the model for the gear pair utilized in this study. Two disks represent the mass moments of inertia of gears and spring as mesh stiffness of value $K(t)$.

First of all, equations of motion for healthy gear were formulated and then spalling defect was incorporated in that equation [21].

3.2.2 Equations of Motion for Spalling Defect

Equation of motion for healthy gear can be written as

$$m_e x'' + c_m x' + K(t)x = F_0 + m_e e''(t) \tag{17}$$

where $m_e = I_a I_b / I_a R_a^2 + I_b R_b^2$, c_m represents mesh damping coefficient, $K(t)$ is mesh stiffness, $F_0 = T/R_a$ represents external torque load.

Equation of motion with incorporation of spalling defect is written as

$$m_e x'' + c_m x' + K(t)x = F_0 + m_e e''(t) + G(t) \tag{18}$$

where $G(t)$ represents time-periodic defect excitation due to deviations in tooth shape and errors. And mathematically, $G(t) = k_0 F(t) D(t) P(\eta)$ with k_0 representing the mesh stiffness per unit contact length, $F(t)$ representing defect morphology,

$D(t)$ representing the instantaneous defect width along the contact line, and $P(\eta)$ represents the defect depth along the contact line.

With the introduction of new parameters $\tau = \omega_0 t$, $\omega_0^2 = k_m/m_e$ (k_m: the mean mesh stiffness of one pair contacted with healthy teeth), the normalized form of Eq. (18) can be given as

$$X'' + 2\mu X' + k(\tau)X = f_0 + F_1(\tau) + \overline{k_0} F(\tau)D(\tau)P(\eta) \tag{19}$$

where $\omega = \omega_e/\omega_0$, $\mu = c_m/2m\omega_0, \overline{k_0} = k_0/k_m$, $f_0 = F_0/m_e\omega_0^2$, $F_1(\tau) = f_1\omega^2\cos(\omega\tau)$

The solution of Eq. (19) gives the comprehensive idea about the effect of spalling defect on the gear system.

4 Conclusion

Dynamic analysis of gears can be utilized to analyze the vibration phenomenon and its effect on gear transmission system. Various authors have analyzed the dynamic models of different types of gears like spur gears, worm gears, etc. Limited studies have been focused on gears with defects. Study in case of spur gear whether it is healthy or defective is limited in two dimensions. On the contrary, worm gear dynamic study is performed in three dimensions. The study of defect model can be utilized not only in improving the efficiency of the system but also in preventing the catastrophic failures in industries.

This study indicates the dearth of literature in case of worm gear dynamics as compared to spur and helical gears dynamic studies. There is a lot of future possibility for work considering defects such as wear, broken tooth, pitting, scoring, etc. in case of helical, worm, and bevel gears.

References

1. Yang DCH, Sun ZS (1985) A rotary model for spur gear dynamics. ASME J Mech Transm Autom Des 107:529–535
2. Özgüven HN, Houser DR (1988) Dynamic analysis of high speed gears by using loaded static transmission error. J Sound Vib 125(1):71–83
3. Özgüven HN (1991) A non-linear mathematical model for dynamic analysis of spur gears including shaft and bearing dynamics. J Sound Vib 145(2):239–260
4. Kahraman A, Singh R (1991) Non-linear dynamics of a geared rotor-bearing system with multiple clearances. J Sound Vib 144(3):469–506
5. Chung MY, Shaw D (2007) Parametric study of dynamics of worm and worm-gear set under suddenly applied rotating angle. J Sound Vib 304(1):246–262
6. Dresig H, Schreiber U, Rodionow P (2007) Stability analysis and simulation of the vibration behavior of worm gears in drive systems. Wuxi, China: Proceedings of ICMEM, pp 24–28

7. Sun W, Chen T, Zhang X (2011) A new method to calculate bending deformation of involute helical gear. UPB, Sci Bull, Series D China 73(3):17–30
8. Kar C, Mohanty AR (2008) Determination of time-varying contact length, friction force, torque and forces at the bearings in a helical gear system. J Sound Vib 309(1):307–319
9. El Badaoui M, Cahouet V, Guillet F, Daniere J, Velex P (2001) Modeling and detection of localized tooth defects in geared systems. J Mech Des, Trans ASME 123:422–430
10. Li CJ, Lee H, Choi SH (2002) Estimating size of gear tooth root crack using embedded modeling. Mech Syst Signal Process 16(5):841–852
11. Choy FK, Polyshchuk V, Zakrajsek JJ, Handschuh RF, Townsend DP (1994) Analysis of the effects of surface pitting and wear on the vibrations of a gear transmission system (No. NASA-E-9014). National Aeronautics and Space Administration Cleveland Oh Lewis Research Centre
12. Choy FK, Mugler DH, Zhou J (2003) Damage identification of a gear transmission using vibration signatures. Trans ASME, J Mech Des 125(2):394–403
13. Chaari F, Fakhfakh T, Hbaieb R, Louati J, Haddar M (2006) Influence of manufacturing errors on the dynamic behavior of planetary gears. Int J Adv Manuf Technol 27(7–8):738–746
14. Parey A, El Badaoui M, Guillet F, Tandon N (2006) Dynamic modelling of spur gear pair and application of empirical mode decomposition-based statistical analysis for early detection of localized tooth defect. J Sound Vib 294(3):547–561
15. Parey A, Tandon N (2007) Impact velocity modelling and signal processing of spur gear vibration for the estimation of defect size. Mech Syst Signal Process 21(1):234–243
16. Wu S, Zuo M, Parey A (2008) Simulation of spur gear dynamics and estimation of fault growth. J Sound Vib 317(3):608–624
17. Chaari F, Baccar W, Abbes MS, Haddar M (2008) Effect of spalling or tooth breakage on gearmesh stiffness and dynamic response of a one-stage spur gear transmission. Eur J Mech-A/Solid 27(4):691–705
18. Chaari F, Fakhfakh T, Haddar M (2009) Analytical modelling of spur gear tooth crack and influence on gear mesh stiffness. Eur J Mech-A/Solid 28(3):461–468
19. Ma H, Pang X, Feng R, Zeng J, Wen B (2015) Improved time-varying mesh stiffness model of cracked spur gears. Eng Fail Anal 55:271–287
20. Simon V (2003) Load distribution in cylindrical worm gears. Trans-ASME J Mech Des 125(2):356–364
21. Ma R, Chen Y, Cao Q (2012) Research on dynamics and fault mechanism of spur gear pair with spalling defect. J Sound Vib 331(9):2097–2109
22. Jiang H, Shao Y, Mechefske CK (2014) Dynamic characteristics of helical gears under sliding friction with spalling defect. Eng Fail Anal 39:92–107
23. Litvin FL, Gonzalez-Perez I, Yukishima K, Fuentes A, Hayasaka K (2007) Design, simulation of meshing, and contact stresses for an improved worm gear drive. Mech Mach Theory 42(8):940–959
24. Večeř P, Kreidl M, Šmíd R (2005) Condition indicators for gearbox condition monitoring systems. Acta Polytech 45(6)
25. Loutas TH, Sotiriades G, Kalaitzoglou I, Kostopoulos V (2009) Condition monitoring of a single-stage gearbox with artificially induced gear cracks utilizing on-line vibration and acoustic emission measurements. Appl Acoust 70(9):1148–1159
26. Waqar T, Demetgul M (2016) Thermal analysis MLP neural network-based fault diagnosis on worm gears. Measurement 86:56–66

Non-Cavitating Noise Control of a Marine Propeller by Optimizing Number and Pitch of Blades

V. Ramakrishna, P. Bangaru Babu, and Ch. Suryanarayana

Abstract The Propeller is a vital component which is delicate for the safe operation of a ship at sea. The propeller noise is very much significant to warship designers and military strategists for many years. By using large eddy simulation (LES) it is possible to simulate the sound radiation of turbulent flows effectively, providing access to resolve turbulent scales at higher accuracy. The main aim of this paper is to study and identify a propeller which is producing less noise. The present paper deals with the prediction of the unsteady non-cavitating marine propeller noise of 5, 6, and 7 blades with +10° pitch, +5° pitch, existing pitch, and −5° pitch, −10° pitch at rotational speed 840 rpm and vehicle speeds of 7.62 m/s, using eddy viscosity model of large eddy simulation (LES) available in computational fluid dynamics fluent software and using Ffowcs Williams–Hawkings (FW-H) formulation. Here solver is chosen as pressure-based, unsteady formulation of second-order implicit. Finite volume method is used to predict the noise in time-domain acoustic analogy. Sound pressure levels are predicted at different receiver positions. The receiver position is 1 m distance in radial direction to the propeller. From this numerical study on these propellers, it is found that the propeller of 6 blades with +5° pitch generates least noise. Finalized the propeller which is producing low noise as well as which generates required Thrust and Torque by changing the diameter of the propeller. Finally, obtain the noise reduction of 9.4 dB with the new redesigned propeller when compared with the baseline propeller for the same thrust and torque.

Keywords Non-cavitating · Marine propeller noise · Computational fluid dynamics (CFD) · Large eddy simulation (LES) · Ffowcs Willams–Hawkings formulation (FW-H)

V. Ramakrishna (✉) · Ch. Suryanarayana
Scientist 'E', Naval Science and Technological Laboratory, Visakhapatnam, India
e-mail: ramakrishna.v@nstl.drdo.in

P. Bangaru Babu
National Institute of Technology, Warangal 506004, Telangana, India

M. Singh and Y. Rafat (eds.), *Recent Developments in Acoustics*, Lecture Notes in Mechanical Engineering, https://doi.org/10.1007/978-981-15-5776-7_19

1 Introduction

Control of propeller noise is an important task in designing the propeller for warship designers and stealth designers. The radiated noise from the propeller depends on Propeller Geometry, Propeller Wake Inflow, and Propeller Isolation. The most effective way of reducing the radiated noise is by reducing pressure oscillations on the propeller. Different geometrical parameters of the propeller that will affect these pressure oscillations are changing the pitch angle, changing the number of blades, changing the diameter of the blade, increase in blade area, increase in skew angle, trailing edge geometrical modifications, and propeller blade finishing fineness. These parameters influence the noise levels produced by the propeller. A major noise source of an underwater vehicle is the propeller. The unsteady force on the propeller generates the discrete tonal noise which is the major source of the propeller noise. Therefore in this paper, it is aimed essentially to reduce the noise caused by the unsteady force on the propeller for the survivability of the underwater vehicle for military applications. The present problem focused on the reduction of propeller noise by varying the number of blades and pitch angle of the propeller.

The first time acoustic analogy was proposed by Lighthill [1] in the year 1950. Then it is extended by Ffowcs Williams and Hawkings (FW-H) [2] for the general surfaces in the arbitrary motion. Now almost all authors are using the FW-H equation to predict noise of the complex motion of the solid bodies. Chang [3] developed the methodology to predict flow pattern and performance of marine propeller using k-ε turbulence model. Yang et al. [4] proposed another method to predict the propeller noise by using large eddy simulation. Bertetta et al. [5] proposed another method for numerical analysis of propeller by using a potential panel method and RANSE solver. Zhi-feng and Shi-liang [6] used viscous multiphase flow theories and hybrid grid based on Navier–Stokes equation to study the cavitation performance of a propeller. Song et al. [7] used periodic structure theory for reduction of noise and vibration from an underwater vehicle due to propeller forces. The author analyzed and compared the dynamic properties of various isolators for underwater vehicles. Yu-cun and Huai-xin [8] predicted the noise generated by the underwater propeller using detached eddy simulation (DES) and Ffowcs Williams–Hawkings (FW-H) equation. Muscari et al. [9] predicted the complex flow of propeller using RANS and DES for numerical simulations. Yu-cun and Huai-xin [10] captured the unsteady forces and moments on the propeller by using the DES approach and it is compared with an experiment for validation of DES approach. Ffowcs Williams–Hawkings and Farassat equations are used to predict the far-field sound radiation of propeller.

The present paper focused on prediction of non-cavitating marine propeller noise of 389 mm diameter propeller of 5, 6, and 7 blades with +10° pitch, +5° pitch, existing pitch, and −5° pitch, −10° pitch at rotational speed 840 rpm using large eddy simulation (LES) approach and Ffowcs Williams–Hawkings (FW-H)equation. Numerical simulations are carried on the above propellers to identify low noise generating propeller. Then modify the diameter of the low noise propeller for further reduction on noise for the same thrust and torque generated by the baseline propeller.

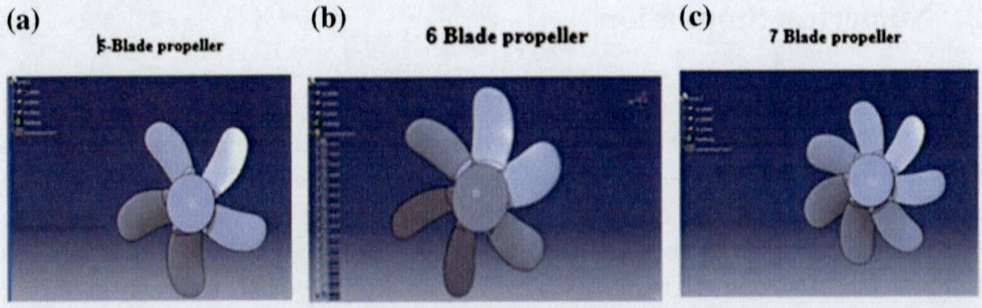

Fig. 1 (**a**) 5-Blade Propeller (**b**) 6-Blade Propeller (**c**) 7-Blade propeller

Table 1 Existing pitch details of 5, 6, and 7 bladed propellers

r/R (r = Pitch circle radius R = Propeller radius)	Pitch angle
0.3	46
0.4	42
0.5	40
0.6	38
0.7	35
0.8	32
0.85	31
0.9	30
1	29

CATIA V5R20 is used to generate a solid model of propeller. Section of the propeller is generated by using non-dimensional geometry data of the propeller. These generated cross section of the propeller which is rotated as per the pitch angle, and finally it is wrapped around cylindrical diameters to get the final sections of the propeller. These sections are connected smoothly by lofted surfaces to generate the final propeller model for simulations. Figure 1 shows the propeller model with 5, 6, and 7 blades with existing pitch, respectively. Table 1 shows the details of the existing pitch. Another 12 propeller solid models are generated having 5, 6, and 7 blades by adding the +10°, +5°, −5° and −10° pitch angles to existing pitch shown in Table 1 to study the variation of propeller noise on the number of blades and pitch angle of propeller blades.

2 Geometric Modeling of Propeller

See Table 1.

3 Numerical Simulation

The propeller model generated in the CATIA V5R20 is imported into ANSYS workbench 14.5 in IGES format for further analysis. The flow and noise simulation on all fifteen propellers are carried out using Fluent. For the prediction of hydrodynamic noise requires the information of the flow domain, and it plays a dominant role in the hydrodynamic noise prediction. For this purpose, flow domain is generated as follows. The flow domain is modeled by using a distance of 4D in downstream for the outlet of the propeller. The inlet distance is 3D (where D is the diameter of the propeller) in upstream from the center of the propeller. The flow domain in radial direction is 4D from the axis of the hub. Figure 2 shows the boundary conditions on the propeller domain. Mesh generation is playing a vital role to capture the flow properties with significant quality. For this purpose, small cell sizes are generated near the blade tip and cell size is increased toward the hub. Figure 3 shows the grid over the surface of the blade and hub. Then convergence of the domain is checked by varying the number of elements in the entire domain. The number of cells generated after convergence in the total grid is 1.2 million.

The fluid and the properties of water are assigned to the entire fluid domain. A rotational velocity of 840 rpm is assigned by using multiple rotating reference frames

Fig. 2 Boundary conditions on propeller domain

Fig. 3 Grid over the surface of the blade and hub

to fluid in the domain. Relative rotational velocity of the wall forming the propeller blade and hub are assigned as zero velocity with respect to the adjacent cell zone. An inlet velocity of 7.62 m/s is assigned at the inlet. Outflow boundary condition is assigned to the outlet of the fluid domain. An absolute rotational velocity of zero is assigned to the far field. These boundary conditions are shown in Fig. 2. The same boundary conditions are applied for the numerical study of unsteady non-cavitating marine propeller with 5, 6, and 7 blades with +10° pitch, +5° pitch, existing pitch, and −5° pitch, −10° pitch at rotational speed of 840 rpm.

Table 2 shows the details of non-cavitating flow details used for present noise simulations using Fluent.

From the noise analysis of 5, 6, and 7 blades with $+10^0$ pitch, +5° pitch, existing pitch, and −5° pitch, −10° pitch propellers at propeller rotating speed of 840 rpm and vehicle speeds of 7.62 m/s, it is observed that 6 blade propeller with +5° pitch generates low noise of **127 dB** compared with all other propellers. Figure 4 shows the noise prediction graph of six blades propeller with +5° pitch at 840 rpm. From these results it is observed that overall sound pressure level at 1 m along radial direction of propeller with 5 blades is 144 dB,138 dB,142 dB,145 dB, 150 dB, with 6 blades it is 132 dB,127 dB,130 dB,139 dB,143 dB, and for propeller with 7 blades it is 149 dB,146 dB,147 dB,152 dB, 156 dB for +10° pitch, +5° pitch, existing

Table 2 Details of non-cavitating flow

Pressure link	Simple
Pressure	Standard
Discretization scheme for convective fluxes and turbulence parameters	Second-order upwind
Turbulence model	Large eddy simulation
Near wall treatment	Standard wall functions
Solver	Unsteady

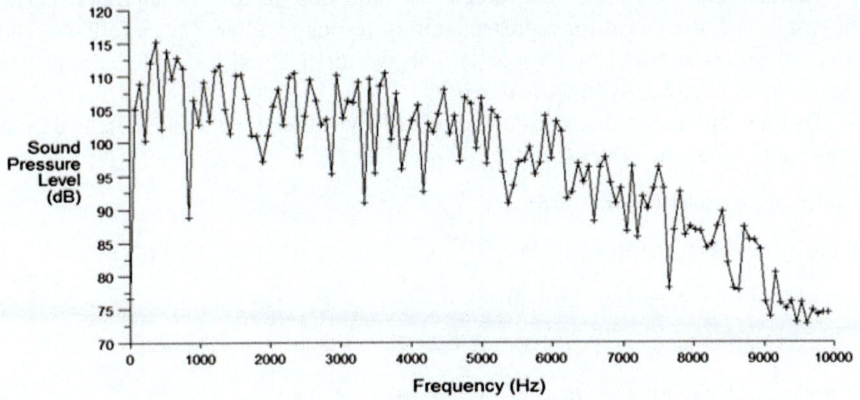

Fig. 4 Noise prediction graph of propeller with six blades +5° pitch at 840 rpm

pitch, and −5° pitch, −10° pitch at 840 rpm. Table 3 shows results of overall sound pressure levels of 5, 6, and 7 blade propellers with +10° pitch, +5° pitch, existing pitch, and −5° pitch, −10° pitch. Reference pressure used for all sound pressure level predictions is 1 μPa.

4 Results and Discussion

4.1 Study the Performance of Low Noise Propeller

Though the 6 blades with +5° pitch propeller generates low noise (127 dB) compared to all other propellers, it should produce basic requirements like thrust and torque as per design. Hence in this paper, we studied the performance of low noise propeller and baseline propeller in terms of total force and total moment. Tables 4 and 5 show the total force and moment on the baseline propeller. The co-ordinate system used for representation of results in Table 4 is as follows. The *x*-coordinate is aligned to propeller's hub symmetrical axis, *y*- and *z*-coordinates are normal to the hub symmetrical axis.

Thrust coefficient and torque coefficients of baseline propeller are obtained from the results shown in Tables 4 and 5.

Speed of propeller (N) = 14 rps

Density of seawater (ρ) = 998 kg/m^3

Diameter of the propeller (D) = 0. 389 m

Thrust Coefficient $Kt = T/(\rho N^2 D^4) = 0.354$

Torque Coefficient $Kq = Q/(\rho N^2 D^5) = 0.03099$

Tables 6 and 7 show the total force and moment on the low noise propeller. The co-ordinate system used for representation of results in Table 7 is as follows. The *x*-coordinate is aligned to the propeller hub symmetrical axis, *y*- and *z*-coordinates are normal to the hub symmetrical axis.

Thrust coefficient and torque coefficients of low noise propeller are obtained from the results shown in Tables 6 and 7.

Speed of propeller (N) = 14 rps

Density of seawater (ρ) = 998 kg/m^3

Diameter of the propeller (D) = 0.389 m

Thrust Coefficient $Kt = T/(\rho N^2 D^4) = 0.455$

Torque Coefficient $Kq = Q/(\rho N^2 D^5) = 0.0993$

Table 3 Noise of the propeller with 5, 6, and 7 blades

Overall sound pressure level (dB) @ 840 rpm														
5 blades					6 blades					7 blades				
+10° pitch	+5° pitch	Existing pitch	−5° pitch	−10° pitch	+10° pitch	+5° pitch	Existing pitch	−5° pitch	−10° pitch	+10° pitch	+5° pitch	Existing pitch	−5° pitch	−10° pitch
144	138	142	145	150	132	**127**	130	139	143	149	146	147	152	156

Table 4 Total normal force on baseline propeller

Pressure force (N)	Viscous force (N)	Total force (N)	Pressure coefficient	Viscous coefficient	Total coefficient
1542.2	43.9	1586.1	2517.9	71.7	2589.6

Table 5 Total moment of the baseline propeller

Pressure moment (N-m) (x, y, z)	Viscous moment (N-m) (x, y, z)	Total moment (N-m) (x, y, z)
34.7, 54.1, 52.4	−8.0, −0.26, −0.30	26.70, 53.9, 52.1

Table 6 Total force on low noise propeller

Pressure force (N)	Viscous force (N)	Total force (N)	Pressure coefficient	Viscous coefficient	Total coefficient
1989.7	47.8	2037.6	3248.6	78.1	3326.8

Table 7 Total moment on low noise propeller

Pressure moment (N-m) (x, y, z)	Viscous moment(N-m) (x, y, z)	Total moment(N-m) (x, y, z)
(58.6, −172.8, −172.8)	(−8.7, 0.44, −0.56)	(49.85, −172.3 −173.4)

The thrust coefficient and torque coefficient are increased for the low noise propeller as compared with baseline propeller, but the motor power is constant. So to maintain motor power constant, it is required to reduce the diameter of the low noise propeller for maintaining the same thrust and torque equal to the baseline propeller. Therefore, it is required to calculate the diameter of the low noise propeller to maintain same thrust and torque equal to the baseline propeller.

4.2 *Finding the Diameter of the Low Noise Propeller to Maintain Same Thrust and Torque Equal to BaseLine Propeller*

Thrust Coefficient (New) $Kt = T\,(old)/(\rho N^2 D^4)$

$0.455 = 1586.1/(998 \times 14^2 \times D^4)$

$D = 0.365$ mm

Torque Coefficient (New) $Kq = Q/(\rho N^2 D^5)$

$0.0993 = 53.9/(998 \times 14^2 \times D^5)$

$D = 0.308$ mm

Out of these two diameters of the propeller, **0.365 mm** diameter is taken as its higher diameter which is considered for modeling of the low noise propeller used for further noise simulations.

4.3 *Modeling and Noise Analysis of Low Noise New Propeller*

Modeling and noise analysis is carried on 365 mm diameter low noise new propeller as per previous boundary conditions explained in the Sect. 3. The overall sound pressure level at 1 m along the radial direction of low noise new propeller is 120.6 dB (ref.pr is 1e-6). It is the least noise observed with all other models used in the noise simulations. From Fig. 5, it is observed that the overall sound pressure level is

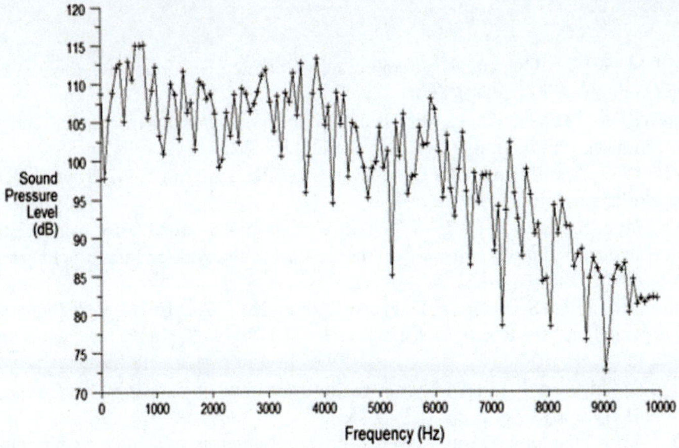

Fig. 5 Noise prediction graph of 6 blade 365 mm dia propeller from simulations

120.6 dB (ref. pr is 1e-6), which is the least noise compared to all other propellers considered in the present noise analysis.

5 Conclusions

1. Developed the methodology for finding the low noise propeller out of fifteen propellers by geometrical modifications of the propeller.
2. Developed the methodology for finding the low noise propeller by maintaining the same thrust and torque of the propeller.
3. Hydrodynamic noise of an underwater vehicle propeller with 5, 6, and 7 blades with +10° pitch, +5° pitch, existing pitch, and −5° pitch, −10° pitch is investigated at 840 rpm and vehicle speed of 7.62 m/s with the large eddy simulation (LES) approach and Ffowcs Williams–Hawkings equation.
4. From these results, it is observed that the propeller with six blades with +5° pitch at speed 840 rpm generates least noise (127 dB) compared to all other propellers.
5. Overall Noise level of redesigned propeller is 120.6 dB (ref. pr is 1e-6) which is least among all the other propellers considered in the present analysis.

Acknowledgements The authors wish to express their sincere gratitude to Dr. O.R. Nandagopan, Scientist 'G' and Director, NSTL Visakhapatnam for permitting to publish this paper. The authors also wish to acknowledge to Sri PVS Ganesh Kumar, Scientist'G', Sri. Ch. Sankara Rao, Scientist'G', NSTL, Visakhapatnam and Dr. S. Rama Krishna, Assoc. Prof., GVP College of Engineering, Visakhapatnam.

References

1. Lighthill MJ (1952) On sound generated aerodynamically: I. General theory. Proc. Royal Society (London) A 211:564–587
2. Ffowcs Williams JE, Hawkings DL (1969) Sound generated by turbulence and surfaces in arbitrary motion. Philos Trans Royal Soc Series 115:1321–1342
3. Chang B (1998) Application of CFD to P4119 Propeller. In: 22nd ITTC Propeller RANS/Panel method workshop, France
4. Yang Q, Wang Y, Zhang M (2011) Calculation of propeller's load noise using LES and BEM numerical acoustics coupling methods. Transactions of the Wessex Institute, Paper. https://doi.org/10.2495/BE110081
5. Bertetta D, Brizzolara S, Canepa E, Gaggero S, Viviani M (2012) EFD and CFD characterization of a CLT propeller. Int J Rotat Mach 12. Article ID 348939, 22 pages. https://doi.org/10.1155/2012/348939
6. Zhi-feng Z, Shi-liang F (2012) Numerical investigation of cavitation performance of ship propellers. J Hydrodyn Ser B 24(3):347–353
7. Song Y, Wen J, Dianlong Y, Liu Y, Wen X (2014) Reduction of vibration and noise radiation of an underwater vehicle due to propeller forces using periodically layered isolators. J Sound Vib 333:3031–3043

8. Yu-cun P, Huai-xin Z (2012) Numerical hydro-acoustic prediction of marine propeller noise 15(6):707–712
9. Muscari R, Di Mascio Andrea, Verzicco Roberto (2013) Modeling of vortex dynamics in the wake of a marine propeller. Comput Fluids 73:65–79
10. Yu-cun P, Huai-xin Z (2013) Numerical prediction of marine propeller noise in non-uniform inflow. China Ocean Eng 27(1):33–42

Noise Control of a Marine Propeller by Modifying the Skew Angle

V. Ramakrishna, D. A. Rao, Ch. Sankara Rao, P. V. S. Ganesh Kumar, T. Gunasekhar, and V. Mani kumar

Abstract Any underwater vehicle like submarine, torpedo and AUV uses a propeller for propulsion. The significance of propeller is to develop thrust which propels the vehicle at its operational speed. The propeller geometry and design are complicated which involves many controlling parameters. The pressure difference is produced on the propeller blade between the forward and rear surfaces which accelerates water behind the blade. In addition to thrust generation, it creates noise which could be detected by the enemy. From stealth point of view, this noise is to be reduced. The present work is aimed at controlling the noise generated by the propeller by changing the skew angle of the propeller. The aim of this paper is to identify the skew angle of a propeller at which propeller noise is least. Unsteady non-cavitating noise of marine propeller of six blades with $+5^{\circ}$ skew, $+10^{\circ}$ skew, $+15^{\circ}$ skew and $+20^{\circ}$ skew and existing propeller at rotating speed of 780 rpm and vehicle speed of 7.08 m/s was predicted. The methodology adopted in CFD analysis is eddy viscosity model of Large Eddy Simulation (LES) and for acoustic analysis is the Ffowcs Williams–Hawkings (FW-H) formulation. From this numerical study on these propellers, it is found that the propeller having $+15^{\circ}$ skew angle generates the least noise.

Keywords Non-cavitating noise · Marine propeller · Computational Fluid Dynamics (CFD) · Large Eddy Simulation (LES) · Ffowcs Willams–Hawkings Formulation (FW-H)

1 Introduction

A propeller is a rotating fan-like structure used in underwater vehicles like submarines, torpedoes and AUVs for propulsion by using the power generated from the main engine. The transmitted power is converted from rotational motion to

V. Ramakrishna (✉) · D. A. Rao · Ch. Sankara Rao · P. V. S. Ganesh Kumar
Naval Science and Technological Laboratory, Visakhapatnam 530027, India
e-mail: ramakrishna.v@nstl.drdo.in

T. Gunasekhar · V. Mani kumar
JNTU Kakinada-University College of Engineering, Vizianagaram 535003, India

M. Singh and Y. Rafat (eds.), *Recent Developments in Acoustics*, Lecture Notes in Mechanical Engineering, https://doi.org/10.1007/978-981-15-5776-7_20

generate a thrust which imparts momentum to the water, resulting in a force that acts on the vehicle and pushes it forward. A pressure difference is produced between the forward and rear surfaces of the blade, and water is accelerated behind the blade. The blade geometry and its design are more complex involving many controlling parameters.

Stealth technology plays a major role in avoiding detection of underwater vehicles. Vehicles can be detected through radiated noise and hence, there is a requirement to minimize the radiated noise. Radiated noise mainly consists of machinery noise, flow noise and propeller noise. In modern underwater vehicles, emphasis is laid on designing propellers with low noise. Chekab et al. [1] investigated the influence of changing the geometry, wake inflow modification and Propeller Isolation on generated noise. Ji et al. [2] studied simulation of marine propellers with different skew angles while operating in the non-uniform wake. Seol et al. [3] predicted the non-cavitating noise generated by underwater propeller using the Ffowcs Williams–Hawkings equation. Kim et al. [4] conducted experiments to predict broadband noise, and numerical method was developed to predict propeller tonal noise. Bagheri et al. [5] analysed the source of flow field of underwater propellers (FVM) and then the time-dependent flow field data are used as the input for Ffowcs Williams–Hawking's formulation to predict the far-field acoustics. Noise characteristics are presented according to noise sources and conditions.

2 Geometric Modelling of Propeller

This study is carried out by using an existing propeller. The geometry and material details are given below in Tables 1 and 2. The solid model of the existing propeller is shown in Fig. 1.

The solid modelling of propeller was carried out in CATIA V5 R20. The shape of the blade of a propeller is usually defined by specifying the shape of the

Table 1 Dimension of propeller model

Diameter (m)	0.389 m
BAR = A_E/A_O	0.58
No. of Blades	6
Hub ratio	0.245
Series	NACA 66

Table 2 Material specification

Density (g/cm^3)	2.6898
Modulus of elasticity (Gpa)	68.3
Poisson's ratio	0.34
Proof street (MPa)	385

Fig. 1 Solid model of existing propeller

section obtained by the intersection of a blade by coaxial right circular cylinders of different radii. These sections are called radial sections or cylindrical sections. Since all propeller blades are identical, only one blade needs to be defined. The non-dimensional geometry data of the propeller was converted into point co-ordinate data to generate the expanded sections. The detailed design of a propeller consists of designing the expanded sections at a number of radii, usually at r/R = 0.2, 0.3….1.0, where R is the propeller radius.

2.1 Modelling Procedure of Propeller

The expanded sections drawn are moved along the radius of propeller to get respective skew angles and they are rotated to respective pitch angle at each section. The rotated sections are developed on the cylinder to get wrapped section at each radius. The multi-section surface is used to join all the profiles to get the blade model. Finally, the surface model is made into solid by closing all the surface edges and filling the volume between the surfaces with solid. The hub is generated by revolving the profile along the propeller axis and then the blade is joined with the hub. At the intersection of hub and the blade, the Edge fillet is used to create the root around the blade. All the edges of the propeller are filleted to get smooth curved edges. This reduces the trailing edge noise radiated from the propeller. The remaining blades are generated on the hub using circular array so that the equally spaced blades are obtained along the axis of the propeller. This modelling procedure of propeller is depicted in Fig. 2. The solid models of existing, $+5^{\circ}$ Skewed, $+10^{\circ}$ Skewed, $+15^{\circ}$ Skewed and $+20^{\circ}$ Skewed propellers are shown in Figs. 3, 4, 5, 6 and 7.

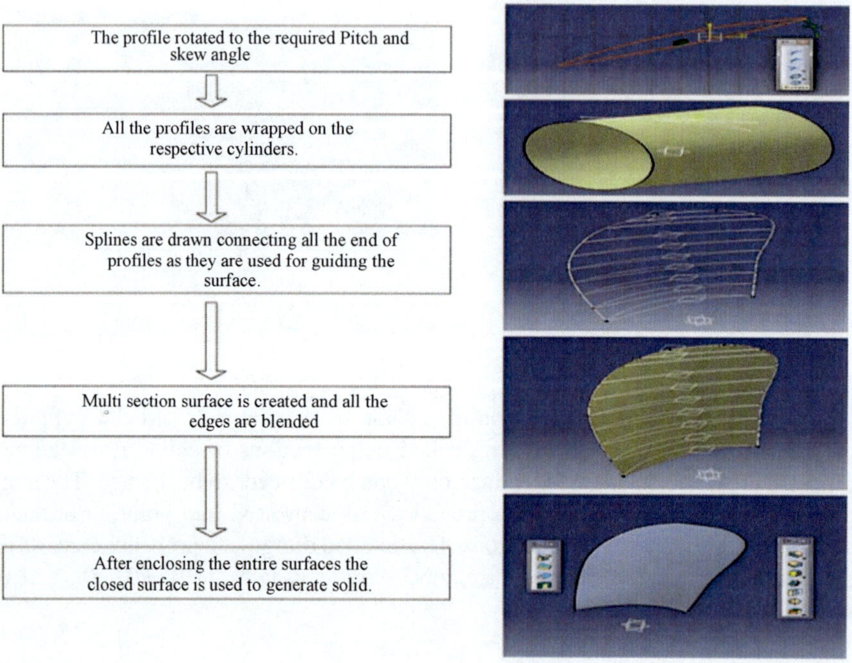

Fig. 2 Modelling procedure of propeller

Fig. 3 Existing propeller

3 Numerical Simulation

The solid model of propeller is imported into ANSYS workbench 15.0. The flow and noise simulation were performed on all propellers using ANSYS-Fluent. The prediction of propeller noise requires the information of the flow domain and it plays a dominant role. Figure 8 shows the boundary conditions imposed on the propeller domain. The inlet is taken at a distance of 3D (where D is the diameter of the propeller) from mid of the chord of the root section at upstream. The outlet is taken

Fig. 4 + 5° Skewed propeller

Fig. 5 + 10° Skewed propeller

Fig. 6 + 15° Skewed propeller

at a distance of 4D from the same point at downstream. In radial direction, domain is considered up to a distance of 4D from the axis of the hub as shown in Fig. 8. The input parameters given are rotational speed of 780 rpm and flow velocity of 7.08 m/s at the inlet. The mesh is generated with tetrahedral cells in such a way that cell sizes near the blade tip are very small and increased towards hub as shown in Fig. 9 to

Fig. 7 +20° Skewed propeller

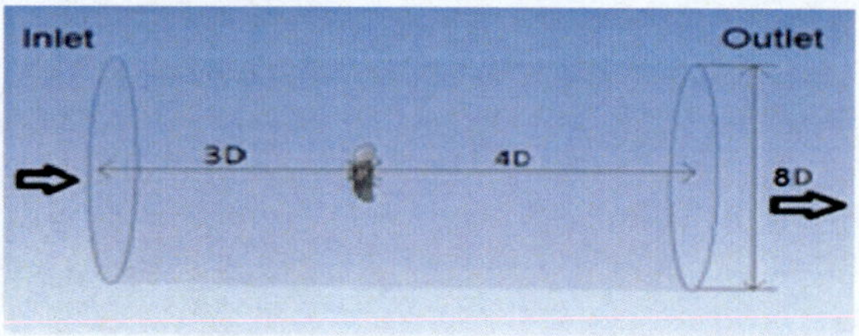

Fig. 8 Computational domain

Fig. 9 Meshed model

capture the flow properties with significant quality. The same boundary conditions are applied for the numerical study of unsteady non-cavitating noise of all propellers with different skews. The turbulence model chosen for the present study was Large Eddy Simulation (LES). Table 3 shows the details of Non-Cavitating flow used for present simulations.

Table 3 Details of non-cavitating flow

Pressure link	Simple
Pressure	Standard
Discretization scheme	Second-order upwind
Turbulence model	Large Eddy Simulation
Near wall treatment	Standard wall functions
Solver	Unsteady

4 Results and Discussion

The output of CFD analysis is given to the input of acoustic analysis using the Ffowcs Williams–Hawkings (FW-H) formulation to obtain the noise levels. The CFD and Acoustic analysis on six-blade propeller with existing skew(41°), +5° skew, +10° skew, +15° skew and +20° skew with a rotating speed of 780 rpm and flow velocities of 7.08 m/s were carried out. It was observed that the propeller with +15° skew generates less noise compared with other propellers. Figures 10, 11, 12, 13 and 14 show the Noise prediction graphs at 1 m distance along the radial direction of the propeller with six blades for existing skew, +5° skew, +10° skew, +15° skew and + 20° skew at 780 rpm and overall sound pressure levels are 130 dB, 124 dB, 121 dB, 119 dB, 123 dB, respectively, referred to//1μ Pa@ 1 m. The propeller of six blades with +15° skew gives the low noise of 119 dB compared with other propellers. The results of all varieties of propeller noise are shown in Table 4.

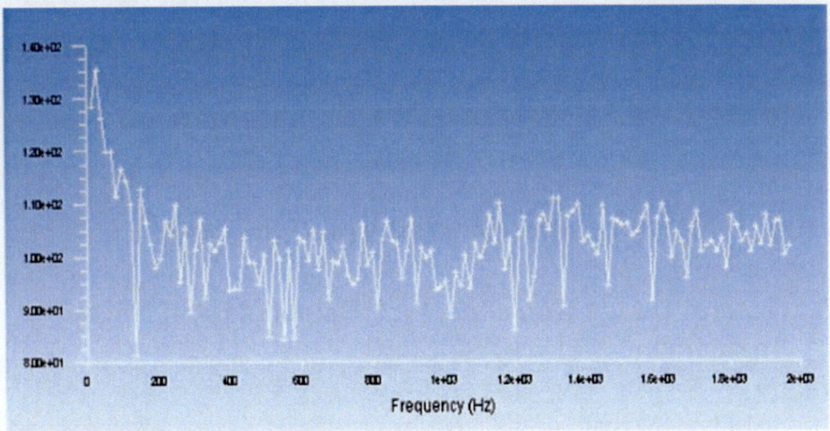

Fig. 10 Noise prediction graph at inlet of the existing propeller from simulation

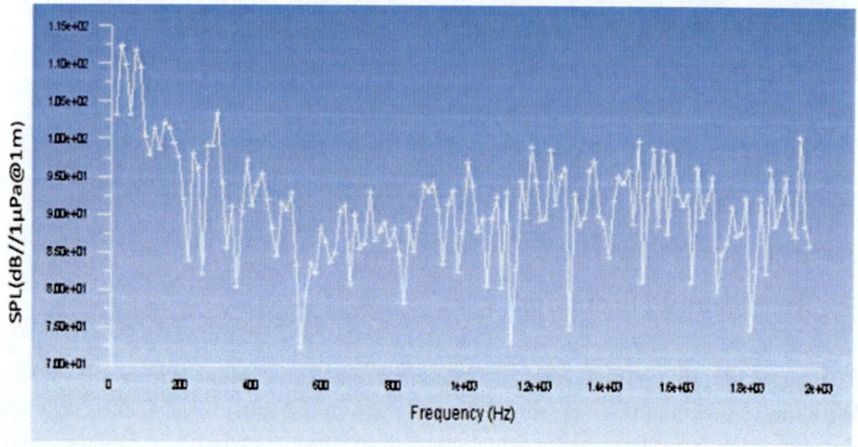

Fig. 11 Noise prediction graph at inlet of $+5^{\circ}$ skew propeller from simulation

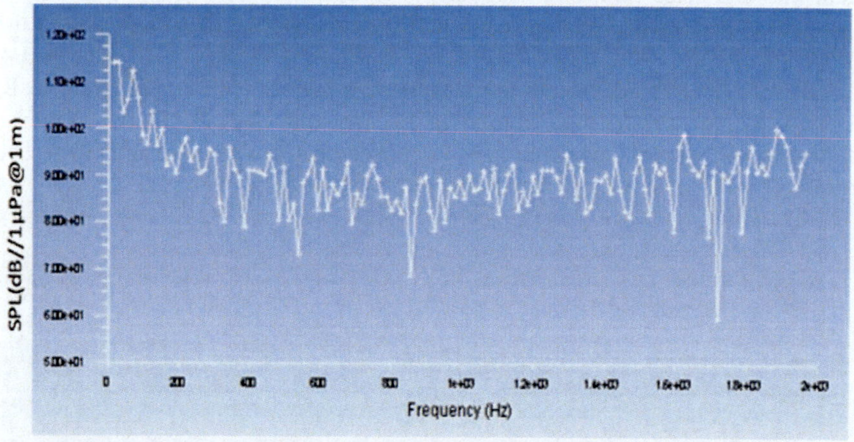

Fig. 12 Noise prediction graph at inlet of $+10^{\circ}$ skew propeller from simulation

5 General Design Guidelines for "Methodology of Noise Prediction of Marine Propeller"

1. Modelling procedure of the propeller-like profile rotation, wrapping, profile generation, multi-section surface creation.
2. Generic Meshing of the propeller used is Tetrahedral meshing.
3. Boundary conditions used: inlet is 3D, outlet is 4D and Radial distance is 4D from the centre of the Propeller.

4. Methodology used in CFD analysis is Large Eddy Simulation (LES).
5. Methodology used in Acoustic analysis is the Ffowcs Williams–Hawkings (FW-H) formulation.

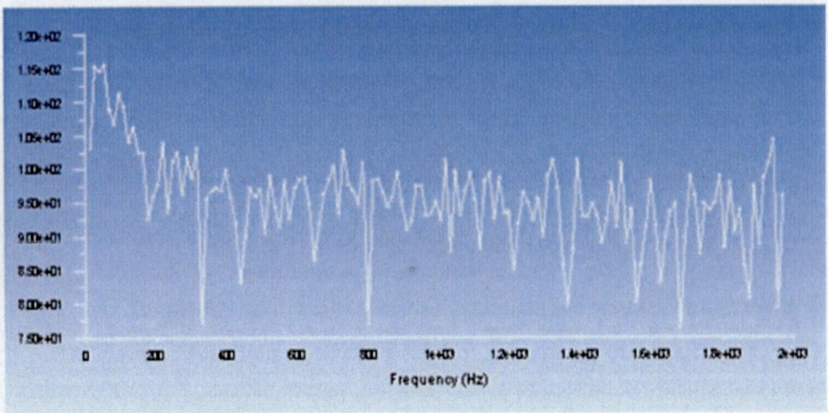

Fig. 13 Noise prediction graph at inlet of +15° skew propeller from simulation

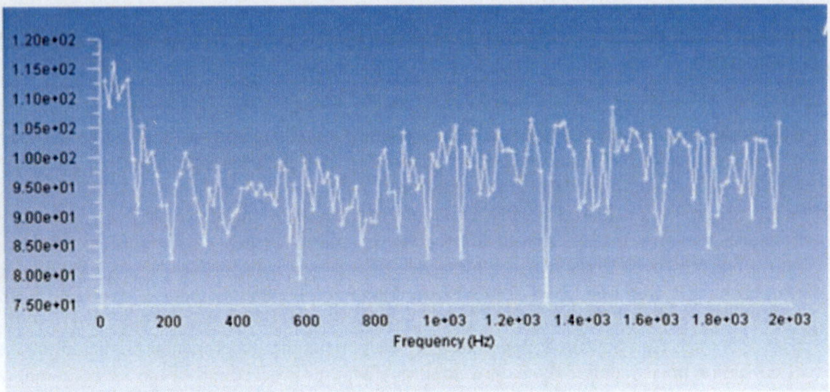

Fig. 14 Noise prediction graph at inlet of +20° skew propeller from simulation

Table 4 Noise of six-blade propeller with different skew

Overall sound pressure level (dB//1μ Pa@ 1 m) @ 780 rpm and flow velocity 7.08 m/s				
Six-blade propeller				
Existing blade	5° Skew	10° Skew	15° Skew	20° Skew
130	124	121	**119**	123

6 Conclusions

1. This paper presents a numerical investigation of unsteady non-cavitating turbulent flow around a full-scale marine propeller of 389 mm diameter with six blades and varying skews, at rotational speed 780 rpm with the eddy viscosity model of Large Eddy Simulation (LES) approach.
2. Propeller behaviour is investigated for its hydrodynamic parameters and then computational acoustic analysis is carried out using the Ffowcs Williams–Hawkings (FW-H) formulation.
3. The optimum values of these design parameters are arrived and presented.
4. It is seen that for the propeller under study, $+15^{\circ}$ skew generates the least noise of 119 dB compared with other skew angles.
5. This methodology of noise prediction can be employed for any propeller to determine the skew angle at which least noise will be generated.

Acknowledgements The authors wish to express their sincere gratitude to Dr. O. R. Nanda Gopan, Director, NSTL Visakhapatnam for Permitting to publish this paper. The authors also wish to express gratitude to Vibration Studies Division of NSTL, Visakhapatnam.

References

1. Chekab MAF, Ghadimi P, Djeddi SR, Soroushan M (2013) Investigation of different methods of noise reduction for submerged marine propellers and their classification. Am J Mech Eng 1(2):34–42. https://doi.org/10.12691/ajme-1-2-3
2. Ji B, Luo1 X, Wu Y (2014) Unsteady cavitation characteristics and alleviation of pressure fluctuations around marine propellers with different skew angles. J Mech Sci Technol 28(4):1339–1348. https://doi.org/10.1007/s12206-013-1166-8
3. Seol H, Jung B, Suh J-C, Lee S (2002) Prediction of non-cavitation underwater propeller noise. J Sound Vib 257(1):131–156
4. Kim T, Jeon J, Chu S, Kim S, Joo W (2016) Numerical and experimental prediction methods of cavitation noise radiated by underwater propellers. In: Proceedings of the 22nd international congress on acoustics underwater acoustics, Paper ICA2016-35
5. Bagheri MR, Seif MS, Mehdigholi H (2014) Numerical simulation of underwater propeller noise. J Ocean, Mech Aerosp 4:1–6

Bibliography

1. Zhu Z (2015) Numerical study on characteristic correlation between cavitating flow and skew of ship propellers. J Ocean Eng 99:63–71
2. Da-Qing L (2006) Validation of rans predictions of open water performance of a highly skewed propeller with experiments. In: Conference of global Chinese scholars on hydrodynamics
3. Ghasseni H, Ghadimi P (2011) Numerical analysis of the high skew propeller of an underwater vehicle. J Marine Sci Appl 10:289–299. https://doi.org/10.1007/s11804-011-1071-4

4. Reese H, Carolus T (2008) Axial fan noise: towards sound prediction based on numerical unsteady flow data—a case study. J Acou Soc Am 123(5):3539–3539
5. Ji B, Luo X, Wang X, Peng X, Wu Y, Xu H (2011) Unsteady numerical simulation of cavitating turbulent flow around a highly skewed model marine propeller. J Fluids Eng 133

Acoustic Signature of Square Cylinder in a Compressible Flow

Mohd Masoom Ali and Syed Fahad Anwer

Abstract The Aeolian tones radiated from structures are important in various engineering fields. For example, in military and defense applications, it is used for detection and survivability. Aeroacoustic noise is increasingly used as a critical design variable in modern engineering designs. This in turn has led to an increase in the research efforts aimed at numerical prediction of aerodynamic noise, often dubbed computational aeroacoustics (CAA). A numerical study of aerodynamically generated acoustic noise was carried out. Simulations of 2D, laminar, and compressible, viscous flow were performed at Reynolds Number (Re) $\in$ [60, 200] and Mach no. (Ma) $\in$ [0.2, 0.4]. Roe Flux-Difference Splitting Scheme was used to solve the governing equations with finite volume differencing. The Least square cell based spatial discretization scheme was used with second-order upwind scheme for calculating convection terms. Acoustic calculations were done using Ffowcs-Williams and Hawkings (FW-H) formulation. The peak in the receiver spectra matches the dominant frequency in the lift force history. Sound Pressure level decreases with decreasing flow velocity and Reynolds number. Acoustic peak frequency shifts toward higher frequency value with increasing Mach number for all cases of Reynolds number. The sound pressure level was observed to be higher downstream of the flow. Wide band noises at frequencies higher than the fundamental frequency were observed to be the subharmonics of the fundamental frequency of sound.

Keywords Aeolian tones · Aeroacoustics · Mach no · Roe Flux-difference · Finite volume · Ffowcs-Williams and Hawkings (FW-H) formulation

M. M. Ali · S. F. Anwer (✉)
Computational Aerodynamics Lab, MED, ZHCET, AMU, Aligarh, India
e-mail: sfahadanwer@zhcet.ac.in

M. M. Ali
e-mail: techmasoomali@gmail.com

M. Singh and Y. Rafat (eds.), *Recent Developments in Acoustics*, Lecture Notes in Mechanical Engineering, https://doi.org/10.1007/978-981-15-5776-7_21

1 Introduction

The Aeolian tone is sound generated by an obstacle in a flow. Research of Aeolian tone has a long history of more than one hundred years. The Aeolian tone that is radiated from the flow around a square cylinder has received far less attention than that from the flow around a circular cylinder, and works from the open literature are rather scarce. From an engineering point of view, Aeolian tones that are radiated from the flow around a square cylinder are important; a typical example is the tones generated by landing gears, pantographs of trains, etc., which has many circular and square cylinders as their structural components.

1.1 Literature Review

Fujita [1] conducted experiments on cylindrical objects and discussed the characteristics of aerodynamic sound radiated from two-dimensional models, such as a circular cylinder, square cylinders with or without rounded corners, and a cylinder with a modified square cross-section. They also investigated experimentally Aeolian tone generation and its relation with surface pressure fluctuation on a circular cylinder at moderate to high Reynolds number. The relation between velocity fluctuations in the wake and the Aeolian tone generation mechanism was quantitatively described.

Silva and Ferreira [2] investigated sound generation by a square cylinder placed in a uniform flow at low Mach and low Reynolds numbers by direct solution of the two-dimensional unsteady compressible Navier–Stokes equations by solving the compressible Navier–Stokes equations by using fourth-order numerical schemes. The simulations were carried out for three different Mach numbers, namely Ma = 0.1, 0.2, and 0.3, and also for three different Reynolds numbers (Re = 100, 150, and 200), in a dipolar nature of the generated sound was observed with main propagation direction toward upstream. The frequency of the pressure waves is the same as that of vortex shedding as well as the fluctuation frequency of the lift force. The results also showed that the Mach number has little effect over the generation mechanisms of the sound, but clearly influences the propagation of sound waves to mid and far fields. The Doppler effects showed to play an important role at a very small Mach number. It was observed that when the Reynolds number increases, the positive and negative sound waves have clearly different propagation angles. This effect has been attributed to a significant increase in drag dipole when the Reynolds number increases.

Ali et al. [3] investigated flow over various bluff body shapes: four different bluff body shapes, i.e.; Triangle, Ellipse, Circular, and Square at Re = 150 and Mach 0.2. The information regarding the flow features and noise generation mechanism was identified. It was found that all bluff bodies exhibited the same physics features, only the magnitudes were different, in which the triangle cylinder was the highest and square cylinder was the lowest. Inspection on the time histories of the sound pressure level found a sinusoidal pattern in the fluctuation. Therefore, it is possible for a downstream flat plate to be applied in the sound-cancelation mechanism for all types of the bluff bodies being investigated. In a later study, Ali et al. [4] investigated numerically the case of a rigid flat plate placed in the wake of a square cylinder for passive sound control at a Reynolds number of 150 and a Mach number of 0.2. Two flow regimes were observed, and they showed to affect the radiated sound. An overall sound reduction at an observer directly above the cylinder was observed for a small gap distance G between the cylinder and the plate ($0 \leq G \leq 2.3D$, in which D is the cylinder height), where a 2.9 dB reduction in the sound pressure level was obtained when there is no gap between the two bodies. In contrast, the emitted sound pressure level increases by at least 8.0 dB for large gap distances ($2.4D \leq G \leq 7D$). Despite this, a 6.3 dB reduction in the sound pressure level is obtained due to a sound-cancelation mechanism when the plate length was reduced to 0.26D and placed 5.6D downstream of the cylinder. The nonlinear unsteady stall process restricts the maximum sound reduction attained.

Most of the studies have been done in low Reynolds number and low Mach number range using incompressible flow equations, while extensive studies have been carried out on the bluff body flows, most of these were primarily focused on the dynamics of the instabilities. The Aeolian tone that is radiated from the flow around a square cylinder has received far less attention as compared to the flow around a circular cylinder. In this work, we will study the effect of Mach number on Aeolian tones to have an understanding of the effects of compressibility on the acoustics emissions. For this purpose, we are going to consider flow past a square cylinder as a canonical problem. Uniform compressible flow past a square cylinder is solved. The flow is in the vertical direction as shown in Fig. 1a. The range of the Reynolds Number selected for the simulations is from 60 to 200, keeping the flow in the 2D regime. Whereas Mach numbers selected for the study are $M = 0.2$, 0.3, and 0.4. This is to have an insight of the compressibility effect on the aerodynamically generated noise.

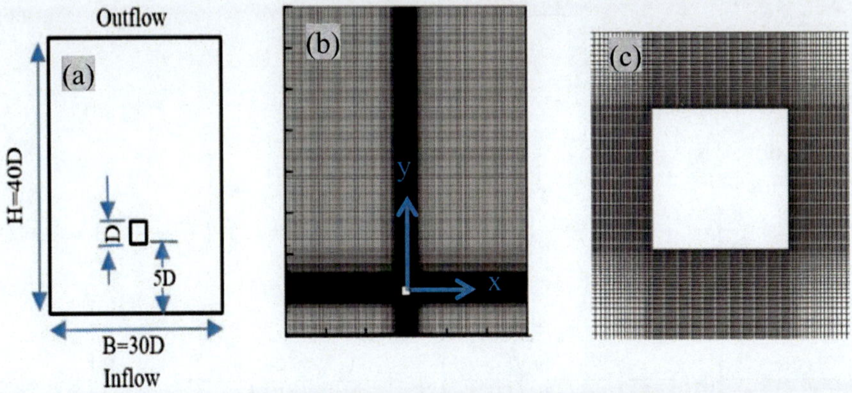

Fig. 1 **a** Geometrical representation of cylinder geometry **b** Block structured Cartesian Grid **c** Grid near cylinder

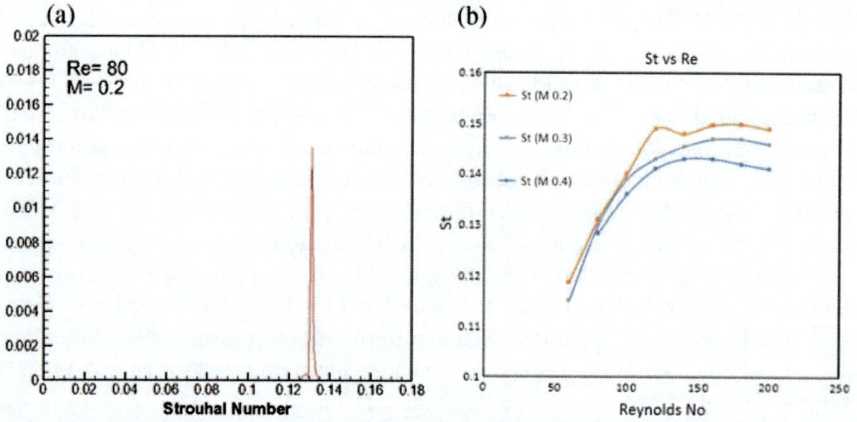

Fig. 2 **a** Fast Fourier transform to calculate the Strouhal number **b** Variation of the Strouhal number with the Reynolds number

2 Mathematical Model and Numerical Methodology

2.1 *Governing Equations*

The flow is taken as compressible, hence the flow is governed by the compressible form of Navier–Stokes and energy equations. Ideal gas law and Sutherland's law is used to complete the model. The non-dimensional form of the governing Eqs. (1)–(4) are given below [5]:

Continuity Equation:

$$\frac{\partial \rho}{\partial t} + \frac{\partial (\rho u)}{\partial x} + \frac{\partial (\rho v)}{\partial y} = 0 \tag{1}$$

Momentum Equation:

x-direction:

$$\frac{\partial (ru)}{\partial t} + \frac{\partial}{\partial x}\left[ru^2 + \frac{1}{gM^2}p - \frac{2m}{\mathrm{Re}}\left\{\frac{\partial u}{\partial x} - \frac{1}{3}(\tilde{N}\cdot\vec{V})\right\}\right] + \frac{\partial}{\partial y}\left\{ruv - \frac{m}{\mathrm{Re}}\left(\frac{\partial v}{\partial x} + \frac{\partial u}{\partial y}\right)\right\} = 0 \tag{2}$$

y-direction:

$$\frac{\partial (\rho v)}{\partial t} + \frac{\partial}{\partial x}\left\{\rho uv - \frac{\mu}{\mathrm{Re}}\left(\frac{\partial v}{\partial x} + \frac{\partial u}{\partial y}\right)\right\} + \frac{\partial}{\partial y}\left\{\rho v^2 + \frac{1}{\gamma M^2}p - \frac{2\mu}{\mathrm{Re}}\left\{\frac{\partial v}{\partial y} - (\nabla\cdot\vec{V})/3\right\}\right\} = 0 \tag{3}$$

Energy Equation:

$$\frac{\partial pE}{\partial t} + \frac{\partial}{\partial x}\left\{\rho u h_T - \frac{k}{\mathrm{Re\,Pr}}\frac{\partial T}{\partial x} + \frac{\gamma(\gamma-1)M^2\mu}{\mathrm{Re}}D_F\right\} + \frac{\partial}{\partial y}\left\{\rho v h_T - \frac{k}{\mathrm{Re\,Pr}}\frac{\partial T}{\partial y} + \frac{\gamma(\gamma-1)M^2\mu}{\mathrm{Re}}D_G\right\} = 0 \tag{4}$$

where

$$D_F = \left[\frac{2}{3}u\left(\frac{\partial v}{\partial y} - \frac{2\partial u}{\partial x}\right) - v\left(\frac{\partial v}{\partial x} + \frac{\partial u}{\partial y}\right)\right],\ D_G = \left[\frac{2}{3}v\left(\frac{\partial u}{\partial x} - \frac{2\partial v}{\partial y}\right) - u\left(\frac{\partial v}{\partial x} + \frac{\partial u}{\partial y}\right)\right]$$

$\mathrm{Re} = \frac{PU_\infty L}{RT\mu}$ is the Reynolds number, ρ and μ are density and dynamic viscosity of the fluid, respectively, while $\mathrm{Pr} = \frac{\nu}{\alpha}$ is the Prandtl number. In the present work, freestream condition is used as initial condition and all the flow properties are initialized with their respective freestream values. At the surface of the cylinder, the flow is governed by no-slip and no penetration condition for the velocity, the cylinder is also considered as adiabatic.

2.2 Computational Model

The computational domain used for the two-dimensional flow simulations over the square cylinder is shown in Fig. 1. The domain is taken from Sharma and Eswaran [6]. The two-dimensional flow field around the bluff body is taken to be unsteady, viscous, and laminar. The non-dimensional width of the square cylinder is equal to unity, D = 1. The leading edge of the cylinder is located 5D units from the inlet ensuring negligible boundary effects. Here 'H' is the height and 'B' is the width of the domain, which is taken as 40D and 30D, respectively. Block Structured grid generated in ICEM CFD is used with a stretching to capture wake–wall interactions.

A grid, time, and domain independence test was performed but for the sake of brevity, we are not including detailed results in this article. The grid Independence tests suggested that the grid having a downstream length greater than or equal to 30D and having near boundary spacing equal to 0.005 is optimum for the numerical calculations by the solver. According to the Time independence test, to get correct numerical results, the time step is chosen to be 0.001. So our domain is having a downstream length of 35D, the near boundary spacing of 0.005 with cell stretching of 7%, and the number of nodes in the computational domain is 93510. Figure 1b, c shows the computational geometry and grid being used.

Table 1 Validation of flow parameters with existing studies at Re 100 M = 0.3

Reference	Model	C_D	St
Present	Compressible	1.60	0.139
Silva and Ferreira [2]	Compressible	1.45	0.141
Franke et al. [9]	Compressible	1.62	0.139
Davis and Moore [7]	Incompressible	1.64	0.152
Sohankar et al. [8]	Incompressible	1.45	0.145

2.3 Numerical Methodology and Validation

ANSYS FLUENT 14.0 is used to solve in this work. The preconditioned Roe Flux-Differencing Splitting Scheme (Roe FDS) a density-based solver is employed to solve equations. The density-based solver solves the governing equations of continuity, momentum, and energy Eqs. (1)–(4). This work employs a projection method for solving the compressible Navier–Stokes equation.

Spatial discretization is done using finite volume formulation, while time discretization is done using an implicit scheme. Spatial discretization is second-order accurate. In order to validate the numerical schemes that have been used in ANSYS Fluent 14.0, numerical simulations were performed over, and the results were compared with the published. Computations are carried out for a single cylinder at Re = 100. The mean drag coefficient and Strouhal number are compared with available data of Silva and Ferreira [2], Franke et al. [9], Davis and Moore [7], and Sohankar et al. [8]. Table 1 shows a comparison of present work with the above-mentioned works.

2.4 Acoustic Scheme

The governing equations for acoustics are indeed the same as the ones governing fluid flows. The main challenge in numerically predicting sound waves comes from the well-recognized fact that sounds have much lower energy than fluid flows, typically by several orders of magnitude. This poses a great challenge to the computation of sounds in terms of difficulty of numerically resolving sound waves, especially when one is interested in predicting sound propagation to the far field. Another challenge comes from the difficulty of predicting the very flow phenomena (e.g., turbulence) in the near field that are responsible for generating sounds. Works in the field of computational aeroacoustics (CAA) can be categorized into three groups depending on the method to use: hybrid method, acoustic/viscous splitting method, and direct numerical simulation (DNS) method. The first group (hybrid method) makes use of an acoustic analogy, under the assumption of a compact source, to predict the far field sound. The source terms are evaluated using the near-field flow quantities, which are obtained by solving the Navier–Stokes equations for low-Mach-number flows. This method saves computational time as well as memory storage compared with DNS, because the flow in the far field is assumed to be stationary or uniform and thus not solved numerically. The second group (acoustic/viscous splitting method) assumes that flow quantities are represented, under the assumption of low Mach number, by an incompressible mean flow and a perturbation about the mean. In the far field, the perturbation quantities are equivalent to acoustic quantities. This method may possibly be a convenient method of predicting sound field resulting from low-Mach-number, non-compact source region. So far the results obtained by this method are qualitative, and detailed descriptions of sound fields have not yet been given. The third group makes use of DNS, where both the fluid motion and the sound which it generates are directly computed. Recent development of a high-performance supercomputer and highly accurate numerical schemes makes it possible to simulate sound field by directly solving the compressible Navier–Stokes equations over the entire region from near to far-fields. This method does not suffer from restrictions such as low Mach number and compactness of the source region, but requires a large amount of computer resources; the studies using DNS are very few. We have used Ffowcs-Williams and Hawkings model for this. The source terms are evaluated using the near-field flow quantities. This method is often called the hybrid method, the flow in the far field is assumed to be stationary or uniform and thus not solved numerically.

3 Result and Discussion

The two-dimensional flow field around the bluff body is taken to be unsteady, viscous, and laminar. The non-dimensional width of the square cylinder is equal to unity, D = 1. The Strouhal number (St) and sound pressure level (SPL) were determined for the airflow across the square cylinder by performing simulations of flow having the Reynolds Number (Re) $\in$ [60, 200] and Mach no. (Ma) $\in$ [0.2, 0.4]. This is to have an insight of the compressibility effect on the aerodynamically generated noise. The Strouhal number (St = fD/U) was calculated by performing a Fast Fourier Transform of the lift coefficient time history.

The Strouhal number decreases with increasing Mach number for a particular Reynolds number. It has also been observed that the Strouhal number first increases with the Reynolds number and reaches a maximum at Re = 160 after which it slowly decreases with a further increase in the Reynolds number. The flow at Re = 60 and Mach 0.4 did not undergo vortex shedding; this may be attributed to the fact that compressibility suppresses turbulence.

Acoustics calculations were performed when the lift and drag histories became statistically stationary. This is the time during the simulation process when the receivers along with the acoustics reference pressure were set up. The acoustic reference pressure is used to convert the acoustic pressure into the Sound Pressure level in dB. Acoustic reference pressure (P_{ref}) is the international standard for the minimum audible sound of 2.10^{-5} Pa [5].

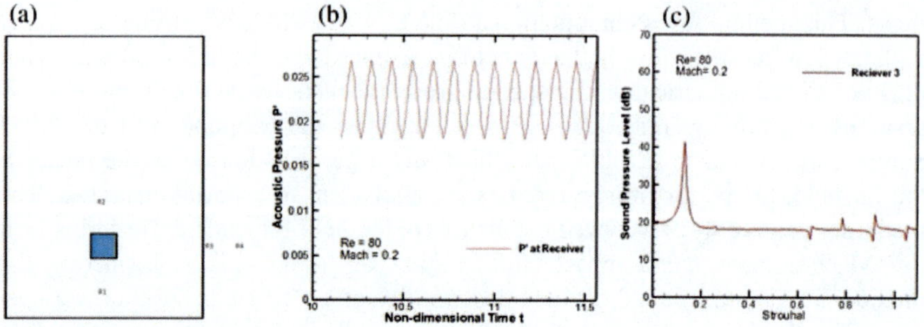

Fig. 3 **a** Geometrical representation of receivers **b** time variation of acoustic pressure level **c** SPL obtained from FFT of acoustic pressure

- Sound Pressure Level, $\text{SPL}_m(\text{dB}) = 20\log\left(\frac{P'_m}{P_{\text{ref}}}\right)$

Receivers were placed in upstream, downstream, and transverse directions as shown in the figure. Receivers were placed at R1(0, –4D), R2(0, 4D), R3(15D, 0), and R4(50D, 0). The centroid of the cylinder lies at (0, 0) as shown in Fig. 3a. In the direction of flow, receivers were placed equidistant both in the upstream and downstream direction to study the effects due to the flow. Due to symmetry in the transverse direction, receivers were placed only in one direction. Figure 3b shows Acoustic Pressure signals at the receiver. It has been observed that there exists a time lag in the receiver signals, this is because of the gap between the receivers. So the peak pressure of the pressure pulse reaches the nearest receiver at the earliest and then the next receiver and so on. There is also a reduction in the magnitude of the pressure pulse with the distance of the receiver. This is because of the dispersion of the wave. The amplitude of the pressure wave is observed to be increasing with the Mach number and Reynolds number. It is also observed that with increasing Mach number, the gap between the acoustic pressure levels at receiver 3 and receiver 4 also decreases for all cases of Reynolds number under study.

Sound pressure levels for all the receivers were generated by spectral analysis (Fig. 3c). When compared with the power spectral density of the lift force history (Fig. 2a), it was observed that peaks in the receiver spectra match the dominant frequency in the lift force history. This indicates that the sound is generated because of vortex shedding. Therefore, this peak corresponds to the shedding off frequency of the vortices.

From the data of the overall sound pressure level (SPL) shown in Figs. 4 and 5, it is also observed that the SPL increases with the Mach number for any particular Reynolds number, and it also increases with the increasing Reynolds number for any particular receiver in a fixed Mach number flow. The pattern indicates a well-defined SPL peak at a frequency which decreases with decreasing flow velocity. When compared for a particular Reynolds number, it is observed that the Acoustic Peak frequency shifts toward the right (toward higher frequency value) with increasing Mach number for all cases of Reynolds number. As shown in Fig. 4, the peak value

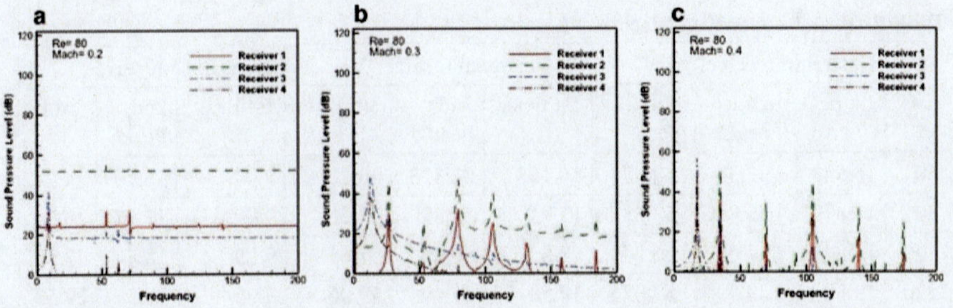

Fig. 4 SPL of all Receivers at a fixed Re = 80 and at **a** Ma = 0.2 b Ma = 0.3 **c** Ma = 0.4

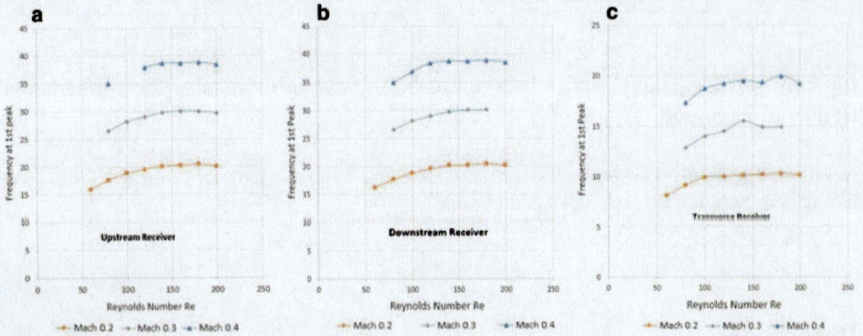

Fig. 5 Dominant frequency at different Re and Ma for receivers placed in **a** Upstream **b** Downstream and **c** Transverse directions

shows the sound pressure level's dominant frequencies. The sound spectrum consists of one strong peak as the fundamental frequency, a weaker peak at double that frequency, and wide band noise at frequencies higher than the fundamental. The value of the acoustic peak frequency for all the three configurations has been tabulated below (Table 2). It was found that the weaker peaks are actually multiples of the fundamental frequency. This shows that the wide band noises are actually the subharmonics of the fundamental noise. The wide band Noise at the frequencies above the fundamental appears to be quadruple from the wake.

4 Conclusions

It can be concluded that the peak in the receiver spectra matches the dominant frequency in the lift force history. It was also shown that Sound Pressure level decreases with decreasing flow velocity and Reynolds number, while Acoustic peak frequency shifts toward higher frequency value with increasing Mach number for all cases of Reynolds number. Wide band noises at frequencies higher than the

Table 2 Peak frequency at Mach = 0.2

Re	Upstream receiver			Downstream receiver			Transverse receiver		
	1st peak	2nd peak	3rd Peak	1st Peak	2nd Peak	3rd Peak	1st Peak	2nd Peak	3rd Peak
60	15.848	32.34	48.873	16.1587	32.323	48.2298	8.09	16.16	24.2425
80	17.677	35.866	53.537	17.71	35.89	53.66	9.091	17.57	26.96
100	18.848	37.64	56.24	18.83	36.52	57.61	9.89	18.84	27.55
120	19.58	39.49	59.07	19.58	39.49	59.06	9.96	19.58	29.53
140	20.2	40.41	60.22	20.2	40.4	60.41	10.1	20.2	30.29
160	20.302	61.44	81.73	20.28	40.59	60.89	10.14	20.29	30.44
180	20.514	40.705	61.22	20.507	40.701	61.21	10.25	20.51	30.45

fundamental frequency were observed to be the subharmonics of the fundamental frequency of sound.

Acknowledgements The Authors would like to acknowledge the help Dr. Yasir Rafat has given through encouraging tea table talks in this work.

References

1. Fujita H (2010) The characteristics of the Aeolian tone radiated from two-dimensional cylinders. Jpn Soc Fluid Mech 015002
2. Silva TDA, Ferreira FGT (2010b) Direct sound computation of a flow past a square cylinder at low mach and low reynolds numbers (150)
3. Ali, Doolan CJ, Wheatley V (2013) Aeolian tones generated by a square cylinder with a detached flat plate. AIAA J 51(2):291–301. Available at: http://arc.aiaa.org/doi/abs/10.2514/1.J051378
4. Ali MMS et al (2013) Aeolian tones radiated from flow over bluff bodies. Open Mech Eng J 7:48–57
5. Canonsburg TD (2011) ANSYS CFD-Post User's Guide 15317(November):724–746
6. Sharma A, Eswaran V (2013) Effect of aiding and opposing buoyancy on the heat and fluid flow across a square cylinder. Numer Heat Transf, Part A: Appl: Int J Comput Methodol Pub 37–41
7. Davis RW, Moore EF (1982) A numerical study of vortex shedding from rectangles. J Fluid Mech 453(May):30–36
8. Sohankar A, Davidson L, Norberg C (1995) Numerical simulation of unsteady flow around a square two-dimensional cylinder. In: Twelfth Australasian fluid mechanics conference, pp 517–520
9. Franke R, Rodi W, Schönung B (1990) Numerical calculation of laminar vortex-shedding flow past cylinders. J Wind Eng Ind Aerodyn 35:237–257

Instrumentation and Signal Processing

Direction of Arrival Estimation for Speech Source Using Closely Spaced Microphones

Mohd Wajid, Arun Kumar, and Rajendar Bahl

Abstract The direction of arrival (DoA) of an acoustic source is a salient parameter whose applications are diversified ranging from the defense industry to the entertainment industry. The conventional method of DoA estimation uses sensor (microphone or hydrophone) array with different computationally intensive algorithms, however, the size of sensor-array restricts its deployment in portable devices. The replacement of sensor-array is an acoustic vector sensor (AVS), which is compact in size and can be easily mounted on any small portable device. The AVS can be realized with omnidirectional microphones and/or particle velocity sensors. We have used different geometrical arrangements of closely spaced omnidirectional microphones (called AVS configurations) for determining the DoA of a speech source for real-time applications. The signals acquired by the closely spaced omnidirectional microphones are used to estimate the acoustic intensity vector, which gives the direction of an acoustic source. In this paper, we have given a method for DoA estimation of a broadband quasi-stationary source (speech signal) and compared the DoA estimation performance of different omnidirectional microphones based AVS configurations with the help of Finite Element Method tool, viz., COMSOL Multiphysics. The speech signals for different angular locations at different microphones of an AVS have been recorded using COMSOL Multiphysics modeling software.

Keywords Acoustic intensity · Acoustic vector sensor · Direction of arrival · Speech source

M. Wajid (✉)
Department of Electronics Engineering, ZHCET, Aligarh Muslim University, Aligarh 202002, India
e-mail: wajidiitd@gmail.com

M. Wajid · A. Kumar · R. Bahl
Centre for Applied Research in Electronics, Indian Institute of Technology Delhi, New Delhi 110016, India
e-mail: arunkm@care.iitd.ernet.in

R. Bahl
e-mail: rbahl@care.iitd.ernet.in

M. Singh and Y. Rafat (eds.), *Recent Developments in Acoustics*, Lecture Notes in Mechanical Engineering, https://doi.org/10.1007/978-981-15-5776-7_22

1 Introduction

Acoustic vector sensor (AVS) is a device that measures acoustic pressure (scalar) and particle velocity (vector). The product of pressure and particle velocity gives the acoustic intensity, which is a vector quantity and its direction is from the sound source to the AVS. Therefore, the acoustic intensity gives the direction of a sound source in free space [1–4]. The particle velocity sensor can be estimated using collocated microphones, where the approximation of the spatial derivative of the pressure signal is used [5, 6]. The direction of arrival (DoA) estimation using different AVS configurations of a narrowband source has been studied earlier in our paper [7] (readers can see this for details). However, we had not examined DoA estimation performance for a broadband quasi-stationary source. In this work, we have devised an approach for DoA estimation of a speech source in the absence of ambient noise and tested its performance for different configurations of the AVS using the Finite Element Method tool, viz., COMSOL Multiphysiscs.

The rest of the paper is organized as follows. Section 2 discusses the acoustic intensity components and DoA estimation using collocated microphones. Different AVS configurations used have been explained in Sect. 3. In Sect. 4, DoA estimation of a speech source and its results are presented, and Sect. 5 concludes the paper.

2 Acoustic Intensity Using Collocated Microphones and Direction of Arrival

If two microphones are located on the x-axis and separation between them is d, assuming separation between microphones is much smaller compared to the smallest wavelength of the source signal and range of the source ($d \ll \lambda$ and $d \ll r$), then the particle velocity along the x-axis is approximated as

$$v_x(t) = -\left(\frac{1}{\rho_0 d}\right) \int_{-\infty}^{t} [p_2(\tau) - p_1(\tau)]\mathrm{d}\tau \tag{1}$$

where ρ_0 is the medium density, and $p_1(t)$ and $p_2(t)$ are the measured microphone signals. It has been assumed that the signals $p_1(t)$ and $p_2(t)$ are stationary signals. The pressure at the midpoint of two microphones can be assumed to be the arithmetic mean of the two microphone signals:

$$p(t) = \frac{p_1(t) + p_2(t)}{2}. \tag{2}$$

The approximated acoustic intensity component along the x-axis is given by

$$I_x(t) \approx \left(\frac{1}{2\rho_0 d}\right)[p_1(t) + p_2(t)] \int_{-\infty}^{t} [p_1(\tau) - p_2(\tau)]\mathrm{d}\tau. \tag{3}$$

The frequency domain representation of the acoustic intensity component is expressed as

$$I_x(\omega) = -\frac{1}{\omega\rho_0 d}\mathrm{Im}ag\left(\Gamma_{p_1 p_2}(\omega)\right) \tag{4}$$

where $\Gamma_{p_1 p_2}(\omega)$ is the cross-power spectral density between the microphone signals $p_1(t)$ and $p_2(t)$, and $Imag\,(*)$ is the imaginary part. The average intensity component is given by

$$I_x = -\frac{1}{\omega\rho_0 d}\int_{0}^{\infty} \mathrm{Im}ag\left(\Gamma_{p_1 p_2}(\omega)\right)\mathrm{d}\omega. \tag{5}$$

Similarly, the intensity component $I_y(\omega)$ and I_y along the y-axis can be determined [4, 5, 8–10] (see these papers for detail). The DoA of an acoustic source in a 2D plane can be estimated as

$$\theta = \arctan\left[\frac{I_x}{I_y}\right]. \tag{6}$$

The DoA is measured with reference to the y-axis in the clockwise direction. This intensity method for DoA estimation can be used for real-time applications as it does not need any search in the space.

3 Acoustic Vector Sensor Configurations

We have considered several AVS configurations for calculating the acoustic intensity due to speech source and hence DoA of a speech source. These AVS configurations are called star, delta, OACφ, square, and sub-configurations of square (AVS1, AVS2, AVS3, AVS4, and AVS5), the geometrical arrangement of these configurations are given in Fig. 1.

The configurations like delta, star, and OACϕ ($\phi = 30°$) do not give orthogonal intensity components (along the x-axis or y-axis). For these configurations, the acoustic intensity components have been calculated along the solid lines. These intensity components have been projected on the orthogonal axes in order to obtain the intensity component along the x and y axes [7]. The DoA estimate of AVS13 has been derived after taking the average of the DoA estimate obtained using AVS1

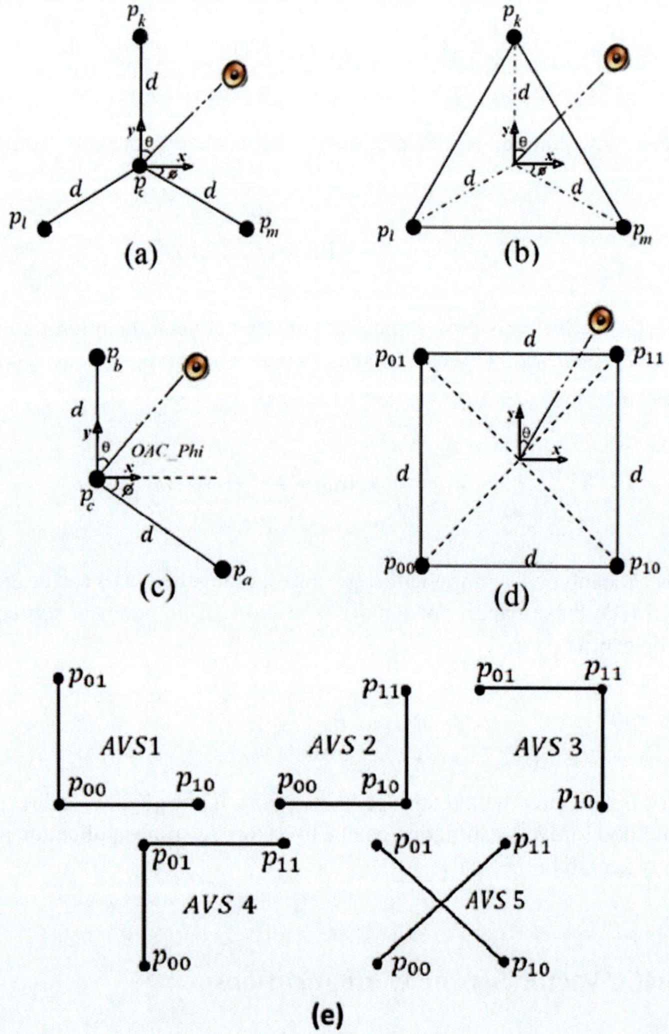

Fig. 1 Microphone configurations for AVS. Circles indicate omnidirectional microphone, speaker symbol indicates the sound source in the far-field. **a** Star configuration, having four measured signals, i.e., p_c, p_k, p_l, and p_m. **b** Delta configuration, having three measured signals, i.e., p_k, p_l, and p_m. **c** Three-microphone obtuse angle configuration (OACϕ), having three measured signals, i.e., p_a, p_b, and p_c, **d** Square configuration, and **e** Sub-configurations of square, having measured signals p_{00}, p_{10}, p_{11}, and p_{01} [7]

and AVS3. Similarly, DoA estimate from AVS1, AVS2, AVS3, and AVS4 have been averaged, and the resultant DoA estimate is called the DoA estimate of AVS1234.

4 Direction of Arrival Estimation of a Speech Source

The speech signal received by microphones of AVS configurations for different angular locations of a speech source has been generated using COMSOL Multiphysics. The distance between speech source and AVS is kept at 1 m and microphone separation d within AVS configurations is fixed at 10 mm. The angular locations of speech source are from 0° to 90° with increments of 15°. It has been assumed that sound source and microphones are of point size, and there is no reflection from the boundary.

The performance has been evaluated in terms of angular error (AE) and average absolute angular error $\overline{AAE}_{0^\circ:15^\circ:90^\circ}$ as defined below:

$$AE = 2\sin^{-1}\left(\frac{||a - \hat{a}||}{2}\right) \tag{7}$$

and

$$\overline{AAE}_{0^\circ:15^\circ:90^\circ} = \frac{1}{7}\sum_{K=0}^{6}|AE(\theta_K)| \tag{8}$$

where $||.||$ is the l_2 norm, $\boldsymbol{a}$ and $\hat{a}$ are the vectors of unit magnitude pointing to the actual direction of the speech source and its estimate, respectively. $AE(\theta_k)$ represents AE for a speech source located at an angle of θ_k ($\theta_k = k * 15^\circ$).

Since speech is generally assumed to be quasi-stationary with a duration of 20 ms; therefore, the speech signal has been analyzed for DoA estimation for successive non-overlapping 20 ms frames. The AE for the speech signal is the average of the estimated angular error from successive 20 ms frame of speech signal.

The DoA for each frame is calculated based on the average intensities (obtained over all frequency bins). The spectrogram of speech source of 3 s duration and sampled at 16 kHz is shown in Fig. 2a. For a speech source with frequency band of 200 Hz–8 kHz, the range of d/λ is from 0.00583 to 0.2332. Figure 2b shows the AE for the speech source of 3 s of 8 kHz band and the average AE spanning the 3 s duration of the signal is given after removal of 10% outliers. The reason for non-zero AE at different DoAs is due to particle velocity estimations. The AE is zero for some DoAs (0°, 45°, 60° and 90° this is due to the cancelation of approximation error (in particle velocity) by AVS configuration geometries. Figure 3 shows $\overline{AAE}_{0^\circ:15^\circ:90^\circ}$ for a speech source.

The AVS configurations, which use three microphones, have higher average error than the case of AVS configuration that used four microphones. It has been observed

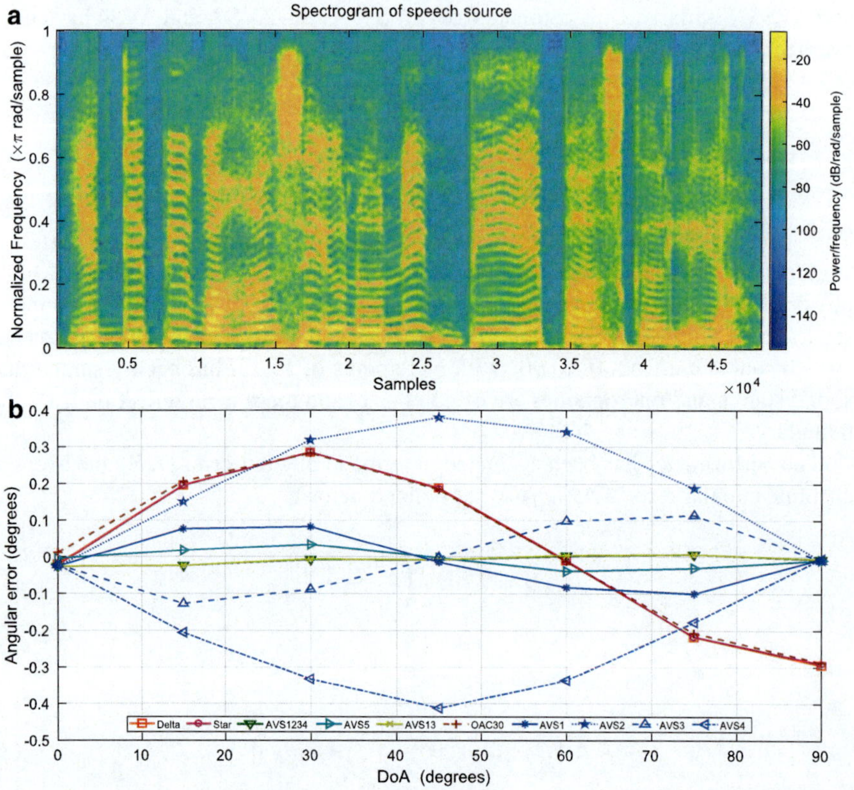

Fig. 2 **a** Spectrogram of the speech source of 3 s duration and sampled at 16 kHz. **b** AE versus DoAs for various AVS configurations for the speech source located at a range of one meter with d = 10 mm

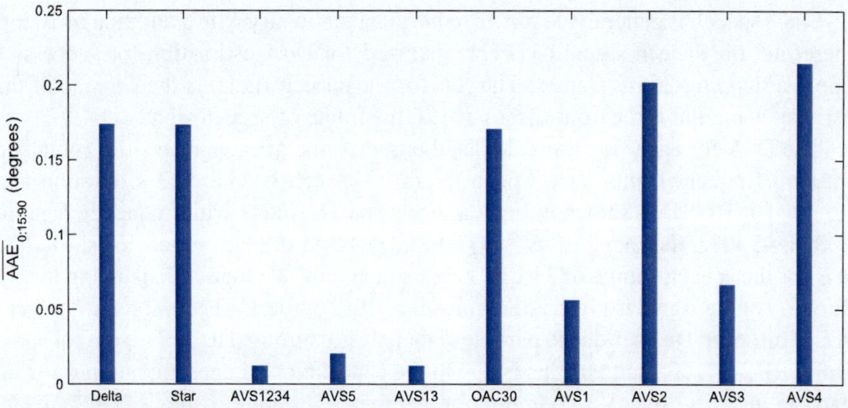

Fig. 3 $\overline{AAE}_{0^\circ:15^\circ:90^\circ}$ for different AVS configurations for speech source of 3 s located at a range of one meter with d = 10 mm

Table 1 Error performance comparison for speech source signal and sinusoidal source signal for different AVS configurations

	$\overline{AAE}_{0^\circ:15^\circ:90^\circ}$ (degree)	
AVS configurations	Speech source	Sinusoidal source (1 kHz)
AVS1234	0.016	0.043
AVS5	0.013	0.078
AVS13	0.016	0.043
AVS1	0.052	0.013
AVS2	0.192	0.208
AVS3	0.072	0.098
AVS4	0.226	0.201

that AVS13 and AVS1234 gives smaller value of $\overline{AAE}_{0^\circ:15^\circ:90^\circ}$ than the other AVS configurations considered, which is also applicable for the narrowband signal as reported in [7]. The presented scheme for DoA estimation of broadband signal using different AVS configurations gives $\overline{AAE}_{0^\circ:15^\circ:90^\circ}$ less than 0.23° . The average error performance for the sinusoidal source has been reported earlier [7], these results are compared with our results for the speech signal and are shown in Table 1.

5 Conclusion

We have presented an approach for DoA estimation of a broadband quasi-stationary source and tested the same for different AVS configurations with the help of the Finite Element Method tool. It has been observed that the average error for speech source using AVS1234 and AVS12 is less than 0.02°, which is better than the case of sinusoidal source. The average error of DoA estimation is larger for the three microphone AVS configurations compared to the four microphone AVS configurations. However, all AVS configurations have angular error below 0.4° and average absolute angular error below 0.23° for the speech source. Future work may include the analysis of DoA estimation of speech source in the presence of ambient noise. In addition, this work may be extended to remove the angular error caused due to particle velocity approximations.

References

1. Bai MR, Juan S-W, Chen C-C (2013) Particle velocity estimation based on a two-microphone array and kalman filter. J Acoust Soc Am 133(3):1425–1432
2. FernáandezComesaña D, Wind J, Grosso A, Holland K (20011) Performance of pp and pu intensity probes using scan and paint. In: International congress on sound and vibrations, ICSV, Rio de Janeiro, Brazil
3. Silvia MT, Richards RT (2002) A theoretical and experimental investigation of low-frequency acoustic vector sensors. OCEANS'02 MTS/IEEE 3:1886–1897

4. Miah KH, Hixon EL (2010) Design and performance evaluation of a broadband three dimensional acoustic intensity measuring system. J Acoust Soc Am 127(4):2338–2346
5. Fahy FJ (1977) Measurement of acoustic intensity using the cross-spectral density of two microphone signals. J Acoust Soc Am 62(4):1057–1059
6. Fahy FJ (1995) Sound intensity. E&FN spon, 2nd ed. London
7. Wajid M, Kumar A, Bahl R (2016) Design and analysis of air acoustic vector-sensor configurations for two-dimensional geometry. J Acoust Soc Am 139(5):2815–2832
8. Fernandez Comesana D, Wind J, Grosso A, Holland KR (2011) Performance of PP and PU intensity probes using Scan and Paint
9. Krishnappa G (1981) Cross-spectral method of measuring acoustic intensity by correcting phase and gain mismatch errors by microphone calibration. J Acoust Soc Am 69(1):307–310
10. Chung JY (1978) Cross-spectral method of measuring acoustic intensity without error caused by instrument phase mismatch. J Acoust Soc Am 64(6):1613–1616

Investigations into Some Parameters of Vibration Signal of Faulty Bearings with Wavelet Transform

Sidra Khanam and N. Tandon

Abstract The vibration signals generated by a rotor bearing system are greatly influenced by the occurrence of a fault on the bearings. Monitoring of parameters like peak, overall rms, kurtosis, crest factor, and power from the time domain signal reveals the status of health of bearings supporting the rotor and serves as the easiest approach. However, the majority of these parameters give an estimation of the overall health of the system and not just the bearings. This work addresses the use of Discrete Wavelet Transform (DWT) to filter the bearing related signal and then monitor the health of bearing. The work also discusses on the selection of appropriate sampling frequency for the collection of time domain signal and demonstrates the dependence of parameters on the same. The results indicate that the use of DWT for a signal with higher sampling frequency has increased the accuracy of prediction of estimation of the defect.

Keywords Vibration monitoring · Time domain signal · Rolling element bearings · Discrete wavelet transform

1 Introduction

Rolling element bearings are widely used in industrial applications and faults therein, unless identified in a timely manner, may lead to malfunctioning of machinery. Condition monitoring has, therefore, become a widely adopted maintenance strategy with monitoring of vibration signal as a common tool. The vibration signals generated by a rotor bearing system are greatly influenced by the occurrence of a fault on the bearings. These faults can be detected from vibration signals through various

S. Khanam (✉)
Department of Mechnical Engineering, Aligarh Muslim University, Aligarh 202002, India
e-mail: sidra.khanam10@gmail.com

N. Tandon
ITMMEC, IIT Delhi, New Delhi 110016, India
e-mail: ntandon@itmmec.iitd.ac.in

M. Singh and Y. Rafat (eds.), *Recent Developments in Acoustics*, Lecture Notes in Mechanical Engineering, https://doi.org/10.1007/978-981-15-5776-7_23

techniques that can be categorized into time domain, frequency domain, or time–frequency domain [1, 2].

Tandon and Nakra [3] have compared the detectability of defect in ball bearings by some vibration and acoustic measurement methods. These measurements included overall vibration acceleration, envelope detected acceleration, overall sound intensity and sound pressure, shock pulse, and acoustic emission ringdown counts and peak amplitude. The authors weighted envelope detected acceleration over other methods. In another research, Tandon [4] contrasted vibration parameters like total RMS, peak, crest factor, power, and cepstrum and found that power is best for defect detectability followed by peak and RMS measurements. Shakya et al. [5] have presented an exhaustive study of various time domain, frequency domain, and time–frequency domain parameters and provided an overall ranking of several parameters. Patidar and Soni [6] have presented a summary of the vibration monitoring techniques and concluded that wavelet analysis is a better tool for fault diagnosis. Because of their flexibility and effective computational execution, wavelet transformation and wavelet decomposition [7–9] have been reported to be a proficient tool for machine failure surveillance. Peng and Chu [7] have stated key aspects for machine fault diagnosis as time–frequency analysis of signals, fault feature extraction, singularity detection, denoising and extraction of the weak signals, compression of vibration signals, and system identification. Researchers in references [10–12] have applied a Discrete Wavelet transform (DWT) for bearing fault detection. Prabhakar et al. [10] have used DWT to identify single and multiple faults and combined faults on ball bearing races, wherein the Daubechies 4 mother wavelet was used to decompose the vibration signals up to level four. Kumar and Singh [11] and Khanam et al. [12] have used Symlet5 wavelet as a mother wavelet, and extracted the bearing defect related feature to predict the defect size on the outer race of taper roller bearing and deep groove ball bearing, respectively.

Monitoring parameters like peak, rms, power, crest factor, and kurtosis from time domain vibration signal serve as one of the easiest approaches of defect detection in bearings. Moreover, these parameters give only an indication of the presence of the defect and require reference data (that of healthy bearing) for the same. In frequency domain monitoring, characteristic defect frequencies are specifically looked for and these lie in the low frequency range. This gives rise to ambiguity for the acquisition of signal using an FFT analyzer, which is more or less user-defined, to be used for monitoring of parameters in the time domain

This work presents the study of these parameters for signal acquired with two sampling frequencies, low (640 Hz) and high (25.6 kHz). DWT has then been applied to observe the improvement in the defect detectability.

2 Experimental Measurements

The vibration signals for a rotor bearing system were acquired from a test rig used in reference [12]. The test facility measures the vibration of radially loaded rolling element bearing for bore diameters ranging from 15 mm to 35 mm. For this study, the test bearing is fixed on the stepped part of the shaft with bore 20 mm. The test bearing SKF BB1B420204 is a deep groove ball bearing with polymer cage. The polymer cage enables in the mantling and dismantling of components for the successive development of defects on the race. Circular defects, to represent spall like fault, have been progressively seeded onto the outer race of bearing with Electric Discharge Machining (EDM) process. The defect size was chosen to vary from very small defect (to represent the signal of defect at early stage) to medium (to represent the signal of reasonably identifiable defect). For this, the diameter of the defect was selected in steps of 250 micron; however, due to experimental limitations, the accurate values in steps of 250 micron could not be achieved. The details of the test bearing and defects created on the bearing's outer race are provided in Tables 1 and 2, respectively.

For carrying out vibration measurements, test bearing was loaded with 100 N and the inner race was rotated with 1500 rpm and the outer race was held stationary. The vibration signals were obtained from the top of bearing housing (load zone) in the form of acceleration with Brüel & Kjær type 4368 accelerometer. The captured signal was stored in ONO SOKKI (CF- 3200) Fast Fourier Transform (FFT) Analyzer through charge amplifier (B&K 2635) and analyzed via post processing in the MATLAB environment. Data acquisition started after 20 minute-run, with 4096 vibration data samples (each sample an average of 100 samples) collected at both low and high frequency sampling rates. Faults like flakes and spalls are represented

Table 1 Details of the test bearing SKF BB1B420204

Bearing bore	0.020 m
Inner race diameter (d_i)	0.02424 m
Outer race diameter (d_o)	0.04164 m
Pitch diameter (D_p)	0.03294 m
Ball diameter (d_b)	0.0087 m
Diametric clearance (P_d)	10×10^{-6} m
Number of balls (Z)	7
Grooves radii	0.00461 m

Table 2 Description of defect seeded on the outer race through EDM

	Stage1	Stage 2	Stage 3	Stage 4	Stage 5	Stage 6	Stage 7	Stage 8
Diameter (mm)	0.35	0.50	0.65	0.90	1.30	1.50	1.77	2.02
Depth (mm)	0.13	0.13	0.25	0.25	0.25	0.25	0.40	0.50

by bearing characteristic frequency in the bearing defect frequency region (zone B) at stage 3 of bearing failure. Bearing defect frequency region is a low frequency region and so, the time signal for this low frequency region (span varying from 2 to 300 Hz) has been studied to investigate the variation of few parameters of this signal. High frequency acceleration signal in the range of 2–20 kHz has also been studied for the variation in vibration parameters and statistical properties like kurtosis.

3 Calculation of Vibration Parameters and Application of Discrete Wavelet Transform

Parameters like peak value, overall rms, power, kurtosis, and crest factor are calculated from the time domain vibration signal. The estimation of these parameters acts as the easiest way to inspect any change in the signal and is therefore categorized among the easiest technique for condition monitoring using vibration signal. These parameters are described again for the sake of completeness using references [4, 6]. The peak value is the maximum value that a time frame of vibration signal attains. Root mean square (rms) value gives the effective value of a varying vibration signal and is expressed as

$$\mathrm{RMS}(x) = \sqrt{\sum_{i=1}^{N} x_i^2 \Big/ N} \tag{1}$$

where N is the total number of samples. The value of overall rms acceleration level is compared with that of healthy bearing to monitor any change; however, this parameter is insensitive to detect a fault at the early stage.

In signal processing, the power of a discrete time domain signal is defined as

$$\text{Power } P = \frac{\sum_{i=1}^{N} |x_i|^2}{N} \tag{2}$$

Crest factor is defined as the ratio of peak acceleration over overall rms, and is mathematically expressed as

$$\text{Crest factor} = \frac{\mathrm{Peak}(x)}{\mathrm{RMS}(x)} \tag{3}$$

Crest factor is a non-dimensional parameter, and Igarashi et al. [14] have reported a value of approximately 5 for a good bearing. With the progression of bearing damage, rms increases and crest factor decreases.

Kurtosis gives a measure of peakedness in the signal. The examination of statistical moments of the data reveals quite useful information [1]. Kurtosis is nothing but the fourth moment normalized with respect to the fourth power of standard deviation, and is expressed as

$$k = \frac{\sum_{n=1}^{N} (x(n) - \mu)^4}{\sigma^4} \quad (4)$$

where μ is the mean and σ is the standard deviation. Bearing with good surface finish has a theoretical kurtosis of 3, and with the deterioration in surface finish value of kurtosis increases, but kurtosis is insensitive to loads and speeds [6].

All the above parameters are obtained for both the signals; one with a low sampling rate, i.e., 640 Hz and the other with a high sampling rate, i.e., 25.6 kHz.

3.1 Application of Discrete Wavelet Transform (DWT)

Discrete wavelet transform is an execution of the wavelet transform using a discrete set of the wavelet scales and translations obeying some defined rules. This work uses DWT as the derivative of the continuous wavelet transform obtained by adopting the dyadic scale and translation to reduce the computational time and is expressed after [10, 12] by the following equation:

$$\mathrm{DWT}(j, k) = \frac{1}{\sqrt{2^j}} \int_{-\infty}^{\infty} s(t)\psi * \left(\frac{t - 2^j k}{2^j}\right) dt \quad (5)$$

where j, k are integers, 2^j and $2^j k$ represent the scale and translation parameter, respectively, ψ is the 'mother' wavelet, and ψ^* is the complex conjugate of ψ. Conceptually, DWT uses filter banks for the study of signal wherein filter banks contain wavelet filters that extract the frequency content of the signal in various sub-bands. The original signal $s(t)$ goes through a set of low pass and high pass filters resulting in low frequency (approximations, a_i) and high frequency (details, d_i) signals at each stage 'n' of decomposition. Therefore, the original signal $s(t)$ can be written as [11, 12]

$$s(t) = a_n + \sum_{i=1}^{n} d_i \quad (6)$$

In this work, the mother wavelet chosen for decomposition of signal is the Symlet wavelet. The Symlets are compactly supported wavelets with least asymmetry and highest number of vanishing moments for a given support width [13]. The perfect reconstruction and cancelation capability allows them to be used in both continuous wavelet transform and discrete wavelet transform. The use of sym5 wavelet has been reported [11, 12] for defect quantification in rolling bearings because of its shape. This work also makes the use of sym5 wavelet for decomposing the signal.

4 Results and Discussion

The results of peak value, rms value, and power of the low-frequency content vibration signal (2–300 Hz) is shown in Fig. 1. The actual values are shown with the points of different markers for different parameters and the solid line in the curve denotes the exponential fit curve for the respective parameters. Exponential fit has been chosen from literature where accelerated life test results are reported and the vibration parameters are observed to undergo exponential increase with time/rate of degradation. Figure 2 shows the variation of non-dimensional statistical parameters for the signal acquired with low-frequency content. Figure 2a, b shows the variation of crest factor and kurtosis, respectively. Literature [14] reveals a value of crest factor to be approximately 5 for good bearing and a decrease with the increase of defect, but the crest factor is reported to be a poor indicator of the defect. Figure 2a shows the value of crest factor less than 5 for all the cases, but a definite trend is not observed. Similarly, kurtosis variation in Fig. 2b gives a value of close to 3 and a non-defined trend; however, literature reports a value of greater than 3 for defective bearings [1].

Figures 3 and 4 present the result for vibration signal acquired with high sampling frequency or high-frequency content (2–20 kHz). Figure 3 depicts the variation of peak value, rms value, and power along with exponential fits for the same. The values of power are observed to be very sensitive to defect followed by peak values and rms values. The defect detectability is also reported in the same sequence in literature [4]. Figure 3 also reports a lesser deviation in the experimental values and curve fitted values. Figure 4a, b shows the variation of crest factor and kurtosis for a signal with high-frequency content; the trend is observed to undergo increasing pattern as

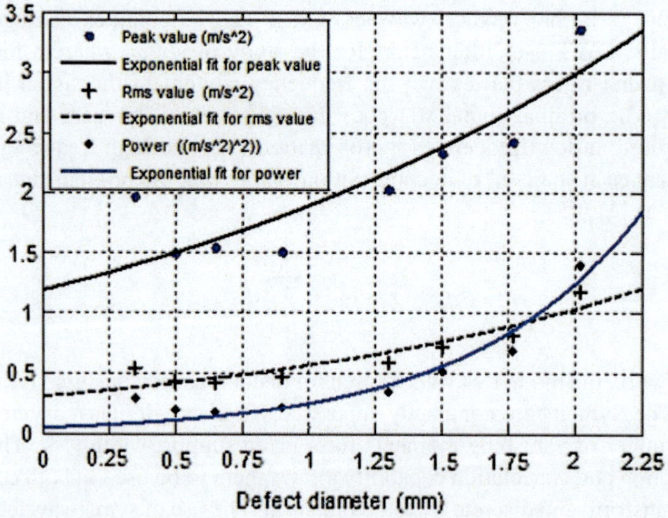

Fig. 1 Variation of peak value, rms value, and power of signal with defect size for a signal with low-frequency content

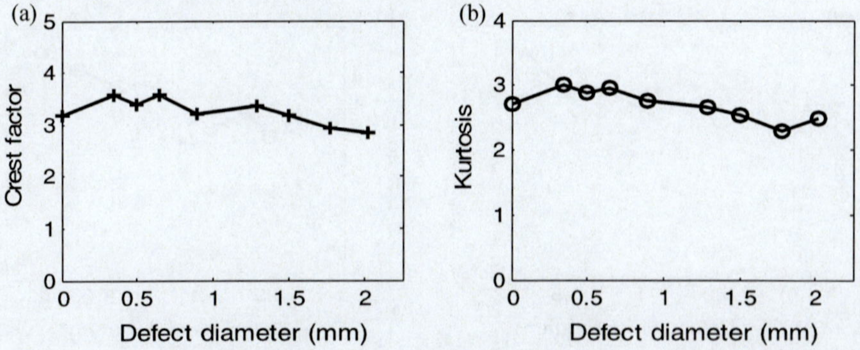

Fig. 2 Variation of non-dimensional parameters with defect size for a signal with low-frequency content **a** crest factor variation, **b** kurtosis variation

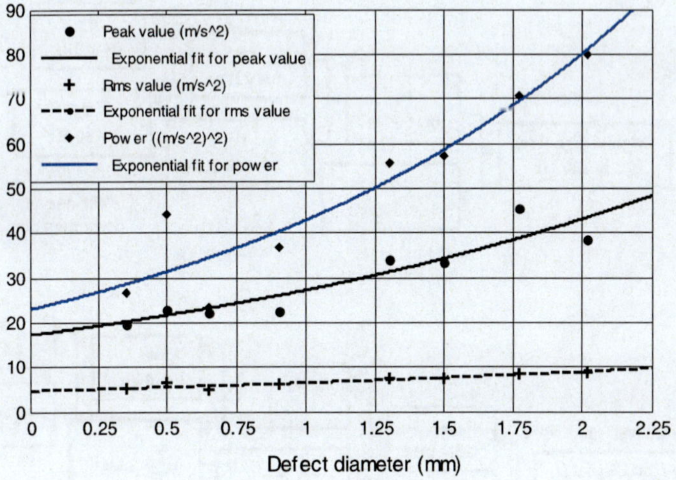

Fig. 3 Variation of peak value, rms value, and power of signal with defect size for a signal with high-frequency content

compared to the one with low-frequency content signal and an improvement in the value of kurtosis (i.e., greater than 3) is observed.

DWT has been applied for both low-frequency content signal and high-frequency content signal. The sampling frequency of low-frequency content signal is 640 Hz; therefore, the highest frequency content of the low frequency according to Nyquist's criterion is 320 Hz. The sampling frequency of high-frequency content signal is 25.6 kHz and hence the highest frequency of this signal is 12.8 kHz. Figure 5 shows the frequency bandwidth of wavelet decomposition for the time signal.

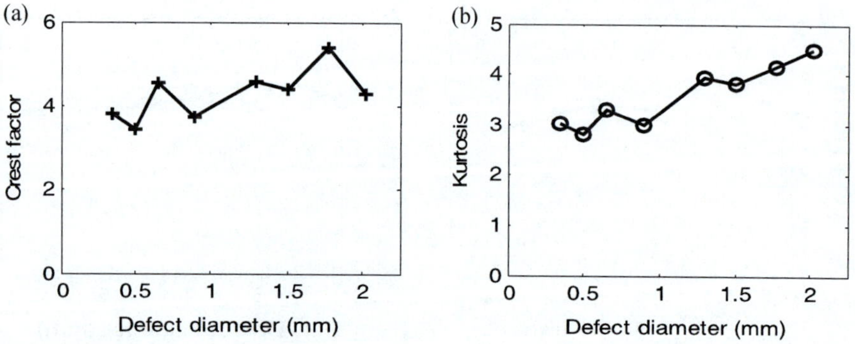

Fig. 4 Variation of non-dimensional parameters with defect size for a signal with high-frequency content **a** crest factor variation, **b** kurtosis variation

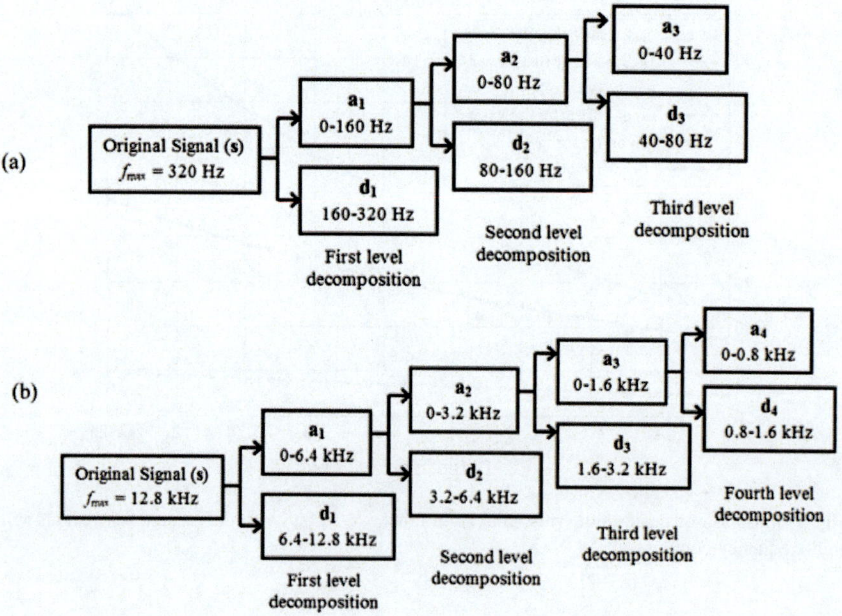

Fig. 5 Frequency bandwidth of wavelet decomposition **a** for low-frequency content signal; **b** for high-frequency content signal

Figure 5a shows the frequency bands for three levels of decompositions for low-frequency content signal and Fig. 5b shows the frequency bands for four levels of decompositions for high-frequency content signal.

It is known that the local defects in bearing excite high frequency vibrations but the characteristic defect frequencies fall in low frequency regime. When a defect is monitored in frequency domain, the presence/increase of/in vibration at the defect frequency is observed. The defect frequency corresponding to outer race defect is given by Eq. 7 following [1] and its value for the given test bearing and shaft rotational speed of 1500 rpm is found out to be 64.38 Hz.

$$f_{od} = \frac{Zf_s}{2}\left(1 - \frac{d_b}{D_p}\cos\alpha_c\right)\ \text{Hz} \tag{7}$$

The signal acquired in the low frequency region is such that it contains four harmonics of bearing characteristic frequency (2–300 Hz). Now, if the feature pertaining to outer race defect frequency is to be picked from the time domain of the above signal, the signal needs to be decomposed up to a maximum of three levels of decomposition and the feature corresponding to outer race defect frequency should be present in detail d_3 (40–80 Hz) of Fig. 5a.

Figure 6 presents the low-frequency content signal for two defect sizes (a) 0.65 mm and (b) 2.02 mm; the top of the figure shows the original signal and the bottom part represents the wavelet decomposed signal at $D3/d_3$. The decomposed signal at $D3/d_3$ depicts the reduction in noise level; however, the clarity in defect-related signal is not prominent.

When looking into the aspect that the presence of defect generates impulses that produce high frequency components, the defect features are supposed to appear in the high frequency range of wavelet decomposition [10, 12]. So, approximation at the fourth level of decomposition ($a_4/A4$) is chosen for this study. Figure 7 presents the high-frequency content signal for two defect sizes. Figure 7a shows the original signal and decomposed signal at $a_4/A4$ for defect size of 0.65 mm and Fig. 7b presents the original signal and decomposed signal at $a_4/A4$ for defect size of 2.02 mm. The periodicity of defect impulses is clearly visible for defect of 2.02 mm and the signal is significantly improved after DWT.

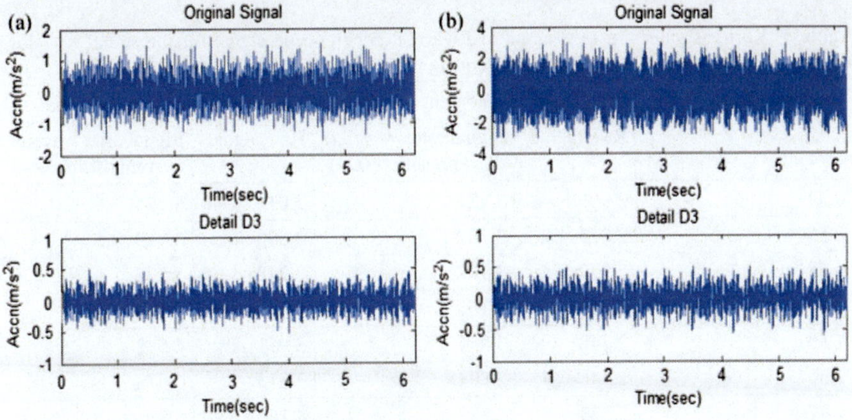

Fig. 6 Original signal and detail at level 3 (low-frequency content signal) for **a** outer race defect diameter of 0.65 mm **b** outer race defect diameter of 2.02 mm

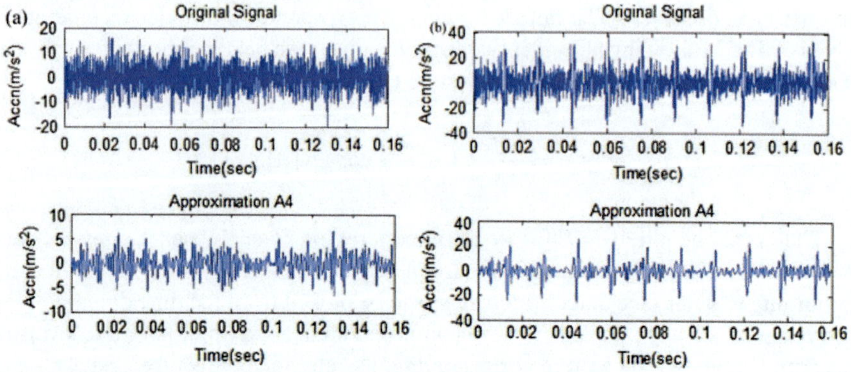

Fig. 7 Original signal and approximation at level 4 (high-frequency content signal) for **a** outer race defect diameter of 0.65 mm **b** outer race defect diameter of 2.02 mm

DWT is also known to reduce the effect of noise, so the selected signal (appropriate bandwidth signal for low and high-frequency content signal) is investigated for defect indicator in the time domain. In this study, only kurtosis has been selected as the parameter for investigation since the trends of other parameters (peak, rms, power) is already established and it would not vary after application of DWT and crest factor is reported as a weak indicator of the presence of a defect. Kurtosis gives a measure of peakedness in the signal which is observed to increase after the application of DWT as evident from the time domain representations in Figs. 6 and 7. Table 3, therefore, represents the kurtosis values of raw signal and wavelet decomposed signal of appropriate bandwidth for vibration signals with low and high-frequency content. The values from the low-frequency content signal give a marginal improvement in the defect indication (value greater than 3). However, the kurtosis values from the

Table 3 Kurtosis values of raw signal and wavelet decomposed signal for the signal with low-frequency content (0–320 Hz) and high-frequency content (0–12.8 kHz)

Defect diameter (mm)	Low-frequency content signal		High-frequency content signal	
	Raw signal	Signal after wavelet decomposition (*D*3)	Raw signal	Signal after wavelet decomposition (*A*4)
0.35	2.996	3.177	3.014	3.274
0.50	2.875	3.167	2.806	3.419
0.65	2.932	3.101	3.288	3.576
0.90	2.766	3.105	2.983	4.274
1.30	2.748	3.234	3.951	4.542
1.50	2.546	2.894	3.813	6.078
1.77	2.333	2.883	4.162	5.538
2.02	2.503	2.693	4.490	6.866

decomposed signal of high-frequency content provide a better picture of the presence of defect as well as a definite increasing trend.

5 Conclusion

This work presents the study of vibration parameters for signal acquired with two sampling frequencies, low and high. DWT has then been applied to observe the improvement in the defect detectability. The conclusions from this work can be summarized as follows:

- Acquisition of signal, which is user-defined, affects the vibration parameters and governs the detection of defect.
- The signal acquired with low-frequency content gives an exponentially increasing trend (with some deviation from experimental values) of peak value, rms value, and power. However, non-dimensional parameters like crest factor and kurtosis are observed to undergo random trends and deviate from the standard values.
- The signal acquired with high-frequency content also gives an exponentially increasing trend but a better trend is observed than that with low-frequency content signal. The sensitivity to defect detection is best represented by power, followed by peak and rms values; a similar trend is reported in the literature as well. Crest factor and kurtosis are observed to depict a definite trend and their values are also acceptable with reference to the standard values.
- The application of DWT for a signal with higher sampling frequency has significantly increased the value of kurtosis. This is because DWT segregates the signal in the appropriate bandwidth and therefore, bearing related impulsive signals are filtered out from the raw vibration signal and hence kurtosis, a measure of peakedness, is significantly improved.

Thus, when using time domain monitoring techniques, one should acquire the signal with high sampling frequency, and the application of DWT enables an improvement in defect detection.

Acknowledgments Authors are thankful to the staff members of Machine Dynamics Laboratory (ITMMEC), IIT Delhi for providing all possible help during experimentations.

References

1. Tandon N, Choudhury A (1999) A review of vibration and acoustic measurement methods for the detection of defects in rolling element bearings. Tribol Int 32:469–480
2. Patil MS, Mathew J, RajendraKumar PK (2008) Bearing signature analysis as a medium for fault detection: a review. ASME J. Tribol 130(1):14001
3. Tandon N, Nakra BC (1992) Comparison of vibration and acoustic measurement techniques for the condition monitoring of rolling element bearings. Tribol Int 25(3):205–212

4. Tandon N (1994) A comparison of some vibration parameters for the condition monitoring of rolling element bearings. Measurement 12:285–289
5. Shakya P, Darpe AK, Kulkarni MS (2013) Vibration-based fault diagnosis in rolling element bearings: ranking of various time, frequency and time-frequency domain data-based damage identification parameters. Int J Cond Monit 3(2):1–10
6. Patidar S, Soni PK (2013) An overview on vibration analysis techniques for the diagnosis of rolling element bearing faults. Int J Eng 4(5):1804–1809
7. Peng Z, Chu F (2004) Application of the wavelet transform in machine condition monitoring and fault diagnostics: a review with bibliography. Mech Syst Signal Process 18:199–221
8. Yan R, Gao RX, Chen X (2014) Wavelets for fault diagnosis of rotary machines: a review with applications. Sig Process 96:1–15
9. Chebil J, Noel G, Mesbah M, Deriche M (2009) Wavelet decomposition for the detection and diagnosis of faults in rolling element bearings. Jordan J Mech Ind Eng 3(4):260–267
10. Prabhakar S, Mohanty AR, Sekhar A (2002) Application of discrete wavelet transform for detection of ball bearing race faults. Tribol Int 35(12):793–800
11. Kumar R, Singh M (2013) Outer race defect width measurement in taper roller bearing using discrete wavelet transform of vibration signal. Measurement 46(1):537–545
12. Khanam S, Tandon N, Dutt JK (2014) Fault size estimation in the outer race of ball bearing using discrete wavelet transform of the vibration signal. Procedia Technol. 14:12–19
13. http://visl.technion.ac.il/documents/wavelet_ug.pdf
14. Ingarashi T, Noda B, Matsushima E (1980) A study on the prediction of abnormalities in rolling bearing. JSLE Int 1:7176

Standard Practices for Acoustic Emission Measurement on Rotating Machinery

Surojit Poddar and N. Tandon

Abstract Acoustic Emission Technique (AET) is a state-of-the-art method to measure faint surface vibration resulting from stress waves when a material is subjected to crack, fracture, corrosion, friction, wear, turbulence, etc. Acoustic emission is a widely used and standardized technique in the field of structural diagnosis. The technique has been extended to machinery condition monitoring in recent years. However, with the non-availability of standards dedicated to the machinery section, the measurements are mostly carried out on personal experiences. This article is a set of standard practices, drawn from already established standards, new methods conceived and applied by researchers in their work and authors' personal experiences, for application of AET in rotating machinery.

Keywords Acoustic emission · Condition monitoring · Rotating machinery · Standard practices

1 Introduction

Acoustic emission as defined by the American Society for Testing and Materials is "the class of phenomena whereby transient stress/displacement waves are generated by the rapid release of energy from localized sources within a material, or the transient waves so generated" [1]. A basic AE system consists of a sensor, a preamplifier, and a data acquisition system as shown in Fig. 1.

The stress wave propagating through the surface induces vibration whose amplitude can be as low as a picometer. The embedded piezoelectric element inside the sensor converts this displacement into an electrical signal [2], which is appropriately processed by the preamplifier and data acquisition system (DAQ) to extract parametric and signal-based features.

A measurement process consists of a series of activities and steps—implementation of which according to standard practices is essential to yield accurate and

S. Poddar (✉) · N. Tandon
Centre for Automotive Research & Tribology (formerly ITMMEC), Indian Institute of Technology Delhi, New Delhi, India
e-mail: surojitpoddar88@gmail.com

M. Singh and Y. Rafat (eds.), *Recent Developments in Acoustics*, Lecture Notes in Mechanical Engineering, https://doi.org/10.1007/978-981-15-5776-7_24

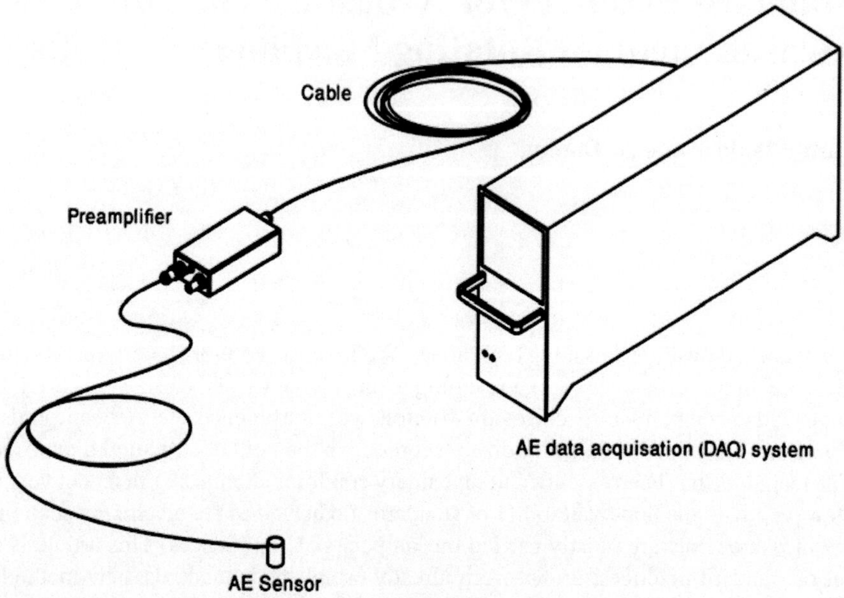

Fig. 1 Schematic diagram showing components of a basic acoustic emission system

repeatable results. Execution of standard practices becomes essentially important for acoustic emission technique as the surface displacement associated with AE is very small, and the probability of noise infestation from surrounding is very high. Added to this, the transmission loss of AE energy from the source to the sensor is also high.

There is plenty of literature, standards, and books pertaining to the use of AE for structural diagnosis. However, the application of AE in machines being sporadic, there is a lack of proper literature and references describing standard practices for the effective and errorless application of AE for condition monitoring of machines. This article fills this gap. The standard practices in the selection of instrumentation settings, measurement parameters, sensor type, couplant, mounting methods, and sensor location have been exhaustively discussed and illustrated.

2 Basic AE Instrumentation Settings

2.1 Threshold Level and Elimination of Noise

The AE threshold is an important setting in acoustic emission measurement as only the signal that crosses the threshold level is processed by the system [3]. A threshold set too low will freeze the system and a threshold set too high would miss useful emissions in case of low noise signals [4]. Usually, the threshold is set above background noise. The threshold should be adjusted in steps to eliminate background noise without blocking useful emissions. Commercial AE systems come with a fixed and a floating threshold setting that can be chosen through a software interface. The floating threshold takes into account the peak dB of background noise and set threshold level [5]. When the noise level is constant throughout the test, the use of a fixed threshold is recommended. The floating threshold is recommended for cases where noise level fluctuates. The floating threshold is suitable for machinery with variable operating speeds and conditions.

Major types of noise encountered in AE measurement are background noise generated by the movement of loose parts, jigs, and contact friction; acoustic noise from fans and hydraulic systems; and electromagnetic noise from electrical and electronic equipment and movement of cables [6]. Mechanical noise due to loose parts can be eliminated by tightening of fasteners. Electrical noise can be eliminated through proper shielding and grounding [6, 7]. Researchers have used nylon film between rotor and bearing, and rotor and coupling in their laboratory setup, to suppress transmission of AE signal from support bearings and motor to test bearing [8]. Sharp bend, stress at connector junction, and movement in cable induce noise and false signal through triboelectric effects [9]. Hence, the cable and sensor should be appropriately taped as illustrated in Fig. 2.

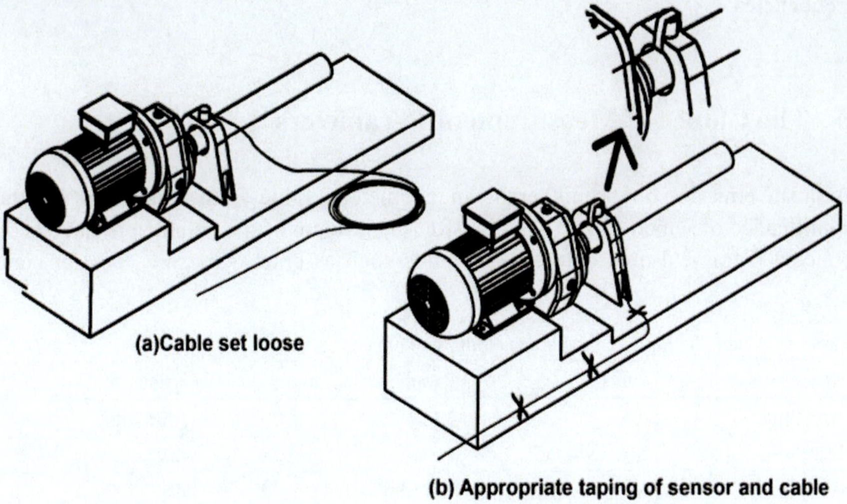

Fig. 2 Appropriate taping of sensor and cable to avoid noise and false signal

2.2 *Preamplifier and Filter Settings*

The voltage output from the sensor is usually very low. The preamplifier converts this low voltage to the usable form. The gain in voltage is expressed in dB. As such, a gain of 20 dB would amplify the signal by ten times, a gain of 40 dB would amplify by a hundred times, and a gain of 60 dB would amplify by a thousand times. Each measurement system has a saturation limit with respect to maximum voltage input it can handle. Above this saturation limit, the signal gets distorted. Hence, the amplifier gain should be chosen wisely as the maximum gain may not be the best choice. In general, strong AE sources are monitored at low gain, and weak AE sources are monitored at high gain [10]. The AE software comes with three digital filters: low pass filter, high pass filter, and band pass filter. A low pass filter passes signal content having a frequency below the set cutoff frequency; the high pass filter passes a signal content having frequency larger than the set cutoff frequency. The band pass filter allows signal having frequency content within a specified frequency range. Choosing the right AE filters for machinery diagnosis applications can be quite tricky, especially when no standards are available. For cases where reference is not available, a preliminary study can be carried out using a wideband sensor to examine the frequency content of an acoustic emission signal originating from a machine under different conditions. Based on knowledge gained from preliminary studies, suitable filters can be set for the remaining set of experiments to avoid external noise frequencies.

3 The Choice of Measurement Parameters

Acoustic emission has around eighteen parameters (Table 1), the definition, and the significance of which can be found in Ref [11]. Some of the parameters have been conceived for real-time detection of faults such as cracks, its size, location, and

Table 1 A table listing AE parameters along with unit

Parameters	Units	Parameters	Units
Amplitude	dB	RMS	Millivolts
Duration	Microseconds	ASL	dB
Rise time	Microseconds	Threshold	dB
Counts	Counts	Peak frequency	kHz
Counts to peak	Counts	Average frequency	kHz
Time of hit	Microseconds	Reverberation frequency	kHz
Energy	μvolt-sec/count	Initiation frequency	kHz
Absolute energy	Joules	Frequency centroid	kHz
Signal strength	Picovolt-sec	Partial power	% of total power

propagation in structural diagnosis. As for rotating machinery, researchers besides using counts, amplitude, energy, and frequency, have also used statistical parameters [12–17]. If an important conclusion cannot be drawn from a particular parameter, including it in a measurement is irrelevant.

4 The Selection of Sensors

AE sensors can be broadly classified as low, standard, and high-frequency sensors depending on their operating regime. Technically, the AE sensor is either resonant or broadband type. While a wide-band sensor has a flat response over a wide range of frequencies, a resonant sensor is most sensitive near its resonance frequency [18] and has a small band of working frequencies. For instance, a typical AE sensor with resonant frequency 150 kHz would have an operating band in the range 50–200 kHz [19]. Consequently, AE signal having a frequency outside the operating band of a transducer may not be reported by the system [20]. A wideband sensor is used for research purposes where frequency-based analysis is required. A resonant sensor is more sensitive than a wideband sensor and is preferred for fieldwork where only parametric analysis is required [18]. The commonly used piezoelectric material such as lead-zirconate-titanate loses its piezoelectric property at around 300 °C [21]. Hence, for high-temperature applications [22–24], special sensors based on lithium niobate piezo-composite are used [25]. AE on rotating machinery elements are monitored through roller type AE sensor (Fig. 3).

In case a roller type sensor is not available, a conventional AE sensor can also be mounted on rotating machinery element using a slip ring [14] as shown in Fig. 4. A slip ring is a device to transmit electrical power or data signal from a rotating component to a stationary component or vice versa.

5 The Choice of Couplant

The acoustic impedance of a medium is the product of material density and speed of sound in that medium. The difference in the acoustic impedance between the two mediums is called an acoustic mismatch. A Higher mismatch results in a larger amount of energy being reflected and a lower amount of energy being transferred [26]. The acoustic impedance of piezoelectric material being very high and that of air being very low, there would be a loss of AE energy as it propagates from its source to the sensor. Couplant is used to effectively transmit AE energy from source to the sensor [27].

The National Physical Laboratory of United Kingdom [27] recommends gel and grease type couplant for rough mounting surfaces and liquid type for smooth surfaces. For very high and low temperature, special couplant that maintains physical and chemical integrity at this extreme temperature is recommended. For cryogenic cases,

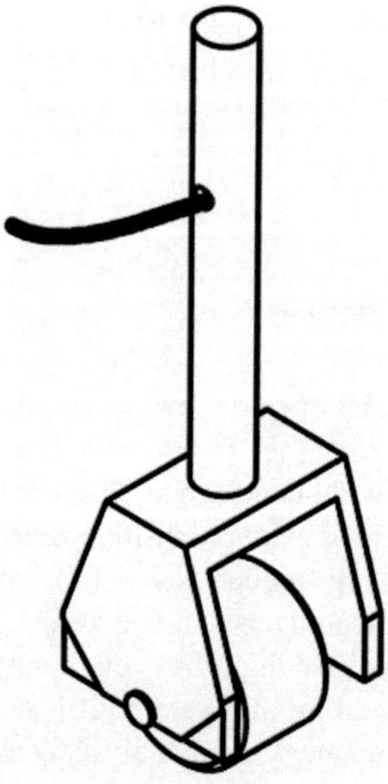

Fig. 3 Sketch of a typical roller type AE sensor

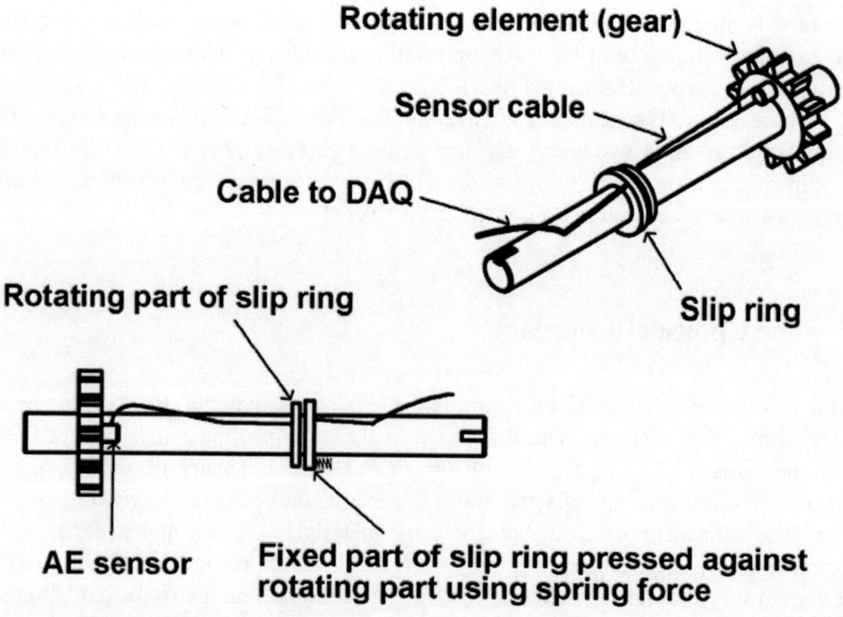

Fig. 4 Schematic diagram showing the use of slip ring for mounting AE sensor on rotating machine element

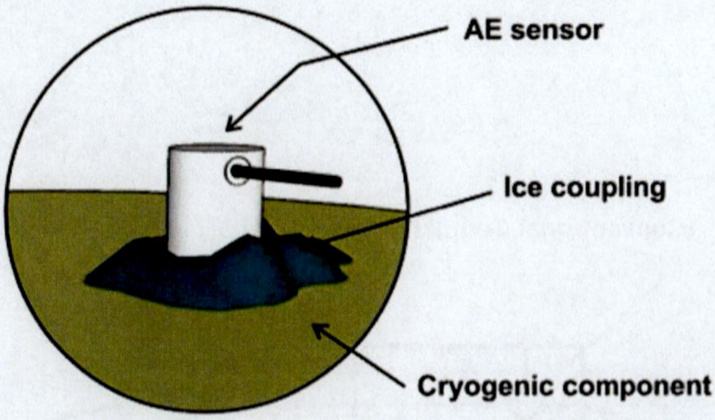

Fig. 5 Schematic diagram showing ice coupling for the cryogenic case. A thin layer of ice is formed beneath the sensor using a drop of water. The coupling is further strengthened using ice layer around the periphery of the sensor

Table 2 A table listing temperature ranges and recommended couplants

Temperature range	Couplant type
Room temperature	Petroleum grease and ultrasonic couplant
−40 to 200 °C	High vacuum stop cock grease
−273 to 100 °C	Nonaq stop cock grease
20 to 700 °C	50% Galium-50% Indium grease

water in the form of ice (Fig. 5) has been used by researchers [28, 29]. Table 2 provides a list of commercial couplants and temperature ranges within which they can be used [19]. As has been stated above, the purpose of a couplant in the acoustic emission technique is to facilitate AE energy transmission to the sensor [30] and not to provide a strong bond with the mounting surface. Hence, before trying something as a couplant, one should find out its suitability based on the criteria of acoustic mismatch. The suitability of a new couplant can also be practically assessed by measuring the response of a lead pencil test [31] carried on a plate with two identical sensors (Sensor 1 mounted with conventional couplant and Sensor 2 mounted with new couplant) placed at equal distances as shown in Fig. 6. The new couplant is suitable when the amplitude of response at both the sensor is the same or when the amplitude at sensor 2 is higher than that at sensor 1.

The thinnest couplant layer with no air void is recommended. This is ensured by mounting the sensor as described in ASTM E650-97 [9]. The method is illustrated in Fig. 7.

6 The Choice of Mounting Methods

The AE sensor should be secured to the surface using a tape or magnetic holder (Fig. 8) to avoid movement at the junction. Sheet metal spiral spring (Fig. 9) has also been used for this purpose.

For cases where direct mounting may not be possible, the use of a waveguide is recommended [32]. Such cases include corrosive, radioactive, and high-temperature

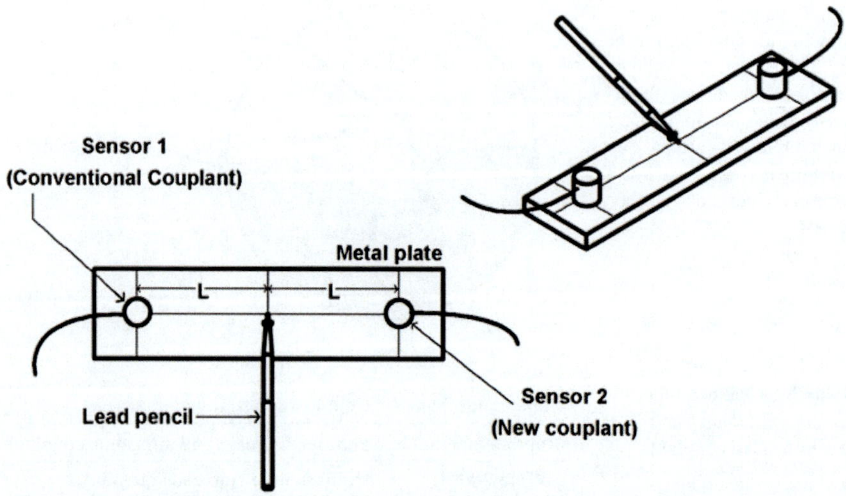

Fig. 6 Setup to measure the suitability of a new couplant

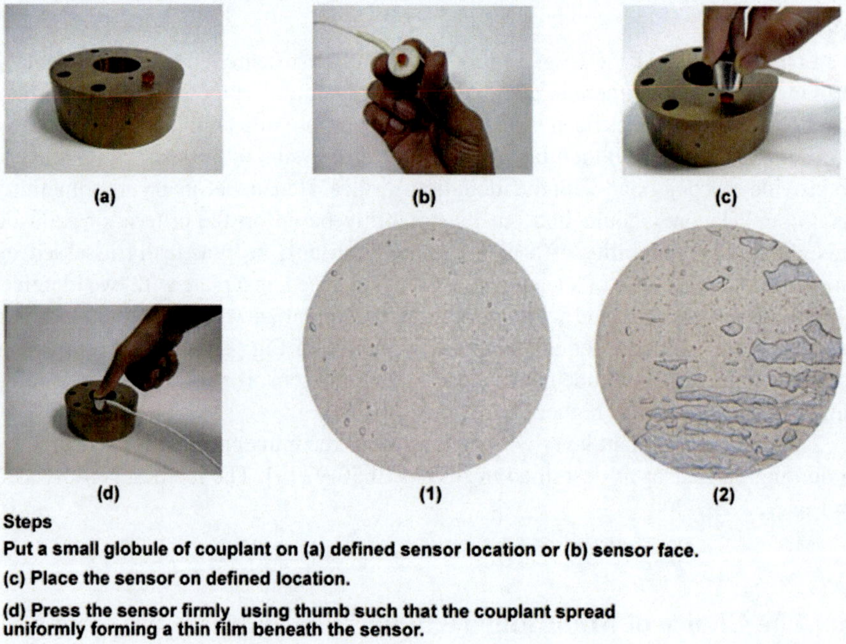

Fig. 7 **(a)–(d)** Figure illustrating the method of applying couplant, (1) Void free couplant layer, (2) Air voids in couplant layer due to improper mounting practice

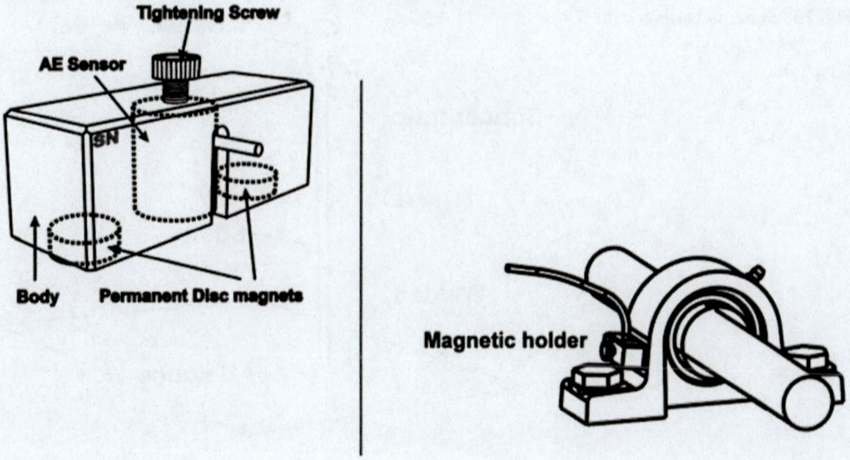

Fig. 8 Parts of magnetic holder used to mount AE sensor

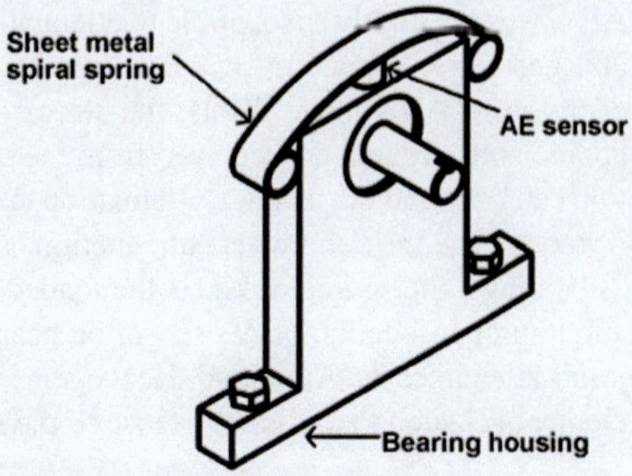

Fig. 9 Figure illustrating the use of sheet metal spiral spring for mounting AE sensor

environment and inaccessible locations. A waveguide not only provides a safe and convenient way to mount AE sensor but also efficiently channelizes AE energy from the source to the sensor. A waveguide in the form of a rod is widely used (Fig. 10).

Steel wire can also be used as a waveguide for special cases [33]. The waveguides are brazed, welded, fastened, or soldered to structure [34]. Some researchers have experimented with nonlinear ball chain waveguides for monitoring high-temperature plasma torch and thermal protection system used in hypersonic spacecraft and aircraft [35].

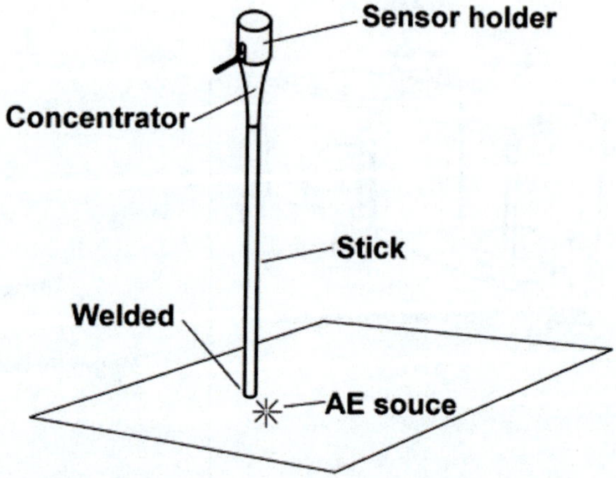

Fig. 10 Schematic diagram of a typical rod type waveguide

7 The Choice of Sensor Location

AE waves are highly susceptible to attenuation resulting from scattering, absorption, and geometric spreading of AE energy. Each decrease of 20 dB means ten folds decrease in AE voltage. Thus, the sensor location should be such that signals of appreciable strength are received from the source [36]. The appropriate sensor location can be found out through attenuation tests using a lead pencil. The sensor should be mounted at the point where attenuation is minimum. For machinery elements such as bearings, the source of AE is the loaded section. Figure 11 illustrates the proper and improper location of AE sensor on bearing housing. A thin layer of air between joints attenuates the AE signal. Hence, the sensor should be mounted on the machine element [2] such that it is very close to the source without any discontinuity.

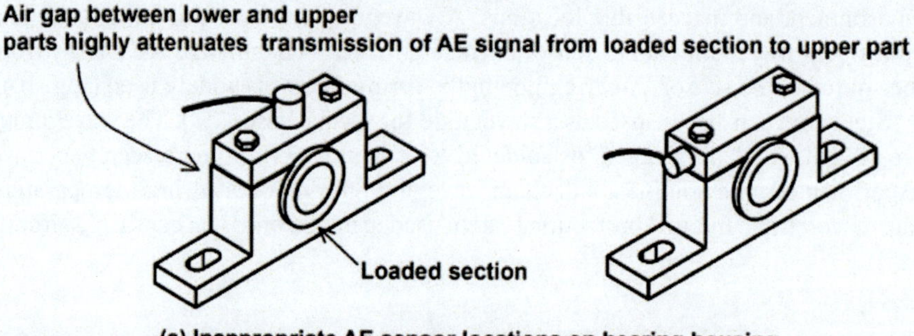

Fig. 11 Inappropriate and appropriate locations of AE sensor on bearing housing

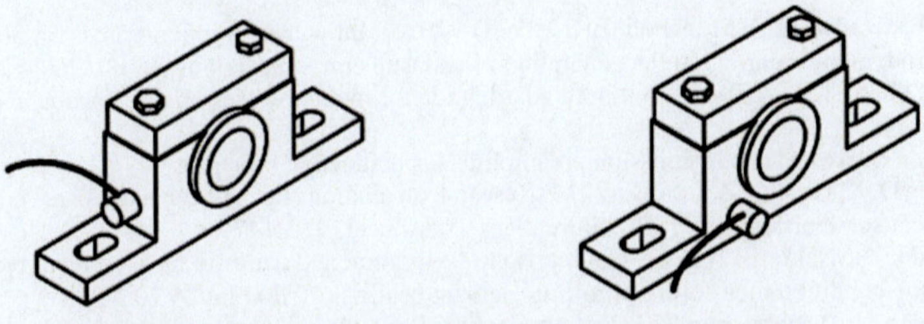

Fig. 11 (continued)

8 Conclusions

Standard practices related to AE measurements on machines have been discussed through illustrations. The key points of this article can be summarized as follows:

- A threshold level is used to eliminate background noise. The threshold should be adjusted in steps to eliminate background noise without blocking useful emissions.
- AE signals of low strength are monitored at high preamplifier gain, and AE signals of high strength are measured at low preamplifier gain.
- Counts, amplitude, energy, and frequency are the most commonly used AE parameters for machinery diagnosis.
- A wideband sensor is used for frequency-based analysis, and a resonant sensor is used for parametric study. The resonant sensor should be appropriately chosen as it is most sensitive around its resonant frequency.
- The sensor should be mounted with proper couplant to facilitate AE energy transmission and secured firmly to the surface using tape or magnetic mounting.
- The sensor should be mounted on the machinery element which is close to the source without any discontinuity.

References

1. ASTM E1316 (2000) 15a. Standard terminology for nondestructive examinations-section B: acoustic emission (AE) terms. Annu B ASTM Stand 03
2. SKF: CMSS 786M SEE/AEE (2012) Sensor mounting for on-line systems
3. Hartmut Vallen JF, Threshold independent AE testing. http://www.ndt.net/article/wcndt00/papers/idn657/idn657.htm
4. Rodgers J (1994) The "Use of a Floating Threshold for Online Acoustic Emission Monitoring of Fossil High Energy Piping." AE Consult
5. Hardy HR Jr (2005) Acoustic emission/microseismic activity: principles, techniques and geotechnical applications, vol 1. CRC Press
6. Fahr A (2013) Aeronautical applications of non-destructive testing. DEStech Publications, Inc.
7. RDSO: BS-104 (2002) Guidelines on use of acoustic emission technique (AET) on railway bridges

8. Mirhadizadeh SA, Moncholi EP, Mba D (2010) Influence of operational variables in a hydrodynamic bearing on the generation of acoustic emission. Tribol Int 43:1760–1767
9. ASTM: ASTM E 650-97 (1997) Standard guide for mounting piezoelectric acoustic emission sensors
10. Vallen (2016) Acoustic emission preamplifiers specification brochure
11. Dong D, Xiangdong Z, Libin L (2014) Research on relationship between parameters correlation of acoustic emission and rock failure. Sens Transducers 183:147
12. Tandon N, Nakra BC (1992) Comparison of vibration and acoustic measurement techniques for the condition monitoring of rolling element bearings. Tribol Int 25:205–212
13. Tandon N, Ramakrishna KM, Yadava GS (2007) Condition monitoring of electric motor ball bearings for the detection of grease contaminants. Tribol Int 40:29–36
14. Toutountzakis T, Tan CK, Mba D (2005) Application of acoustic emission to seeded gear fault detection. NDT E Int. 38:27–36
15. Poddar S, Tandon N (2016) Detection of journal bearing vapour cavitation using vibration and acoustic emission techniques with the aid of oil film photography. Tribol Int 103:95–101
16. Hase A, Mishina H, Wada M (2016) Fundamental study on early detection of seizure in journal bearing by using acoustic emission technique. Wear 346:132–139
17. Miettinen J, Andersson P (2000) Acoustic emission of rolling bearings lubricated with contaminated grease. Tribol Int 33:777–787
18. Arroll C, Aly S, others (2005) Fracture, fatigue, failure, and damage evolution, vol 5. Springer
19. PAC (2005) Physical acoustic corporation's AE system manual, New Jersey, USA
20. Wright AP (2009) A multi-axis capacitive MEMS sensor system for acoustic emission sensing. ProQuest
21. Noma H, Tabaru T, Akiyama M, Miyoshi N, Hayano T, Cho H (2007) High-temperature acoustic emission sensing using aluminum nitride sensor. In: The sixth international conference on acoustic emission (Acoustic Emission Working Group), South Lake Tahoe (USA)
22. Fujicera: Fuji Ceramics corporation. http://www.fujicera.co.jp
23. Vallen: Vallen system GmbH. http://www.vallen.de
24. PAC: Physical acoustic corporation. http://www.physicalacoustics.com
25. Kirk KJ, Scheit CW, Schmarje N (2007) High-temperature acoustic emission tests using lithium niobate piezocomposite transducers. Insight-Non-Destruct Test Cond Monit 49:142–145
26. Gibbs V, Cole D, Sassano A (2011) Ultrasound physics and technology: how, why and when. Elsevier Health Sciences
27. Theobald P, Guide on acoustic emission sensor couplants. http://www.npl.co.uk/acoustics/ultrasonics/research/guide-on-acoustic-emission-sensor-couplants
28. Xu Y, Wang X, Wang Y (2012) Improved or green wave-guide methods in the acoustic emission inspection of high/low-temperature or complicated vessels. In: 2012 Third global congress on intelligent systems (GCIS), pp 323–326
29. Dong S, Wu Q, Xu J, Acoustic emission in-service detection of cryogenic storage tank floors
30. Theobald P, Zeqiri B, Avison J (2008) Couplants and their influence on AE sensor sensitivity. J Acoust Emiss 26:91–97
31. Sause MGR (2011) Investigation of pencil-lead breaks as acoustic emission sources. J Acoust Emiss 29
32. Zakharov DA, Ptichkov SN, Shemyakin VV (2010) Acoustic emission signal attenuation in the waveguides used in underwater AE testing. ECNDT 2010-Acoustic Emiss
33. Ono K, Cho H (2004) Rods and tubes as AE waveguides. J Acoust Emiss 22:243–252
34. Neill IT, Oppenheim IJ, Greve DW (2007) A wire-guided transducer for acoustic emission sensing. In: The 14th international symposium on: smart structures and materials & nondestructive evaluation and health monitoring, p 652913
35. Pearson SH, Huston D (2014) Nonlinear ball chain waveguides for acoustic emission and ultrasound sensing of ablation. In: SPIE smart structures and materials + nondestructive evaluation and health monitoring, pp 90642P–90642P
36. Hellier C (2001) Handbook of nondestructive evaluation. McGraw-Hill

Acoustic Analysis of Timbre of Sarasvati Veena in Comparison to Simple Sonometer

Chandrashekhar Chauhan, P. M. Singru, and Radhika Vathsan

Abstract We characterize the timbre of the Sarasvati Veena, a traditional Indian string instrument, by analysing the sound spectrum and the energy distribution with different frequencies as a function of time. We compare the acoustic analysis with that of a simple sonometer. We find that while the spectrum consists of the fundamental and harmonics of a plucked string under tension, the Veena shows a prominent formant near the resonant frequency of the body. But more notable is the behaviour of the higher harmonics (upper partials) which share substantial energy and show a characteristic revival with time, in contrast to the sonometer spectrum.

1 Introduction

The Sarasvati Veena shown in Fig. 1 is a fretted string instrument of traditional south Indian classical music. The basic sound generation is by plucking stretched metal strings.These pass over a curved brass-alloy bridge on one end and are wound around wooden tuning pegs on the other end.

The timbre of this instrument was first brought to the notice of the scientific world by Raman [4], who argued that the unique sound spectrum is basically attributed to the curved bridge. Studying the nature of vibrations of the plucked string when one end is subjected to extended boundary conditions is essential to characterize the

C. Chauhan · P. M. Singru · R. Vathsan (✉)
BITS Pilani K K Birla Goa Campus, NH17B Zuarinagar, Goa 403726, India
e-mail: radhika@goa.bits-pilani.ac.in

C. Chauhan
e-mail: chandrashekharchauhan378@gmail.com

P. M. Singru
e-mail: pmsingru@goa.bits-pilani.ac.in

M. Singh and Y. Rafat (eds.), *Recent Developments in Acoustics*, Lecture Notes in Mechanical Engineering, https://doi.org/10.1007/978-981-15-5776-7_25

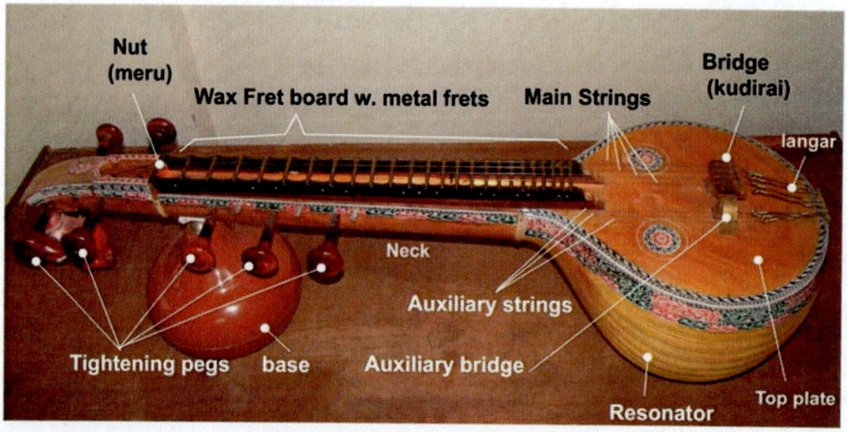

Fig. 1 Sarasvati Veena

timbre of the Veena. The spectrum is also augmented by the nature of the resonant cavity, its shape, material properties and radiation patterns.

While most Western string instruments that have been widely studied have sharp bridges, Indian instruments including the Veena, tanpura and sitar, and the Japanese *shamisen* and *biwa* [7] have extended bridges, some curved to a greater or lesser degree. The nature of the timbre of these instruments is markedly different, with a characteristic richness of harmonics. Theoretical studies and simulations have been performed for sitar and tanpura [1, 2, 8], where the interaction of the vibrating string with the extended bridge is considered from various angles. Siddiq [5] points out the role of dispersion in the sound production. We started studies on the acoustics of the Sarasvati Veena with an examination of the frequency spectrum of each string for different tunings [3, 6]. Variation of FFT with different frets was also studied. Raman's hypothesis of the presence of partial harmonics was experimentally corroborated. The optimum tuning frequency for a given Veena was shown to be related to the resonant frequency of the body.

In this work, we bring out details of difference in timbre of the Veena and the sonometer, by comparing the acoustic spectra. We also found the natural frequency of the body of both instruments through impact tests. For both the instruments, we used strings of the same material, length, mass density and diameter, tuned to the same pitch and plucked at the same position from the bridge using a metallic plectrum. To bring in uniformity, the microphone was also kept at the same location. All the experiments were performed in the semi-anechoic chamber. The time-domain data and FFT were analysed and compared.

This study has brought out a marked difference in the energy distribution among different harmonics in each instrument. The waterfall diagram showed that the higher harmonics get revived in the case of the Veena while they are all uniformly damped in the case of the sonometer. The tonal characteristics, *naad*, expected of a good Veena by accomplished musicians is validated by this scientific study. This paper attempts to quantify the subjective notion of *naad* of the instrument.

2 Experimentation

Experimentation was carried out in two steps: (1) Impact tests on sonometer and Veena were carried out to find the natural frequency of the body of the instruments; (2) Acoustic analysis of plucked string vibrations of Veena and sonometer. Steel string of gauge 31 (diameter 0.3 mm) and linear mass density 0.56 g/m was used on both instruments. Data obtained by these tests were recorded by vibration analyser (Data acquisition system, Spider-81, Crystal Instruments, USA), and the obtained results were then analysed using Engineering Data Management Software.

2.1 Impact Test

Impact test was conducted in Dynamics Lab at BITS Pilani K. K. Birla Goa campus in room ambience. The accelerometer, 352C34 of PCB Electronics, USA, which has a sensitivity of 101.3 mV/g, was attached near the bridge and impact was made at a location found by driving point method of modal analysis. In order to ensure free–free boundary conditions, the instruments were placed on foam at both ends. The plastic-tipped impact hammer with sensitivity of 10.32 mV/lbF is used. The results obtained were recorded.

2.2 Acoustic Study

The acoustic study involves measurement of sound pressure amplitude created near the centre of the vibrating string after plucking. The microphone (1/4") used was model no. 378C01, PCB Electronics, USA, which has sensitivity of 2.29 mV/Pa. The string was plucked using a metallic wire plectrum that is normally used while playing the instrument. We ensured that plucking was of small and almost the same amplitude and in the vertical plane. For a given tuning frequency, eight runs were recorded, normalized and averaged before analysis. We selected four tuning frequencies based on standard tuning pitches in stringed instruments: 146.8 (middle D), 155.5 (D#), 174.6 (F: standard tuning for this Veena) and 185 (F#). This range of tuning was selected since the string tension for lower frequencies than D is too loose for musical effect and for higher frequencies than F#, it is too tight for comfortable playing. The Veena player normally prefers to pluck the playing string at a distance of around 0.17 times the string length. We chose this point to simulate natural playing conditions. For comparison, the sonometer was also plucked at the same position. However, this may restrict the total range of harmonics that are excited. To explore the full range of harmonics in the instruments, we also studied the spectrum by plucking the string at the centre.

3 Analysis of Results

3.1 Impact Test

Impact test FFT graphs are shown in Fig. 2. The FFT spectra for Veena show a peak at 282.5 Hz, which is the natural frequency of the resonant cavity of the particular Veena used. The natural frequency of sonometer was found to be 100 Hz.

3.2 Acoustic Study

The waterfall diagrams for Veena show the evolution of the spectrum with time. We note that the spectra for both instruments are harmonic. Also, the dominant frequency is not always the fundamental, but the fundamental or the third harmonic for sonometer. The Veena spectrum for different tunings shows that the second harmonic dominates. A typical waterfall diagram is shown in Fig. 3. We also note that higher frequencies exhibit revivals at later times, indicating the nature of the typical "ringing" effect in the timbre of the Veena. This prompted us to study the behaviour

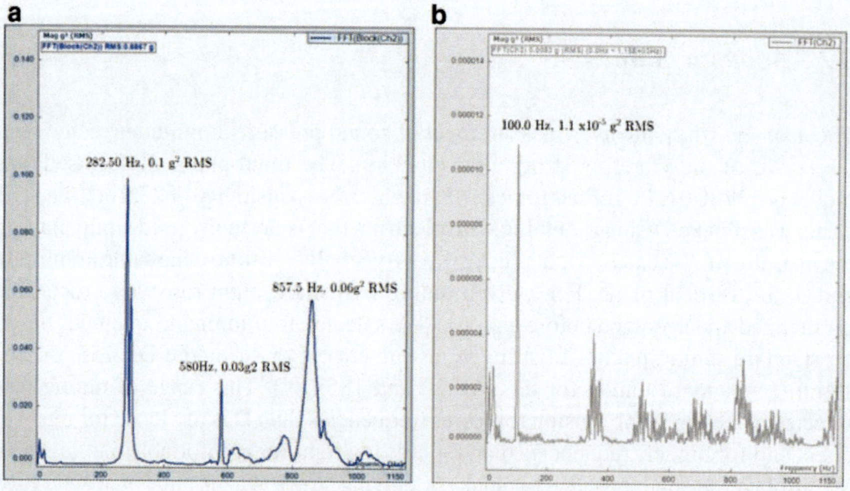

Fig. 2 Impact test for **a** Veena **b** Sonometer

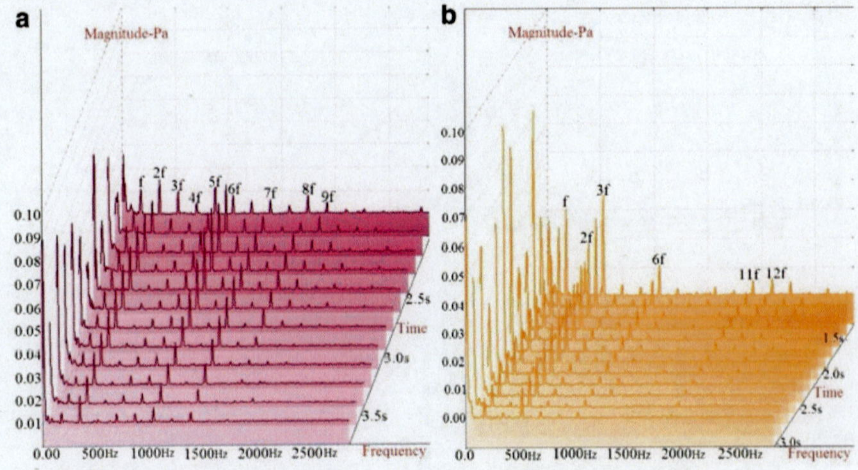

Fig. 3 Waterfall diagrams for plucked string tuned to 174 Hz **a** Veena **b** Sonometer

of FFT amplitude with time for each harmonic. These graphs for each harmonic for both sonometer and Veena are displayed for comparison in Figs. 4 and 5.

These graphs show the following features characterizing the timbre of the Veena:

1. Energy taken by upper partials as compared to the dominant harmonics is much more in the case of Veena than sonometer.
2. In Veena, the upper partials show pronounced revivals, unlike the case of sonometer. We expect this to be a result of the interaction of the string with the curved bridge, and the resonant cavity.
3. The graphs for tuning frequency of 174.3 Hz show that the third harmonic is the strongest. It also shows a stronger presence of upper partials. This indicates richer sound, and is the preferred tuning frequency for this instrument.

3.3 Connection to Music

The graphs of Figs. 4 and 5 give us an interesting connection to the musical content of the string spectrum. The harmonic scale used in Indian classical music has its origin in this quality of string vibrations. Experienced musicians can identify the notes that are heard on plucking a string, and use that identification not only in the improvisation of music, but also in tuning the instrument, which involves placing the metal frets at the right positions on the wax fret board. This method is crucial in the construction of Indian musical instruments, especially the Veena. In contrast, string instruments with fixed key values in Western music, such as the guitar and piano, use the equal tempered scale where the ratios of successive notes are all the same, the twelfth root of 2.

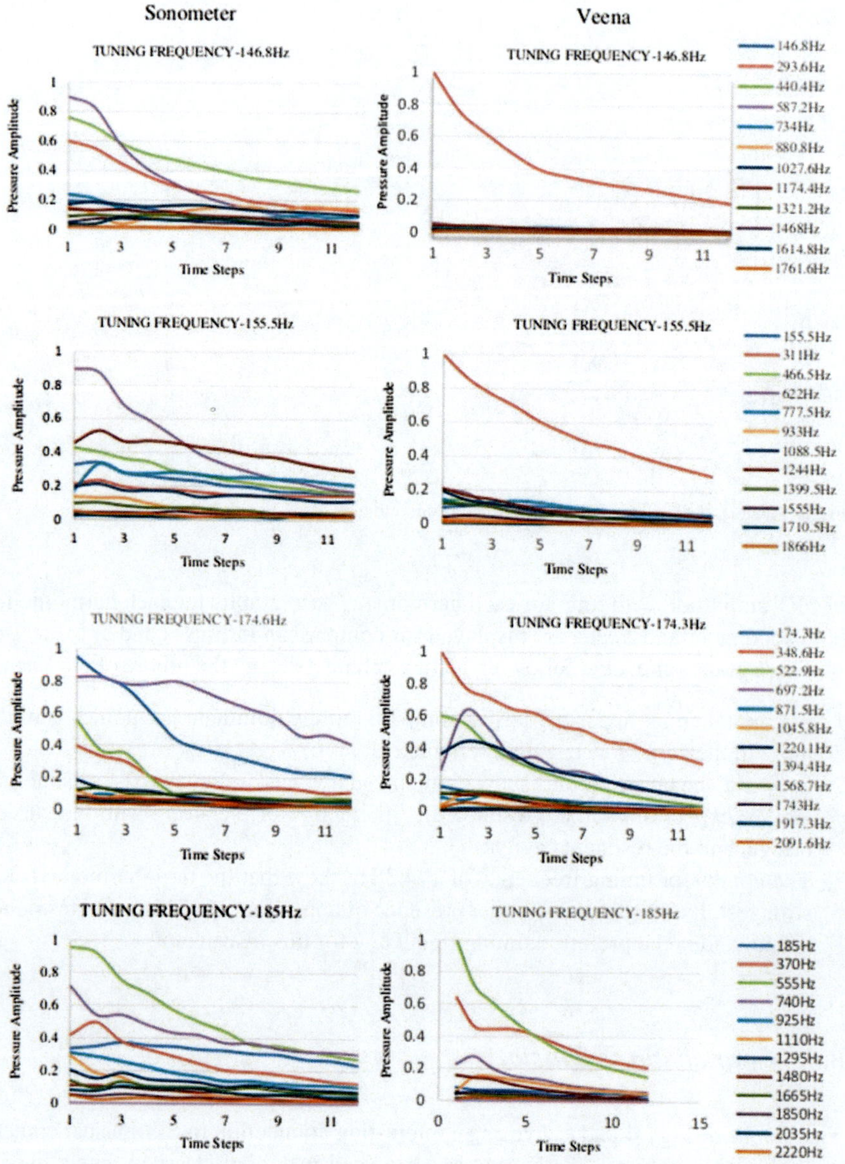

Fig. 4 Comparison of formants for different tuning frequencies for Veena and sonometer plucked close to the bridge

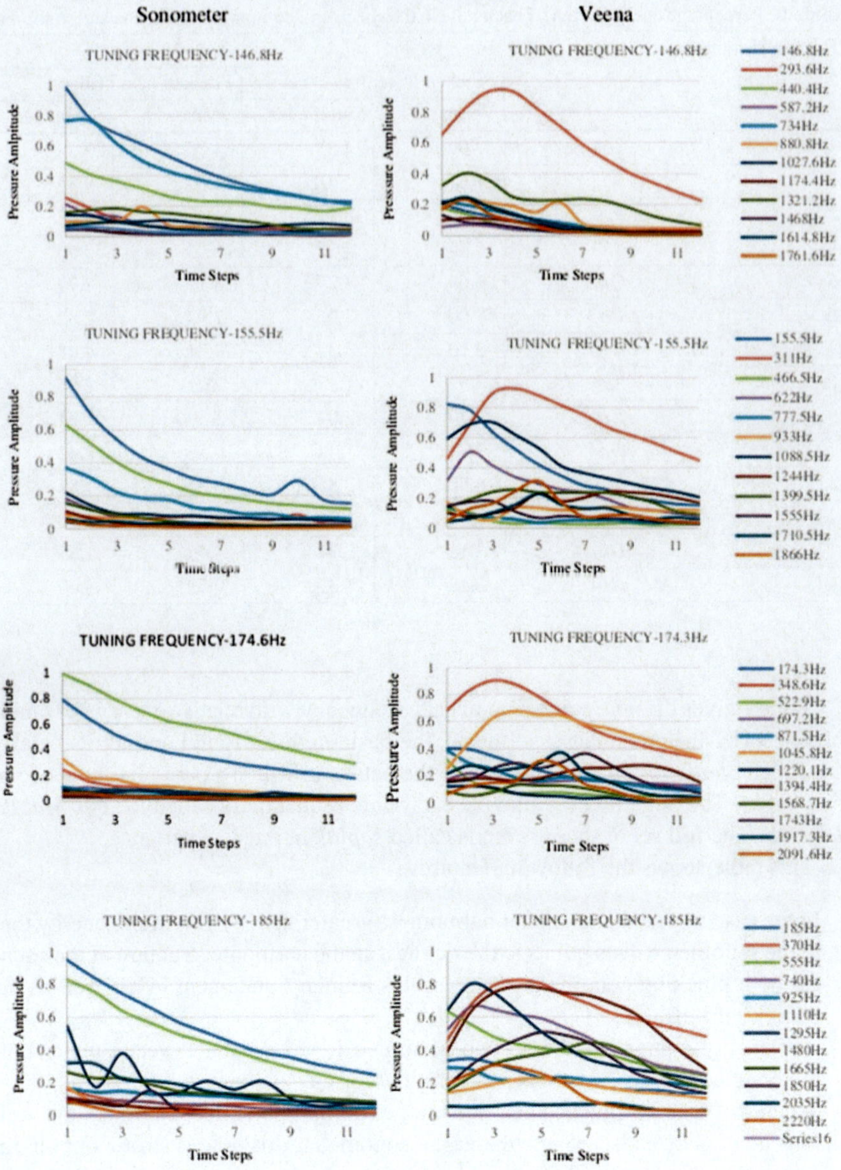

Fig. 5 Comparison of formants for different tuning frequencies for Veena and Sonometer plucked at the centre

Table 1 Percentage energy taken by each note in the spectrum: comparison of Veena and sonometer

Harmonic number	Note	Energy (%)			
		Plucked at centre		Plucked near bridge	
		Veena	Sonometer	Veena	Sonometer
1	*Sa*1	7.99	24.53	4.70	22.70
2	*Sa*2	21.61	5.02	32.37	9.33
3	*Pa*2	3.36	38.12	14.47	8.42
4	*Sa*3	5.64	4.32	16.04	32.60
5	*Ga*3	6.64	3.27	3.11	2.77
6	*Pa*3	18.00	7.06	2.97	2.82
7	*Ni*3	2.10	3.32	14.81	4.48
8	*Sa*4	7.52	1.17	2.21	2.54
9	*Re*4	2.70	2.04	3.50	4.34
10	*Ga*4	7.73	1.44	0.52	1.42
11	*Ma*4	6.80	3.85	0.86	1.87
12	*Pa*4	4.33	1.74	1.52	1.50
13	*Dha*4	3.12	2.74	1.45	1.65
14	*Ni*4	2.45	1.37	0.50	3.60

Table 1 gives the note equivalent of each frequency with relative energy, for tuning of 174.3 Hz. In Indian classical music, the basic note (or tonic) and all its octaves are called *Shadja* or *Sa*. We will denote the octave in which a particular note occurs by a suffix. The other notes within the octave are sequentially called *Re, Ga, Ma, Pa, Dha, Ni*. The full set of these seven is called *saptak* in musical parlance.

This table shows the following features:

- In the case of Veena, the higher harmonics (greater than 4) have more energy than in the sonometer. This points to the richness in the harmonic structure of the sound for both plucking conditions. This verifies Raman's statement [4] as well as our earlier work [6].
- The energy content of the second harmonic is strongest for Veena under both plucking conditions, whereas it is very low for the sonometer.
- When the Veena is plucked near the bridge (normal playing conditions), the tonic (*Sa*) in all octaves is strong, whereas in sonometer, this note is strong only in the first and third octaves. This strengthens the musical quality of Veena.
- We noticed that plucking at the centre shows up considerable energy in all the other notes of the *saptak*, such as *Ni*, *Ma* and *Re*. This is favoured for *tanpura*, the drone instrument for accompanying all *raags*,

3.4 Formants

A peak in the harmonic spectrum of a musical instrument refers to a favoured frequency range, known as a formant. In Fig. 6, we plot the percentage energy for each note of the Veena, under both plucking conditions. We see that the Veena supports three basic groups of frequencies, which we will call its *formants*: The dominant group consisting of the first four harmonics (in musical parlance, Sa, Pa, Sa), the middle formant of the next four harmonics (Ga and Ni) and the upper partials (Ga, Ma, Dha). When plucked at the centre, we find that the relative strengths of these groups are in the ratio 39 : 34 : 27 approximately for Veena whereas for sonometer, the ratio is 36 : 7 : 7. Thus, the Veena timbre shows that the energy content of these formants reduces gradually, whereas in sonometer, the energy reduces drastically after the first formant. When the Veena is plucked near the bridge, which is the normal playing condition, 2/3rd energy is in the first formant, 1/4th in second formant and 1/12th in the third formant. For the sonometer, the ratio of energy content in these formants is the same as in the case of plucking at the centre. This gives a more quantitative characterization of the timbre of the Veena.

Thus, the richness of Veena timbre is most evident when plucked at the centre. However, even under normal playing conditions, there is substantial energy in the second formant. This timbre is still richer than that of the sonometer. This energy distribution among the formants is the scientific basis for the *naad* of the Veena which musicians identify.

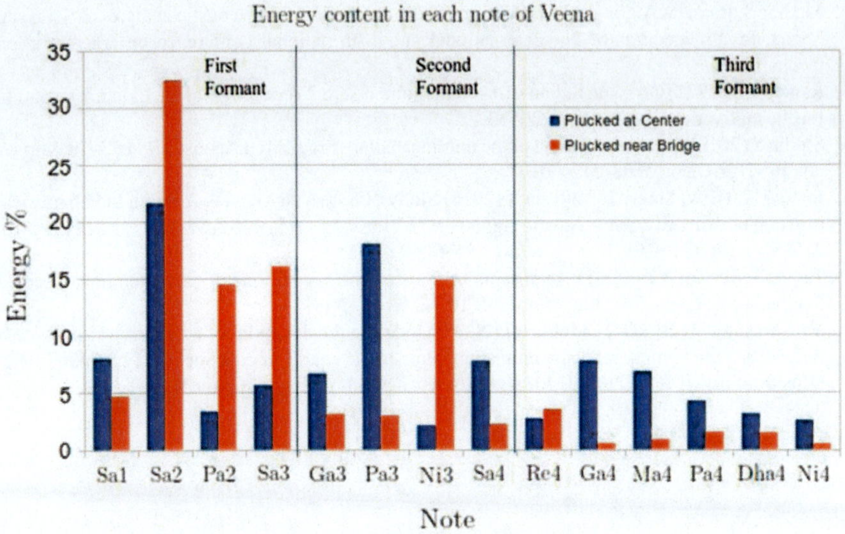

Fig. 6 Energy distribution and formants for Veena, plucked at centre and near bridge

4 Conclusions

This is the first reported work on the experimental determination of the timbre of Veena, in comparison to the sonometer. We find that the typical ringing tone of the Veena, which is observed by accomplished musicians, is due to a revival of the higher harmonics such as the 6th (Pa), 10th (Ga), 11th (Ma) and 13th (Dha). We further studied the formant structure of the Veena, characterized by the strongest group of the first four harmonics, the next four harmonics and the higher harmonics. The relative energy of these three groups is nearly equal for Veena plucked at the centre, while only the first group is strong in the sonometer. Under normal plucking conditions, where the energy is supplied closer to the bridge, the Veena timbre is still richer than that of the sonometer. These features give us a quantitative characterization of the timbre of the Veena, *vis a vis* the sonometer. This explains scientifically the *naad* of the Veena. We expect to substantiate this behaviour by theoretical and experimental studies of the interaction of the vibrating string with the curved bridge and the resonant cavity.

References

1. Burridge R, Kappraff J, Morshedi C (1982) The Sitar string, a vibrating string with a one-sided inelastic constraint. SIAM J Appl Math 42(6):1231–1251
2. Mandal AK, Wahi P (2015) Natural frequencies, modeshapes and modal interactions for strings vibrating against an obstacle: relevance to Sitar and Veena. J Sound Vib 338:42–59. https://doi.org/10.1016/j.jsv.2014.06.010
3. Mathur S, Raizada A, Singru PM, Vathsan R (2015) Experimental acoustic analysis of Sarasvati Veena. In: Proceedings of 2nd international and 17th national conference on machines and mechanisms
4. Raman CV (1921) On some Indian stringed instruments. In: Proceedings of the Indian association for the cultivation of science 33:29–33
5. Siddiq S (2012) A physical model of the nonlinear sitar string. Arch Acoust 37(1):73–79. https://doi.org/10.2478/v10168-012-0010-y
6. Sundar A, HP V, Singru P, Vathsan R (2016) Study of Sarasvati Veena—A South Indian musical instrument using its vibro-acoustic signatures. J Vibroeng 18(5):1392–8716. https://doi.org/10.21595/jve.2016.16930
7. Taguti T, Tohnai YK (2001) Acoustical analysis on the sawari tone of Chikuzen biwa. Acoust Sci Technol 22:199–207. https://doi.org/10.1250/ast.22.199
8. Vyasarayani CP, Birkett S, McPhee J (2009) Modeling the dynamics of a vibrating string with a finite distributed unilateral constraint: application to the sitar. J Acoust Soc Am 125(6):3673–82. https://doi.org/10.1121/1.3123403, http://www.ncbi.nlm.nih.gov/pubmed/19507949

A Simultaneous EEG and EMG Study to Quantify Emotions from Hindustani Classical Music

Uddalok Sarkar, Soumyadeep Pal, Sayan Nag, Shankha Sanyal, Archi Banerjee, Ranjan Sengupta, and Dipak Ghosh

Abstract With the advent of various techniques to assess the bioelectric signals on the surface of the body, it has become possible to develop various Human–Computer interface systems. In this study, for the first time a cross-correlation based data is reported for two different types of bio-signals, viz. Electroencephalography (EEG) and Electromyography (EMG). Whereas EEG refers to the neuro-electric impulses generated in the brain recorded in the form of electric potentials, EMG records the activation potentials of the muscle cells when they contract or relax. The ability of Hindustani Music (HM) to evoke a wide range of emotional experience in its listeners is widely known. For this study, we took simultaneous EEG and EMG data of 5 participants while they listened to two Hindustani *ragas* of contrast emotions namely *Chayanat (*corresponding to happy/joy*)* and *Darbari Kanada (*corresponding to sad/pathos*)* emotion. We make use of two latest signal processing algorithms—Wavelet-based power spectra and cross-correlation coefficient to assess the arousal based activities in response to the acoustic clips in the two different bio-signals. For the first time, an attempt is being made to quantify and categorize musical emotions using EMG signals and an attempt to correlate that with the EEG signals obtained from the brain. The alpha, theta, and gamma frequency range in the frontal and parietal electrodes is found to be the most responsive in case of musical emotions. The EMG response has been studied by segregating the entire signal into different frequency ranges as is done in case of EEG frequency bands. Interestingly, the response in case of EMG data is strongest in the same frequency bands as that of EEG signals. Novel pitch detection algorithm has also been applied to EMG signals

U. Sarkar · S. Pal · S. Nag
Department of Electrical Engineering, Jadavpur University, Kolkata, India
e-mail: uddaloksarkar@gmail.com

S. Sanyal (✉) · A. Banerjee · R. Sengupta · D. Ghosh
Sir C.V. Raman Centre for Physics and Music, Jadavpur University, Kolkata, India
e-mail: ssanyal.sanyal2@gmail.com

S. Sanyal
School of Languages and Linguistics, Jadavpur University, Kolkata, India

A. Banerjee
Rekhi Centre of Excellence for the Science of Happiness, IIT Kharagpur, Kharagpur, India

M. Singh and Y. Rafat (eds.), *Recent Developments in Acoustics*, Lecture Notes in Mechanical Engineering, https://doi.org/10.1007/978-981-15-5776-7_26

to ratify the rationale behind the separation of frequency bands. This is the first of its kind study which looks for categorization and quantification of musical emotions using simultaneously two different bio-signals with the help of robust mathematical analysis. The results and implications have been discussed in detail.

Keywords EEG · EMG · Hindustani classical music · Musical emotion · Wavelet transform

1 Introduction

Human emotions play a direct role in our everyday life. Over the past decades, the interaction between computer and human has resulted in the recognition of human emotions to be of great importance. Thus, the automated detection of human emotions taken utmost importance. Speech, gesture, facial expressions, etc. have been analyzed to recognize such emotions. However, outer manifestations like speech or facial expressions may not reflect the correct human emotion because these can be controlled voluntarily. Thus, biomedical signals like Electroencephalography (EEG), Electromyography (EMG), and Electrocardiography (ECG) have been another important source of such information about human emotions. These signals delve into such depths of human emotions that they cannot be voluntarily controlled. For these reasons, the quantitative assessment of human emotions is being done with different types of bio-signals and they provide a unique platform for automated emotion recognition.

Researchers classify emotion based on a two dimensional plane with orthogonal components-valence and arousal. Classification is done based on the position of the emotion recognized in this plane. Lin et al. [1] classified four basic emotional states: joy, anger, sadness, and pleasure induced from music listening. Features based on spectral power were extracted and classified using multilayer perceptron and Support Vector Machine (SVM). Lin et al. [2] extracted the hemispheric asymmetry alpha power indices of brain activation as feature vector and classified the four basic emotions using multilayer perceptron classifier. Some other methods and classifiers have also been used by Lin et al. [3, 4] to classify emotions from EEG signals. Fractal dimension based approach was used by Liu et al. [5] for real-time EEG based classification of emotions induced from music pieces. Murugappan et al. [6] stimulated subjects through audio-visual signals and identified emotions by extracting features from wavelet decomposition coefficients, i.e., alpha, beta, and gamma and classified using Linear Discriminant Analysis (LDA) and K- Nearest Neighbor (KNN) technique. Musha et al. [7] used the cross-correlation of the powers of EEG wave bands for the classification of emotions. Chopin [8] stimulated human emotions by making the subjects play video games and computed different linear EEG features like Power Spectral Density (PSD), coherence, peak frequency, and trained a neural network to classify emotions from these features. Frantzidis et al. [9] showed different pictures to subjects and classified the emotions evoked on seeing them by classifying

EEG features using Mahalanobis distance classifier and SVMs. Different methods are present in the literature to classify emotions using EMG signals [10, 11, 12]. Yang and Yang [13] classified EMG signals for emotion recognition and compared between standard Back propagation (BP) neural network classifier and BP neural network classifier improved by the Levenberg–Marquardt(L-M) algorithm, with the feature vector being constructed by wavelet coefficients. The emotions were elicited with selective emotional music by Johannes Wagner. Kim and André [14] took four bio-sensor signals, namely EEG, EMG, skin conductivity, and respiration. Music induced emotions were classified using features based on statistical measures of frequency domain, time domain, geometry analysis, etc. and classifying them using probabilistic LDA algorithm. Naji et al. [15] collected forehead bio-signals, which contain information of EEG, EMG, and Electrooculography (EOG), from subjects stimulated by music and extracted features using a fuzzy rough model feature evaluation criterion and sequential forward floating selection algorithm. Emotion classification was done using two parallel cascade-forward neural networks. Hamedi et al. [16] used time-domain analysis of EMG signals to extract features and classified emotions using very fast Versatile Elliptic Basis Function Neural Network (VEBFNN) and compared it with conventional classifiers like SVM and neural network. In [17], Hamedi et al. classified emotions from EMG signals using active data of root mean square (RMS) of the signal at specific time intervals, classifying them using Fuzzy-C-means classifier.

In the domain of emotion analysis from Hindustani Music (HM) clips, research have been scarce, but Hindustani music is unique in its approach of imbibing different *Rasas* (or emotions) that the different *ragas* convey. A number of works tried to harvest this immense potential by studying objectively the emotional experiences attributed to the different *ragas* of Hindustani classical music [18, 19, 20, 21]. The *raga* is a sequence of musical notes and the play of sound which delights the hearts of people. The word Raga is derived from the Sanskrit word "Ranj" which literally means to delight or please and gratify [22, 23]. In this work, we look to quantify emotions from a pair of Hindustani raga taking the help of simultaneous EEG and EMG data using robust analysis techniques. This is the first of its kind work which looks objectively into correlations between brain response obtained from EEG signals and the muscular response acquired from the EMG signals.

2 Data Acquisition

2.1 Subjects

Five naive listeners ($M = 3$, $F = 2$) voluntarily participated in this study whose average age was 23 (SD = 1.5 years) years. All participants were made to sign a consent as per the guidelines of Ethical Committee of Jadavpur University (**Approval No. 3/2013**). All experiments were performed at the Sir C.V. Raman Centre for Physics and Music, Jadavpur University, Kolkata. The experiment was conducted in the afternoon in a normally conditioned room sitting on a comfortable chair and performed as per the guidelines of the Institutional Ethics Committee of SSN College for Human volunteer research.

2.2 Experimental Details

EEG and EMG responses for the two Hindustani Classical *Ragas - Chayanat and Darbari Kanada* were recorded simultaneously with five participants. The emotional contents of every music clip were standardized by a listening test previously performed with 30 respondents. The results have been reported from a previous hysteresis-based research [18], which focuses on the retention of emotional arousal in the human brain (75% of those who participated found Chayanat cheerful, 80% discovered Darbari to express pathos/sad emotion). Amplitude standardization has been performed for both signals to prevent loudness indications. All of these sound signals were sampled at the rate of 44.1 kHz, with 16 bit resolution and in a mono channel. A sound system (Logitech R _ Z-4 speakers) with high S/N ratio was used for the measurement.

2.3 Experimental Protocol

The brain electrical and muscular responses of five participants were recorded simultaneously by EEG and EMG. An EEG cap with 19 electrodes (Ag/AgCl sintered ring electrode) placed in the global 10/20 scheme has been developed for each participant.

Figure 1 depicts the positions of the EEG electrodes while Fig. 2 depicts the positioning of EMG electrodes.

Impedances were checked below 5 kΩ. The EEG recording system (Recorders and Medicare Systems) was operated at 256 samples/s recording on customized software of RMS. The data was band-pass-filtered between 0.5 and 35 Hz to remove DC drifts and suppress the 50 Hz power line interference. Each subject was seated comfortably in a relaxed condition in a chair in a shielded measurement cabin. They were also asked to close their eyes. On the first day after initialization, a 14 min recording period was started, and the following protocol was followed:

1. 2 min No Music
2. 2 min With Drone
3. 2 min With Music 1 (*Chayanat*)
4. 4. 2 min No Music
5. 2 min With Music 2 (*Darbari Kannada*).

2 min No Music. The drone signal was used as a basis for an emotional arousal that corresponds to other musical clips [18].

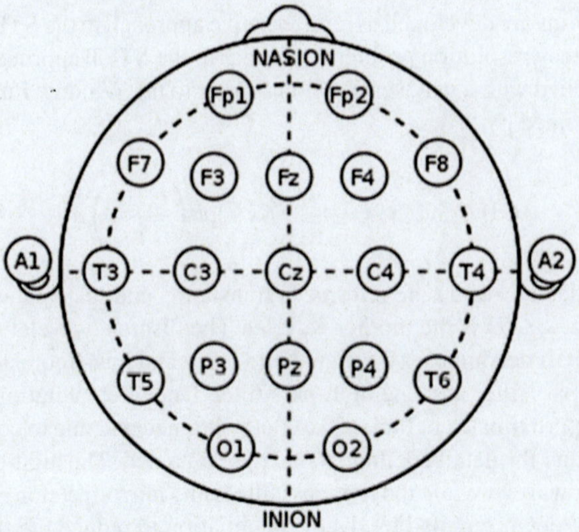

Fig. 1 EEG electrodes

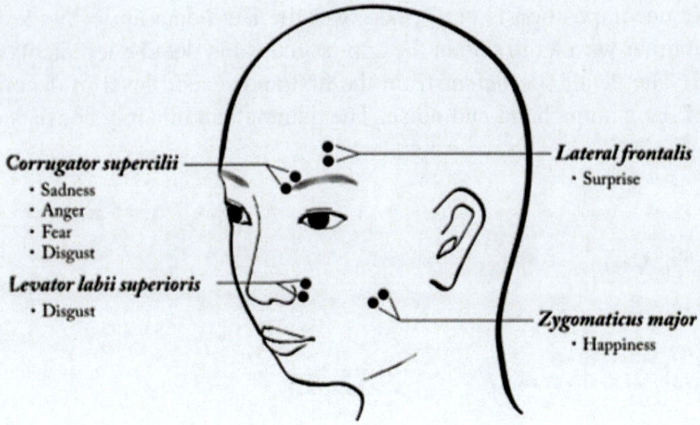

Fig. 2 EMG electrodes

3 Methodology

3.1 Wavelet Decomposition of EEG Signal

The EEG and EMG samples obtained from the experiment were decomposed into distinct frequency bands by wavelet decomposition [24]. Wavelet decomposition on a signal decomposes a signal into a family of functions called wavelets which are derived from the generating function called the mother wavelet. The continuous wavelet transforms are developed as an alternative approach to the STFT to overcome the time-frequency resolution problems. Similar to the STFT approach here also the signal is multiplied with a wavelet function similar to the Window function in STFT. The CWT is defined by

$$CWT_x^{\psi}(\tau, s) = \frac{1}{\sqrt{|s|}} \int x(t)\, \psi\left(\frac{t-\tau}{s}\right) dt$$

Here the variables τ and s are referred to translation and scale parameters, respectively, and $\psi((t-\tau)/s)$ is the mother wavelet. The discrete wavelet transform used for discrete signals decomposes the signal into high- and low-frequency components using one low-pass filter and one high-pass filter. Due to convolution with the low-pass filter the approximate coefficients cA1 are obtained and due to convolution with the high-pass filter the detail coefficients cD1 are obtained. The high-pass filter is the discrete mother wavelet while the low-pass filter is its mirror version. Those cA1 and cD1 vectors are down sampled by dyadic decimation to reduce the time resolution. The next step splits the cA1 vector into cA2 and cD2 using the same scheme followed by the dyadic decimation. Hence, the wavelet decomposition of the signal S at level i obtains [cAi, cDi … cD2, cD1] (Fig. 3).

Wavelet decomposition is performed over the EEG data up to five levels using symlet 5 mother wavelet to obtain the approximate and detail coefficients as shown in Table 1. The detail coefficient from the first and second level of decomposition consists of the gamma band and noise. The gamma band mainly consists of higher

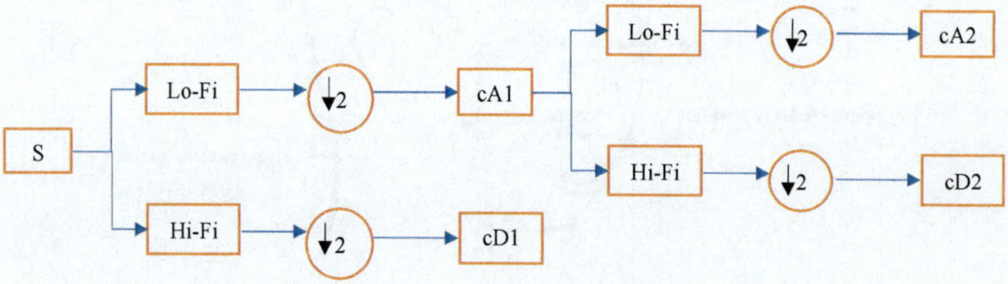

Fig. 3 Wavelet decomposition up to two levels using high- and low-pass filter

Table 1 EEG wavelet decomposition

Wavelet coefficients	Frequency (Hz)	Band name
cD1	64–128	Higher gamma and noise
cD2	32–64	Lower gamma
cD3	16–32	Beta
cD4	8–16	Alpha
cD5	4–8	Theta
cA5	0–4	Delta

frequency range above 32 Hz. Since the response in this frequency range was not significant, this was rejected with an intention to minimize the noise present in the data.

3.2 *EMG Signal Decomposition from Signal Zero-Crossings*

As is the case of EEG, EMG signal can also be subdivided into roughly five frequency bands. The ranges of these bands in EMG almost correspond to that in EEG. To determine these frequency bands an attempt has been made to find the zero-crossings of EMG signal. The zero-crossings are manifestation of pitch in a signal. We calculated the zero-crossings in an interval of one second for an EMG signal of 2 min. We get a plot of a number of zero-crossings versus the signal intervals (here there will be 120 intervals which are of 1 s duration each, for a signal of 2 min) as shown in Fig. 4. The zero-crossing number gives an idea of the frequency. Say for the plot shown, if we consider the zero-crossing number 8, for example, then we can count the number of occurrences of 8 throughout the length of the signal interval-wise. That number is stored as $N(i)$ where i represents the zero-crossing levels, which can also be termed as frequency levels. Then we calculated the probability of occurrence of each level, the probability $P(i)$ be defined as the ratio between $N(i)$ and the sum of occurrences of all the zero-crossing levels throughout the signal. The probability gives an idea of the frequency bands' limits. These probabilities are maximized for certain frequencies thus representing demarcating boundaries between different frequency bands. The above approach has also been tested on EEG signals which clearly indicates the five levels, namely delta, theta, alpha, beta, gamma levels of frequencies for EEG signals. Thus, this approach is verified.

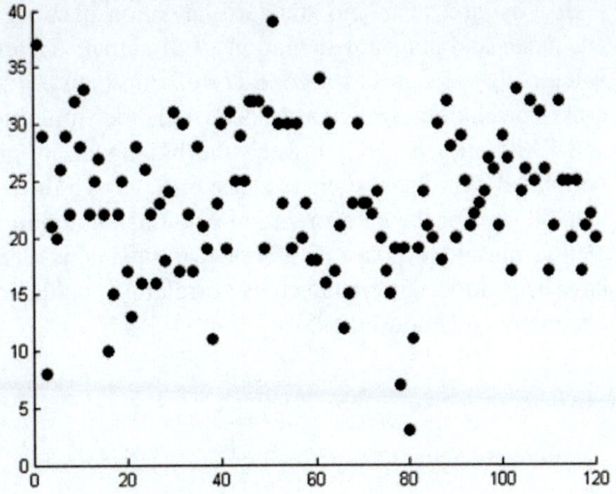

Fig. 4 Plot of zero-crossings of EMG signal versus signal time intervals

3.3 Relative Power of Frequency Bands

The relative power of a signal over a selected frequency range from its bandwidth is defined as

$$\text{relative power} = \frac{\sum \text{energy in the selected frequency range}}{\sum \text{energy in the total frequency range}}$$

The relative power for the beta, theta, alpha, and delta bands of the EEG signals and the frequency ranges of the EMG signals is calculated.

3.4 Cross-Correlation

Cross-correlation coefficients give a measure of similarity between two series of samples as a function of their relative displacement. The cross-correlation coefficient is defined as

$$(f * g)(\tau) = \int_{-\infty}^{\infty} f^*(t)g(t+\tau)dt$$

f^* is the complex conjugate of f and τ is the displacement called lag. For the time series bivariate data, the cross-correlation coefficient is given by,

$$\rho_{XY} = \frac{E[(X_t - \mu_X)(Y_{t+\tau} - \mu_Y]}{\sigma_X \sigma_Y}$$

μ_X, σ_X are mean and standard deviation of X signal, respectively, and μ_Y, σ_Y are mean and standard deviation of Y signal. Clearly $-1 \leq \rho_{XY} \leq 1$. A high value of $|\rho_{XY}|$ signifies a good correlation between the two signals. We used this cross-correlation coefficient to determine the similarity measures between the EEG and EMG signals, i.e., between the bio-signals originated from autonomous and peripheral nervous system as a response to two different music clips. This gives a general cue for the categorization of human emotion simultaneously from brain as well as muscle response data. Since simultaneous recording of EEG and EMG data have been done, so zero lag cross-correlation coefficients have been evaluated for all experimental conditions.

4 Results

In the following section, the simultaneous EEG and EMG responses have been studied for the same set of music clips. The difference in the relative power has been computed for beta, theta, alpha ,and delta bands of the EEG and 4–8, 8–16, 16–32, 32–64 Hz frequency bands of the EMG signals. The relative power difference for these frequency bands has been plotted in the following figures (Figs. 5, 6, 7, 8, and 9) for *Drone*, *Chhayanat*, and *Darbari Kanada.*

The responses for both the frontal electrodes are found to be very similar in both the musical clips (Fig. 5). A maximum change of relative power occurs in case of Drone music which in fact signifies the onset of tranquility. For a happy clip, there is a suppression in relative power, and for a sad clip, there is an arousal in relative power for alpha band. In the beta frequency range, again maximum response is obtained

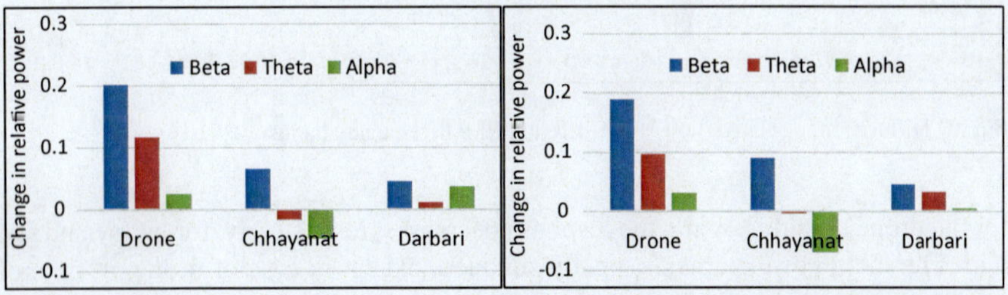

Fig. 5 Difference in relative power for different EEG frequency bands in frontal lobes

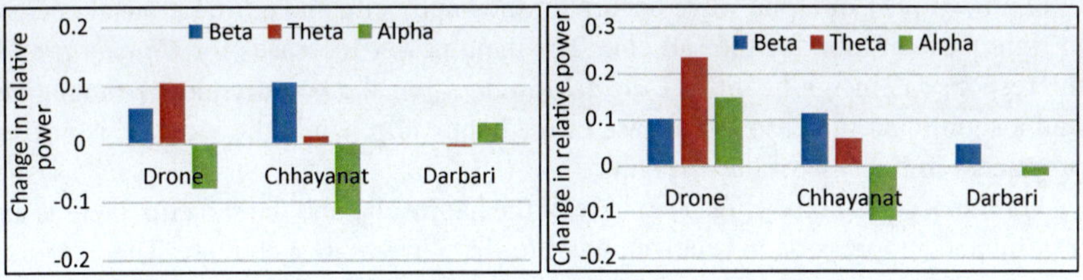

Fig. 6 Difference in relative power for different EEG frequency bands in parietal lobes

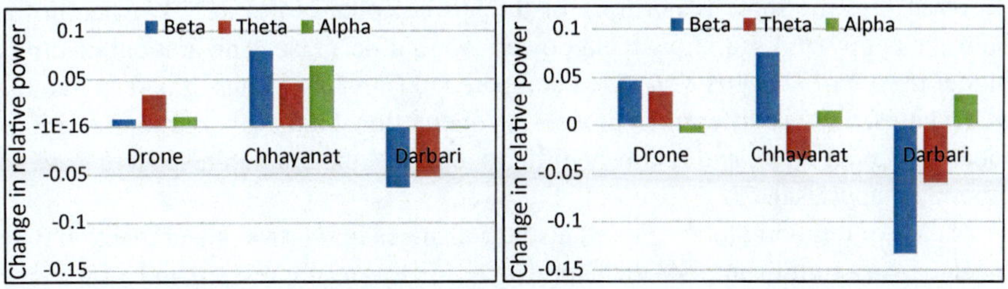

Fig. 7 Difference in relative power for different EEG frequency bands in occipital lobes

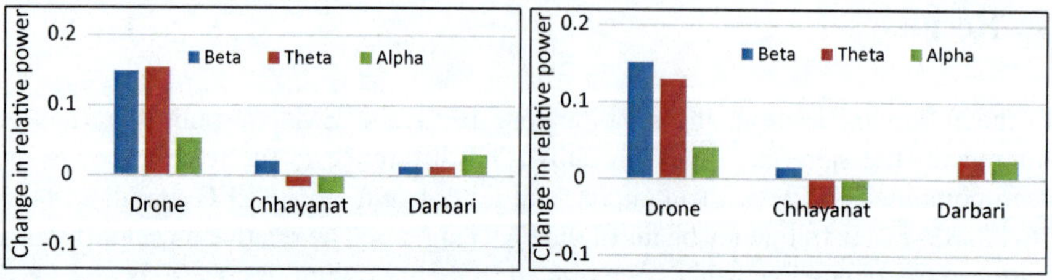

Fig. 8 Difference in relative power for different EEG frequency bands in temporal lobes

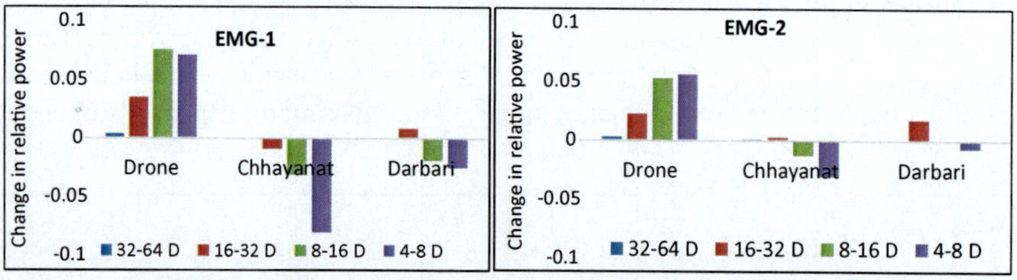

Fig. 9 Difference in relative power for different EMG frequency bands

for the drone stimulus, while the response decreases, respectively, for happy and sad clip. The theta power remains mostly subdued, except in case of drone, in case of the frontal lobe as is evident from the figures.

For parietal lobe (Fig. 6), the responses for Drone music are very significant which inevitably depicts the baseline for other responses. Here also the high suppression of relative power of alpha wave occurs for the happy clip and a little arousal occurs in right parietal lobe for the sad clip. The theta power increases for *Chayanat* and the response is insignificant in *Darbari Kanada.* For the beta frequency range, we find a significant increase in the case of the happy clip, while the increase is not so prominent in the case of the sad clips.

For the occipital lobe (Fig. 7) in case of the happy clip and the sad clip, there is an arousal and suppression in relative power for beta waves, respectively. The baseline response is much lower compared to the other lobes. The right occipital electrode O2 gives a clear indication of emotion categorization in regard to theta response. While a happy clip results in a rise of theta spectral power, a sad clip results in its decrease. But the same is not true for the left occipital electrode O1, wherein for both the happy and sad clips, theta power shows a decrease. The amount of dip is higher in case of *Darbari Kanada* as compared to *Chayanat.* Thus, it is seen in most of the cases, there is differential processing of emotional musical clips in the various lobes of human brain, and it is very difficult to model the brain in respect to specific emotional appraisal.

In case of temporal lobe (Fig. 8) also the changes in relative power due to Drone music is very significant. For the happy and sad clips a suppression and arousal are noticed, respectively, for both the alpha and theta waves. The significant changes in the drone show the sensitivity of temporal lobe. For both the electrodes there is enough evidence for emotion categorization for the happy and sad clip.

In case of EMG electrodes (Figs. 9, 10), significant changes in relative powers of the corresponding frequency bands occur in the case of Drone music. The considerable change in the similar frequency ranges depicts a good correlation or similarity with the frontal, parietal, and temporal lobe electrodes. The change in relative power is most prominent in the 4–8 Hz EMG frequency range which is essentially the theta frequency range of EEG signal. Also, significant response is obtained in the 8–16 Hz EMG frequency range with respect to drone as well as happy/sad stimulus; this range is essentially the alpha frequency range for EEG signals. For the first time, a frequency-based response has been obtained in case of EMG electrodes and surprisingly it has been found to be at par with the EEG frequency bands. The changes are very low in case of the sad clip which also is very similar to the frontal, parietal, and temporal lobe responses.

In the next section, the cross-correlation coefficients between EEG electrodes and EMG electrodes have been reported in the following figures (Figs. 10, 11, 12, and 13). This unique study is the first of its kind which tries to correlate the bio-signals originating from autonomous and peripheral nervous systems.

The cross-correlation coefficient essentially gives the degree of correlation between the two physiological signals originating from brain and muscle. A high correlation coefficient signifies a good correlation between the two signals while a low correlation coefficient signifies a weak correlation between the two. It is seen that for happy music, the cross-correlation coefficient between EMG electrodes and the frontal electrodes, F3 and F4 become significantly high while the same for the sad clip becomes much lower compared to the happy clip. In case of drone signal,

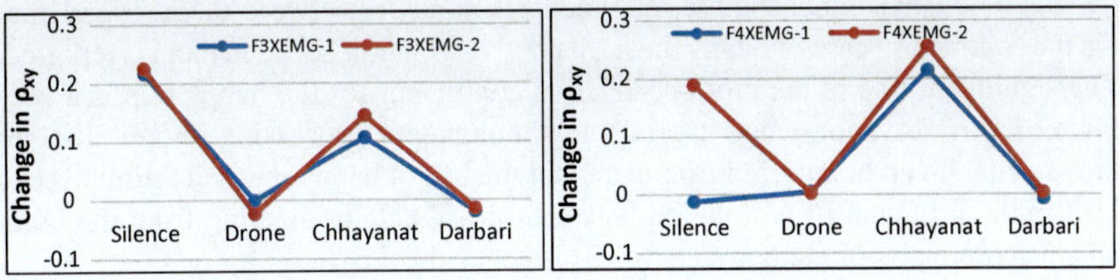

Fig. 10 Cross-correlation coefficients between EMG and EEG of frontal lobes

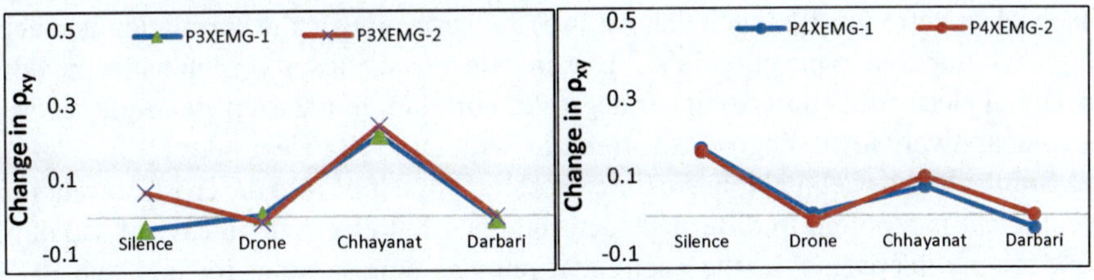

Fig. 11 Cross-correlation coefficients between EMG and EEG of parietal lobes

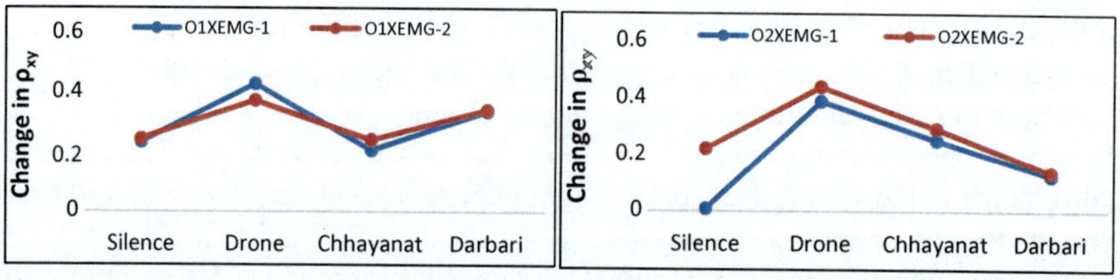

Fig. 12 Cross-correlation coefficients between EMG and EEG of occipital lobes

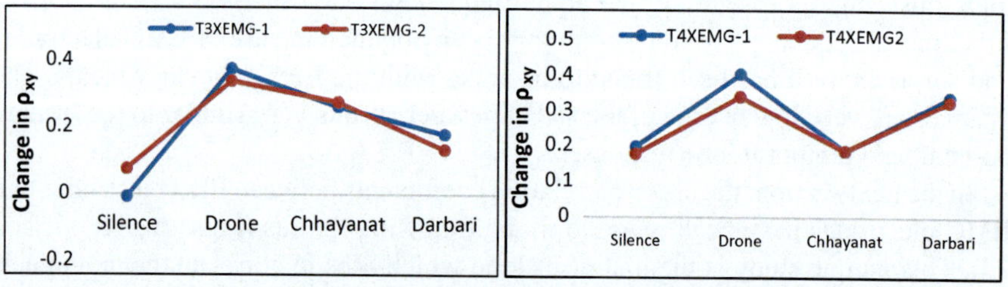

Fig. 13 Cross-correlation coefficients between EMG and EEG of temporal lobes

the correlation becomes low in both the cases. This gives a general cue for the categorization of human emotion simultaneously from brain as well as muscle response data.

In (Fig. 11), for happy music again the cross-correlation coefficient between EMG electrodes and parietal EEG electrodes, P3 and P4 is reported to be high and for the sad music it is very low. An almost zero correlation coefficient between EMG and parietal lobe EEG signal signifies an independent nature between them for sad clip. On the contrary, in case of happy music, a good correlation between them is reported. The significant rise in the cross-correlation coefficient for the happy clip is a very interesting observation when it comes to simultaneous processing to two different bio-signals. It can be inferred from here that the use of happy music as stimulus can lead to the pronounced correlation between bio-signals originating from the brain and different parts of the body.

In case of Occipital lobe, the cross-correlation coefficient with EMG electrodes is reported to differ a lot between right occipital lobe and left occipital lobe in case of the sad clip. For the happy clip, the coefficients remain almost equal for the left and right electrodes. It is seen that here for the Drone music the correlation is much high as compared to the other lobes. But, in case of sad clips, it is seen while the odd occipital electrode reports a higher degree of correlation; the even electrode reports a comparatively lower degree of correlation with the EMG electrodes.

Similar to the occipital lobe, in the case of temporal lobe also the cross-correlation coefficient is reported to differ between right and left electrode in case of sad clip. But for the happy music, the coefficient remains almost equal for both the EEG electrodes. In case of sad music, we find contradictory result compared to the occipital lobe. While the degree of correlation remained lower for the odd electrode, it became significantly high for the even electrode. For Drone music, the coefficient is reported to be very high than sad and happy clip which is in fact very similar to the case of occipital lobe. Thus, for the first time, we report various interesting features of EEG and EMG signal processing in developing a robust emotional categorization model using these specialized techniques.

5 Conclusion

The study presents first of its kind data on simultaneous EEG and EMG processing for emotion quantification from Hindustani classical music stimuli. The relative power corresponding to EEG alpha, beta, and theta waves has been computed for the two clips of contrast emotion and the same have been compared with EMG waves. The cross-correlation coefficients have also been computed for EEG and EMG data for the different experimental conditions. The study presents the following interesting conclusions:

1. For the parietal and frontal electrodes, the cross-correlation coefficient between EEG and EMG electrodes shows a distinct categorization for happy and sad clips; while for the happy clip the cross-correlation coefficient rises significantly, for sad clip, there is a considerable decrease in the cross-correlation coefficient. The rise from non-correlated time series during the silence period to a strong correlation is a revelation of this study.
2. The EMG frequency bands have been verified in this study with the help of zero-crossing rate technique and it is seen that there is a strong response in the same frequency bands as in the case of EEG. The 4–8 Hz frequency (theta in case of EEG) and 8–16 Hz (alpha in the EEG) band has the strongest response in the EMG domain.
3. The frontal, parietal, and occipital lobes show the strongest response in the EEG study with a significant increase in relative power corresponding to the happy clip while a decrease is noted for sad clip.

Thus, in relation to the classic Hindustani music analysis, with a varied range of other music stimuli and a higher number of respondents we propose a new algorithm for quantifying and categorizing emotional stimulus that would produce more statistically important and conclusive outcomes. This is a pilot study to this end.

Acknowledgments SS acknowledges the JU RUSA 2.0 Post Doctoral Fellowship (**R-11/557/19**) and Acoustical Society of America (ASA) to pursue this research AB acknowledges the Department of Science and Technology (DST), Govt. of India for providing (**SR/CSRI/PDF-34/2018**) the DST CSRI Post Doctoral Fellowship to pursue this research work. This study has been conducted under the guidelines of Ethical Committee of Jadavpur University (**Approval No. 3/2013**).

References

1. Lin YP, Wang CH, Jung TP, Wu TL, Jeng SK, Duann JR, Chen JH (2010) EEG-based emotion recognition in music listening. IEEE Trans Biomed Eng 57(7):1798–1806
2. Lin YP, Wang CH, Wu TL, Jeng SK, Chen JH (2007) Multilayer perceptron for EEG signal classification during listening to emotional music. In: TENCON 2007–2007 IEEE Region 10 Conference. IEEE, pp 1–3
3. Lin YP, Wang CH, Wu TL, Jeng SK, Chen JH (2009) EEG-based emotion recognition in music listening: a comparison of schemes for multiclass support vector machine. In: 2009. ICASSP 2009. IEEE international conference on acoustics, speech and signal processing. IEEE, pp 489–492
4. Lin YP, Wang CH, Wu TL, Jeng SK, Chen JH (2008) Support vector machine for EEG signal classification during listening to emotional music. In: 2008 IEEE 10th workshop on multimedia signal processing. IEEE, pp 127–130
5. Liu Y, Sourina O (2014) Real-time subject-dependent EEG-based emotion recognition algorithm. Transactions on Computational Science XXIII. Springer, Berlin, Heidelberg, pp 199–223
6. Murugappan M, Ramachandran N, Sazali Y (2010) Classification of human emotion from EEG using discrete wavelet transform. J Biomed Sci Eng 3(04):390
7. Musha T, Terasaki Y, Haque HA, Ivamitsky GA (1997) Feature extraction from EEGs associated with emotions. Artif Life Robot 1(1):15–19
8. Choppin A (2000) EEG-based human interface for disabled individuals: emotion expression with neural networks. Unpublished master's thesis
9. Frantzidis CA, Bratsas C, Papadelis CL, Konstantinidis E, Pappas C, Bamidis PD (2010) Toward emotion aware computing: an integrated approach using multichannel neurophysiological recordings and affective visual stimuli. IEEE Trans Inf Technol Biomed 14(3):589–597
10. Cheng B, Liu G (2008) Emotion recognition from surface EMG signal using wavelet transform and neural network. In: Proceedings of the 2nd international conference on bioinformatics and biomedical engineering (ICBBE), pp 1363–1366
11. Murugappan M (2011) Electromyogram signal based human emotion classification using KNN and LDA. In: 2011 IEEE international conference on system engineering and technology (ICSET). IEEE, pp 106–110
12. Nakasone A, Prendinger H, Ishizuka M (2005) Emotion recognition from electromyography and skin conductance. In Proceedings of the 5th international workshop on biosignal interpretation, pp 219–222
13. Yang S, Yang G (2011) Emotion recognition of EMG based on improved LM BP neural network and SVM. JSW 6(8):1529–1536
14. Kim J, André E (2008) Emotion recognition based on physiological changes in music listening. IEEE Trans Pattern Anal Mach Intell 30(12):2067–2083
15. Naji M, Firoozabadi M, Azadfallah P (2015) Emotion classification during music listening from forehead biosignals. SIViP 9(6):1365–1375
16. Hamedi M, Salleh SH, Astaraki M, Noor AM (2013) EMG-based facial gesture recognition through versatile elliptic basis function neural network. Biomed Eng Online 12(1):73
17. Hamedi M, Rezazadeh IM, Firoozabadi M (2011) Facial gesture recognition using two-channel bio-sensors configuration and fuzzy classifier: a pilot study. In: 2011 International conference on electrical, control and computer engineering (INECCE). IEEE, pp 338–343
18. Banerjee A, Sanyal S, Patranabis A, Banerjee K, Guhathakurta T, Sengupta R, Ghose P et al (2016) Study on brain dynamics by non linear analysis of music induced EEG signals. Phys A: Stat Mech Appl 444:110–120
19. Nag S, Biswas S, Sengupta S, Sanyal S, Banerjee A, Sengupta R, Ghosh D (2017) Can musical emotion be quantified with neural jitter or shimmer? A novel EEG based study with Hindustani classical music. In: 2017 4th international conference on signal processing and integrated networks (SPIN). IEEE, pp 358–363

20. Sanyal S, Banerjee A, Patranabis A, Banerjee K, Sengupta R, Ghosh D (2016) A study on improvisation in a musical performance using multifractal detrended cross correlation analysis. Phys A 462:67–83
21. Sanyal S, Nag S, Banerjee A, Sengupta R, Ghosh D (2019) Music of brain and music on brain: a novel EEG sonification approach. Cogn Neurodyn 13(1):13–31
22. Ghosh D, Sengupta R, Sanyal S, Banerjee A (2018) Musicality of human brain through fractal analytics. Springer Singapore
23. Martinez JL (2001) Semiosis in Hindustani music, vol 15. Motilal Banarsidass Publ
24. Akin M (2002) Comparison of wavelet transform and FFT methods in the analysis of EEG signals. J Med Syst 26(3):241–247

Studies Different Structure of Atmospheric Boundary Layer Using Monostatic SODAR

Nishant Kumar, Kirti Soni, Ravinder Agarwal, and Mahavir Singh

Abstract The Atmospheric Boundary Layer (ABL) was continuously monitored using Monostatic SODAR, which was conducted at CSIR-NPL, New Delhi. In this paper, an effort has been made to review the studies in the field of the structure of ABL using SODAR echograms in Delhi during a special observation period of the experiment. The change in weather conditions has found to have definite marks on the SODAR observed the structure of ABL. In particular, it has been seen that dot echoes and spiky top layered structures are different typical structures of the different seasons, which are associated with temperature, wind speed, and relative humidity. Also, the knowledge of these structures is great to use in air pollutant studies and dispersion models.

Keywords SODAR · Atmospheric boundary layer · Different ABL structure

1 Introduction

This paper represents the processes governing the different types of structure in the Atmospheric Boundary Layer (ABL) and discusses relevant uses of Sonic Detection And Ranging (SODAR) information in the field of ABL. It is written from the practical point view of a meteorologist involved in consulting and advisory work for industry and government agencies, rather than from SODAR researcher.

Factors that govern the dilution, rise, and spread of pollutants include wind speed and direction, turbulence, temperature lapse rate, and ABL [1, 2]. As a result, to understand local pollutants transfer and dispersion over an area, researchers should know the ABL, the different types of structure of ABL. During recent years, the capacity of SODAR to map the different types of ABL structure is increased. The knowledge of these structures is of great use in air pollutant studies [2, 3]. SODAR

N. Kumar (✉) · R. Agarwal
Thapar University, Patiala 147004, India
e-mail: nishant.kumar@thapar.edu

K. Soni · M. Singh
CSIR-National Physical Laboratory, New Delhi 110012, India

M. Singh and Y. Rafat (eds.), *Recent Developments in Acoustics*, Lecture Notes in Mechanical Engineering, https://doi.org/10.1007/978-981-15-5776-7_27

is a comparatively cost-effective continuously operating remote sensing technique, and can be thus usefully employed to deal with hazardous situations of air pollution [4].

1.1 Acoustic Wave in the Atmosphere

When the flexible diaphragm of a speaker moves, it creates small pressure fluctuations traveling outward from the speaker [3]. These pressure fluctuations are sound waves. The speed, *c*, at which these waves travel can be expected to depend on the mechanical properties p_{atm} (atmospheric pressure) and ρ (air density) [3]. Therefore

$$c \propto \sqrt{\frac{p_{\text{atm}}}{\rho}}$$

and, as already noted, the temperature and density are inversely related to each other at constant pressure through the gas equation

$$p_{\text{atm}} = \rho R_d T$$

where $R_d = 287\ \text{J kg}^{-1}\ \text{K}^{-1}$. This means that

$$c \propto \sqrt{T}$$

Allowing for T being the temperature in K, and that the speed of sound at 0 °C is 332 m s^{-1}, therefore

$$c(T) = 332(1 + 0.00166\Delta T)\ \text{ms}^{-1}$$

where ΔT is the temperature in °C. For air containing water vapor, the air density is the sum of the dry air density, ρ_d, and the water vapor density, ρ_v, or

$$\rho = \rho_d + \rho_v = \frac{p_{\text{atm}} - p_v}{R_d T} + \frac{p_v}{\left(R_d/\varepsilon\right)T}$$

where $\varepsilon = 0.622$ is the ratio of the molecular weight of water to molecular weight of air, and individual gas equations have been used for dry air and for water vapor. A simpler expression is obtained in terms of the water vapor mixing ratio, $w = \varepsilon p_v/(p_{\text{atm}} - p_v)$, which is the mass of water vapor divided by the mass of dry air per unit volume. The adiabatic sound speed is

$$c = \sqrt{\frac{\gamma RT}{M}}$$

where $R = 8.31\ \text{J mol}^{-1}\ \text{K}^{-1}$ is a universal gas constant, γ is the ratio of specific heats for the gas, and M is the average molecular weight. This sound speed does not allow for the effect of air motion (i.e., wind) in changing the speed along the direction of propagation.

1.2 *Reflection and Refraction*

At the point, when an acoustic signal meets an interface where the signal speed changes, some signals are reflected and some proceeds over the crossing point thus far with an alteration in direction [3]. This has been pictured using the Huygens rule, which expresses that each point on a signal goes about as a point source of spherical wavelets, and taking the unrelated twist to the wavelets before long gives the position of the proliferated signal. Generally, acoustic signal traveling through the air, there has no separate interface but rather a continuous change in acoustic signal speed due to a temperature gradient or wind shear [2].

On account of acoustic travel-time tomography where the stimulating way is a pair of meters on the ground, ground reflections has been a major supposed. In this case, the reflection from the ground has combined with the direct line-of-sight signal, causing much-reduced signal amplitude [2, 3]. For this reason, continuous encoded-signal systems have experienced difficulties, therefore short pulses are used. The acoustic signal speed increases by 0.17% for every degree rise in air temperature [3].

2 Data and SODAR

SODAR is an all-around perceived acoustic remote-detecting method [2, 4] that ceaselessly screens ABL warm structures up to statures in the scope of 340–3400 m. ABL height has been measured utilizing monostatic SODAR, which was designed at Council of Scientific and Industrial Research (CSIR)—National Physical Laboratory (NPL) and worked at different frequencies as per its particular requirement. The details of CSIR-NPL monostatic SODAR are described (Table 1).

The antenna is encased in an acoustic shield to constrict the outside noise and has been described at anechoic chambers at CSIR-NPL for the transmitting and accepting proficiency and directional qualities. The transducer yield productivity is measured to be 1.32 Pa V − 1 in the anechoic chamber utilizing a sound level analyzer and is in this way traceable to the national models of sound weight acknowledged at CSIR-NPL, India [4]. Highly directional short bursts of sound energy are radiated into the atmosphere, which after scattering from atmospheric fluctuations of eddy sizes within the inertial subrange of 0.1–10 m are being received back by the receiving antenna, conditioned through a preamplifier, and fed an analog input signal at the microphone input terminal of the computer. Each gained information is perused with an 8-bit determination and put away in the information record with a pre-doled out

Table 1 Characteristics of CSIR-NPL (New Delhi) Monostatic SODAR

Transmitted power (electrical)	90 W
Transmitted power (acoustical)	15 W
Pulse width	100 ms
Pulse repetition period	6 s
Operational range	1000 m
Receiver bandwidth	50 Hz
Frequency of operation	2250 Hz
Acoustic velocity	340 m/s (average)
Receiver gain	80 dB
Transmit–receive antenna	Parabolic reflector dish surrounded by conical acoustic cuff
Receiver area	2.5 m^2
Pre amplifier sensitivity	Fraction of a micro-Volt

document name contingent on the date and time toward the start of the information securing. The dynamic scope of the acquired signal 0–5 V is separated into eight stages, and each progression is pre-allocated an uncommon shading code. Contingent on the time passed t after transmission of tone burst, and digitalization of individual information focuses, the gained information point is doled out a stature estimation of

$$h = \frac{ct}{2}$$

where c is the speed of sound noticeable all around and t is the time passed measured in seconds. Each data point with assigned color is displayed as a two-dimensional image in time versus height graphics, on the computer monitor in real time. Line by line joining of various sweeps creates a pictorial perspective of the SODAR echograms [4]. A tight band channel is utilized as a part of the equipment, hardware, to stifle the commotion at undesirable frequencies and enhance the flag to clamor proportion. The signals are processed to produce an online facsimile display of the dynamics of ABL thermal structures. The SODAR framework has been adjusted utilizing a calibration detailed by Danilov et al. [5] in anechoic chambers at CSIR-NPL. SODAR echograms are reflex pictures of the turbulence in the lower climate. This turbulence is accountable for the dispersion of pollutants. Thus, the measured height of thermals plumes in the daytime and of shear echoes for the duration of night time. The height of the thermal plumes by SODAR during the daytime will always give an underestimated value unless they are capped by a low-level elevated shear echo layer [4]. The below empirical relation has given the mixing height during the daytime:

$$y = 4.24x + 95$$

where y is the blending stature (m) for temperamental ABL and x is the profundity of the SODAR measured thermal plume (m).

3 SODAR Structure and Characteristics of the Different Season ABL

Five thousand eight hundred fifty (approx. one year) SODAR echograms that have been seen at the examiner area (CSIR-NPL, New Delhi) for various seasons were analyzed and classified into different structures. Figure 1 shows the different structures (Table 2) under SODAR echograms [6].

The convection layer structure, that is, the characteristic highlight of the daytime unstable conditions, disseminates in the evening after dawn because of sun-powered go down of the ground and the temperature profile changes its shape [3, 4]. The time is taken for a convection layer to disperse after dawn changes from every day and season to season contingent on the nearness of overcast cover, the quality of the reversal layer framed amid the night, the flood of sun based warmth, and the nearness of rising layer or haze layer. Figure 1a, b, c, d, e displays the thermal plume with tall spikes which were observed during the daytime. They are more common in the month of May due to the high temperature (or solar radiation). Due to high solar radiation, it uncovers that the height of thermal plumes is generally 470 m and can go up to 625 m. In this period, ABL height is more; because of this, the dispersion area of pollutants is more. Therefore, this is a more unstable class and more favorable condition for the industry and human being.

Figure 1k, l displays the multilayer which was observed during nighttime. Due to more variation in temperature, wind speed, wind direction, and relative humidity, it uncovers that the height of the inversion layer is average 100 m and can go up to 600 m. In this period, ABL height has varied and as a result, the dispersion area of pollutants is varied. This is a stable class and unfavorable condition for the industry and human being.

Waves structure (Fig. 1j) is another class of ABL structures, which have been watched and grouped specifically on the SODAR echograms. The wave structure, for the most part, is found in the April, May, and June months, that is, in the pre-monsoon season. This type of structure is essential in the investigation of wave movement [2]. This type of structure forms due to high wind speed and change in direction.

The inversion layer structure (Fig. 1g, h, i, j, k), that is, a characteristic highlight of the nighttime stable conditions, disseminates in the morning after dawn because of sun-powered warming of the ground and the temperature profile changes its shape. The time is taken for an inversion layer to disperse as follows the above convection layer after dawn changes from every day and season to season. Contingent on the

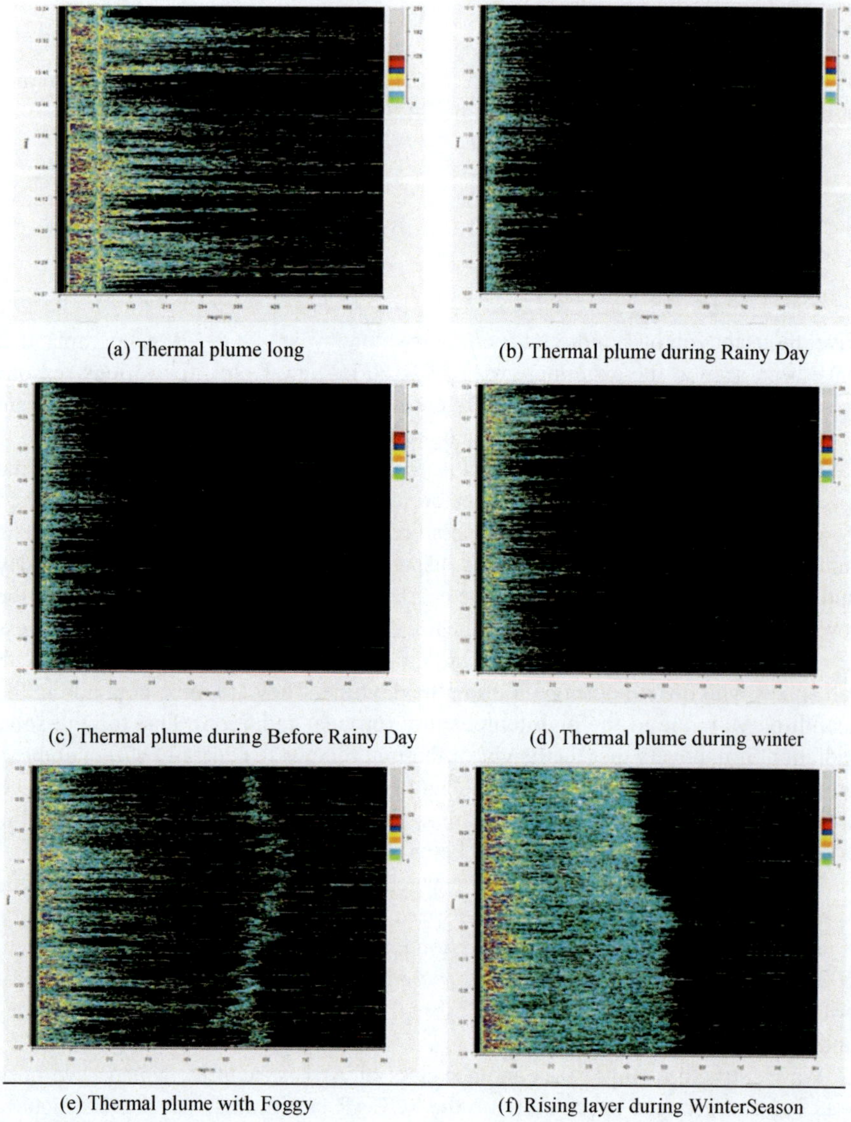

(a) Thermal plume long

(b) Thermal plume during Rainy Day

(c) Thermal plume during Before Rainy Day

(d) Thermal plume during winter

(e) Thermal plume with Foggy

(f) Rising layer during WinterSeason

Fig. 1 Different structures of ABL

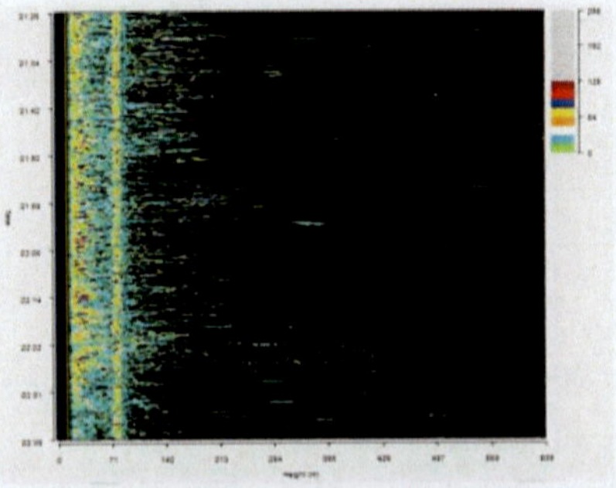

(g) Inversion with long Spikes

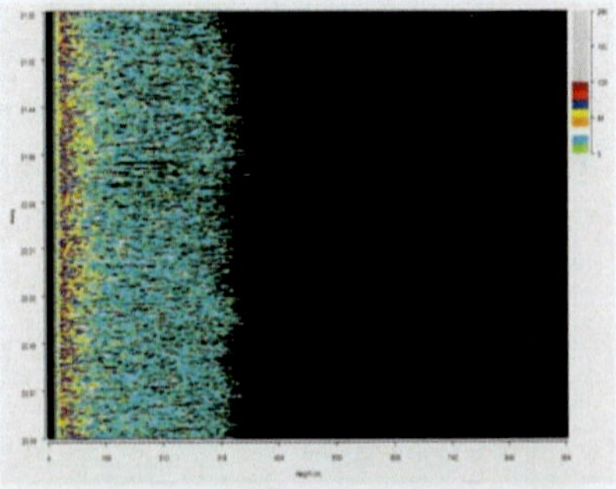

(h) Winter Inversion

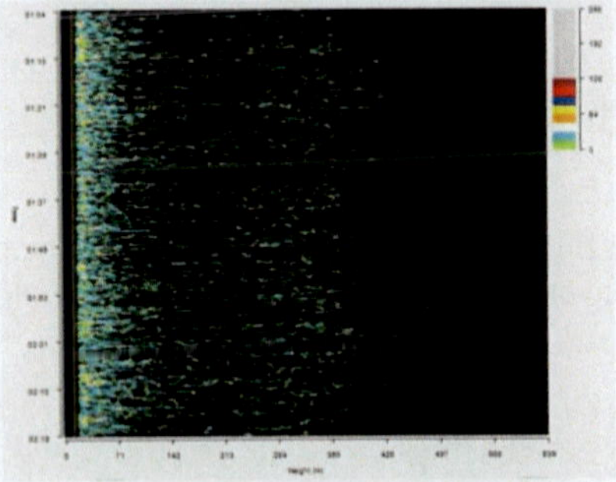

(i) Inversion with one or two elevated layer in motion

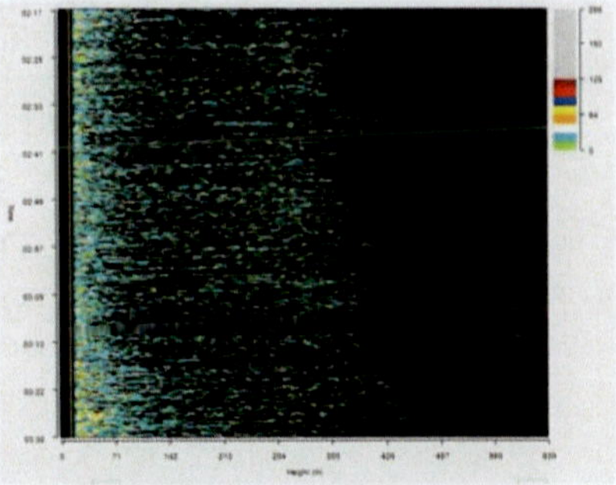

(j) Inversion with wave motion

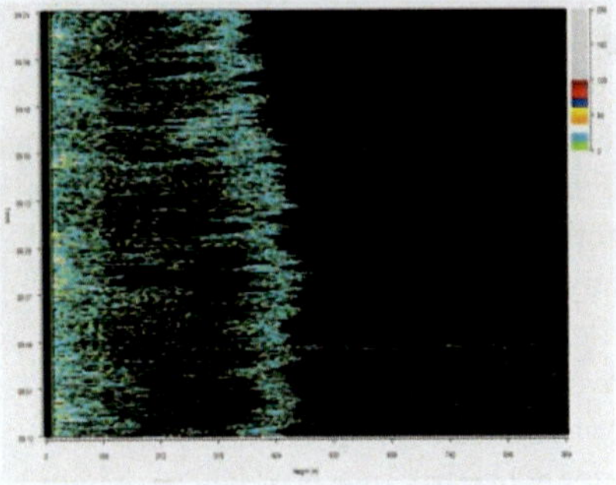

(k) Inversion With Fog

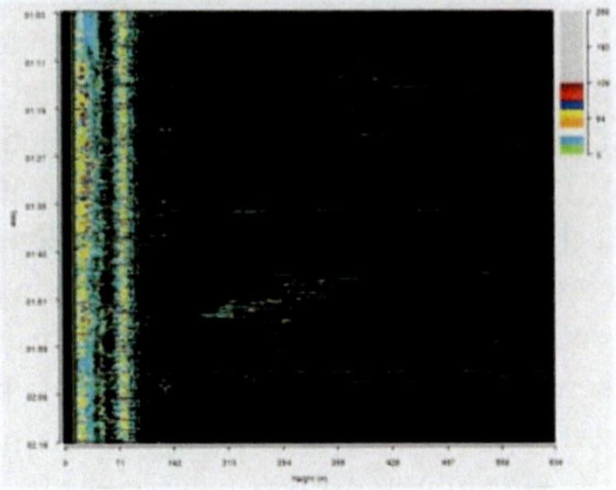

(l) Multilayer

Fig. 1 (continued)

nearness of overcast cover, the quality of the reversal layer framed amid the night, flood of sun based warmth, and the nearness of rising layer or haze layer. In the rising layer (Fig. 1f), the rising rate is quicker in the pre-monsoon season than in the winter and post-monsoon season because of the adjustment in the temperature profile. The night-time inversion while scattering demonstrates its essence as a rising layer, which ranges to a specific most extreme height and after that vanishes. This kind of structure

Table 2 Description of SODAR Structure and Class Numbers

Class	Description of SODAR structure
1.	Inversion with force convection
2.	Inversion with a tall spike
3.	Inversion with a single elevated layer
4.	Inversion with two elevated layers
5.	Inversion with wave motion
6.	Inversion with one or two elevated layers in motion
7.	Inversion with a small spike
8.	Stratified layer
9.	Diffuse thermal plumes
10.	Thermal plumes with normal days
11.	Thermal plumes with a foggy layer
12.	Transition structure

is not common in the morning time. During rainy and cloud cover days, the rising layer does not exist and inversion directly converted into the convection. As a result, less rising layer is found in the monsoon season.

4 Discussion and Conclusion

SODARs have been successfully used to study microclimates and to provide an integrated approach to the meteorological classification of pollutants concentration in major city. The inversion and convection height have a very important parameter for control of air pollution and predication. For the reason that the inversion height has below 300 m [2], and this has been perfectly measured using SODAR. SODAR echogram information is a good option to obtain atmospheric information as well as a real-time picture of the spatial distribution of the ABL structural features. It has been used for prediction of parameters like inversion height, average height of ABL, therefore this information has been helpful for authorities of environmental services. In predication, the direct view on the backscatter echo structure is much more helpful than code information, because the pictorial view of ABL provides more information of atmosphere. The SODAR echogram has provided many valuable ABL structures. On the other hand, SODAR echograms have given the lots of other atmospheric information, i.e., carry capacity of atmosphere, convection condition, fog, rain condition and other seasonal information. As it has been found, SODAR becomes necessary in operation of modern air-quality monitoring systems, which are responsible for operations of far-reaching economic and ecological consequences [2, 3]. Author is deeply convinced that SODARs should be considered as mandatory element of big air-monitoring systems and that this fact should be taken into account in national and international regulations for meteorological and environmental services.

References

1. Singal SP, Gera BS (1982) Acoustic remote sensing of the boundary layer. Sadhana 5(2):131–157
2. Singal SP (ed) (2006) Acoustic remote sensing applications, vol 69. Springer
3. Bradley S (2007) Atmospheric acoustic remote sensing: principles and applications. CRC Press
4. Kumar N, Soni K, Garg N, Agarwal R, Singh DSM, Singh G (2017) SODAR pattern classification and its dependence on meteorological parameters over a semiarid region of India. Int J Remote Sens 38:3466–3482
5. Danilov SD, Gur'yanov AE, Kallistratova MA, Petenko IV, Singal SP, Pahwa DR, Gera BS (1992) Acoustic calibration of sodars. Meas Sci Technol 3(10):1001
6. Chandara ND, Pal S, Chandra Patranabis C, Dutta HN (2010) A neurocomputing model for SODAR structure classification. Int J Remote Sens 31(11):2995–3018